D0053638

NO LONGER PROPERTY OF
SEATTLE PUBLIC LIBRARY

CALL
OF THE
MILD

CALL
OF THE
MILD

CALL OF THE MILD

LEARNING TO HUNT MY OWN DINNER

Lily Raff McCaulou

GRAND CENTRAL PUBLISHING

NEW YORK BOSTON

Names of certain persons have been changed.

Grand Central Publishing
Hachette Book Group
237 Park Avenue
New York, NY 10017

www.HachetteBookGroup.com

Printed in the United States of America

RRD-C

First Edition: June 2012
10 9 8 7 6 5 4 3 2 1

Library of Congress Cataloging-in-Publication Data

Raff McCaulou, Lily.
Call of the mild : learning to hunt my own dinner / Lily Raff McCaulou.—1st ed.
p. cm.
Summary: "A beautifully written and contrarian narrative about what it means to hunt in America today"—Provided by the publisher.
ISBN 978-1-4555-0074-1
1. Raff McCaulou, Lily. 2. Women hunters—Oregon—Bend—Biography. 3. Hunters—Oregon—Bend—Biography. 4. Bend (Or.)—Biography. 5. Bend (Or.)—Social life and customs. 6. Outdoor life—Oregon—Bend. 7. Hunting—Social aspects—Oregon—Bend. 8. Hunting—Social aspects—United States. 9. Hunting—Moral and ethical aspects—United States. 10. Meat industry and trade—Moral and ethical aspects—United States. I. Title.
SK17.R23 A3+
639'.1092—dc23
[B]
2011033189

For Scott: This and everything else.

And for Nathan, who would have been an exceptional hunting buddy or an intolerable one. I wish we'd had time to find out which.

CONTENTS

PROLOGUE

I squint down the long metal cylinder of my shotgun, past the red plastic bump at the tip, and imagine what it will feel like to yank the trigger as a bird soars by. It is not the first time I have envisioned this situation. But it is the first time it could actually come true.

The bits of foam I've squished into my ears drown out the faint, usual sounds like rustling leaves. All I can hear is my heart. Each beat thrums my whole body as if I'm standing too close to the bass speakers at a rock concert.

A brown-and-white dog named Tessa stands frozen about ten yards to my left. A few minutes earlier, she sniffed out a male pheasant, her stubby tail wagging faster and faster as she closed in on its trail, nose to the ground, until suddenly she halted, her tail standing straight. She's still standing there now, which, according to her owner, means her eyes have locked on to the bird's eyes: a primordial staring contest with extraordinarily high stakes, at least for the bird. Tessa's owner, Gerry, is behind me, creeping toward the dog.

I rub the gun's ridged safety switch with my sweaty thumb, hesitating, until finally I shove it forward into the "fire" position, the dangerous one. The rock concert in my chest becomes techno.

"Are you guys ready?" Gerry asks in a flat voice.

"Yes," says Nancy, another novice hunter standing a few yards beyond Tessa, waiting to shoot the same doomed bird. Nancy hasn't hunted in decades, but she grew up shooting guns with her father and brothers. Her husband goes bird hunting every fall, and now, with her children grown and out of the house, she wants to join him. It makes sense that Nancy is here. Nancy should be ready.

But not me. My dad never took me shooting; he walked me to the local scoop shop, Lickety Split, or we rode the subway to downtown Washington, DC, and visited a museum. My husband doesn't hunt, and I only decided about a year ago to give it a try.

Now here I am, on a swampy swath of state-owned land in southern Oregon, with a loaded gun pointed at the sky, waiting for a bird. I'm one of twenty women hunting for pheasants on this foggy Saturday in September. We each paid forty dollars to traipse around with volunteer guides and their well-trained hunting dogs. The pheasants were raised in pens and released one week earlier, for a similar hunt for kids. Our goal is to shoot the birds that the kids missed last week. Years later, I will look back on this event and feel slightly embarrassed by its staged phoniness, as if I went back to the Renaissance fair I attended as a kid and noticed that the women were wearing Nikes under their polyester petticoats. Later, I will hunt for real wildlife, not creatures who have been raised by humans and planted for shooting. I will find myself kneeling before a bull elk, up to my shoulders in blood, as I gut the animal and prepare to pack its meat out of the woods. But for now, here, traipsing through tall grass with a loaded gun for the first time, I feel wild and daring.

The night before the hunt, when I should have been sleeping, I instead compiled a mental list of everything that could go wrong. Around three in the morning, assured that I had exhausted every possible horrible outcome, I began to group and rank them in descending order of tragedy:

1. My own death. Shot by another member of the hunting party probably, or a mistaken rifle hunter hundreds of yards away, or perhaps, in a particularly cruel swoop of irony, my own careless trigger finger.
2. The death of someone else, caused by me. I would then endure crushing, paralyzing guilt for the rest of my life, which would probably take me past the age of ninety and involve frequent run-ins with friends and family of the victim. On second thought, maybe this fear should be number one.
3. Any non-fatal maiming of me or caused by me.

4. Embarrassing failure in one of its smaller, more familiar forms. Perhaps, when the time came, I would feel too guilty to pull the trigger and the other, braver hunters would wonder what was wrong with me.

At eight in the morning, when I pulled up to the concrete building that was our designated meeting point, I noticed that four or five of the pickup trucks parked there had camouflage-covered crates in their beds. Dogs whined and barked from inside.

"Shit," I muttered. I hadn't even thought to worry about the dogs, who would dart close to the birds and in front of my gun and be easier to shoot accidentally than any human.

But by the time I get here, to this point where my weapon is loaded and my safety is switched off and a real, live bird is in my vicinity, I have managed to avoid these disasters for more than five hours. The list, which consumed my thoughts in the early morning, has almost disappeared from my mind completely. Only the last item—good old-fashioned embarrassment—lingers in my mind. As the day lags on, it looks more and more likely that I will return home empty-handed. This once innocuous outcome has already started to morph, unexpectedly, into a new worst-case scenario. I picture all the friends I've told about today's outing. I grit my teeth and practice telling them, one by one, "Nope, I didn't get one. Thanks for asking."

Lori and Debra, two of the four women in my group, have already killed a bird each. They are already using the proud, possessive language of real hunters, who speak as if the purchase of a hunting license includes entitlement to one particular specimen. "I got my pheasant!" They are safe and happy, chatting excitedly about what their husbands and children will say when they arrive home triumphant, with a bird in a cooler.

I don't have my pheasant. I'm tired of hiking through uneven ground in wet socks and boots, tired of carrying this gun that feels twice as heavy as it did in the morning. I'm tired of being on edge, of staying alert in case a bird appears suddenly. Tired of watching Tessa's tail for the fast-wagging signal that a bird is near. Tired of eyeing where the muzzle of my gun is pointed and where Nancy's is. I start to dread the two-and-a-half-hour drive back home to my husband—who remains confused by my nascent interest in hunting—without a feathery trophy of my own.

"When I tell Tessa to release, everything is going to happen in a split second," Gerry whispers. He stops just a couple of feet behind the dog and lets another moment pass before asking, again, "Are you sure you're ready?"

"Mm-hmm." Nancy is getting impatient.

"Ready?" Gerry turns to me. The muzzle is heavy in my left hand. My outstretched arm quivers.

"Yes." I am ready. So ready I can hardly believe it. I cannot wait to kill this thing.

Gerry steps one long stride toward Tessa, who lunges at the bird. I hear a squawk and flapping, thrashing against the grass. I wonder if Tessa has somehow managed to catch the thing in her mouth. Suddenly a dark bird with a long, spiky tail leaps into view. I squeeze my right hand into a fist around the trigger.

Bang.

CHAPTER 1

GOING WEST

You would be hard-pressed to find an unlikelier hunter than me. I'm a woman, and married to a man who does not hunt. I grew up in a city, terrified of guns. I love animals and even entered college on track to become a veterinarian. Yet, at the age of twenty-six, I made the strange decision to pick up a gun and learn to hunt. It was a complicated choice, but it started with one simple thing that almost all of us—hunters and non-hunters, women and men, city dwellers and country bumpkins—have in common: dinner. Not the greens and grains on the sides of the plate, but the hunk of meat in the middle.

Of course, my decision to hunt was also deeply personal. It was a way for me to explore my relationships with animals—the dog for whom I buy Christmas presents, the mice I occasionally trap in my kitchen, the wolves whom I admire in theory but have never met. It made me rethink what it means to be an environmentalist. The experience transformed me from the person I had been just three years earlier. I'll start there, when I'm a few months shy of twenty-four, and nothing could be farther from my mind than hunting:

I live with a girlfriend in a cramped apartment in Manhattan, where I work part-time as a personal assistant to a movie director and screenwriter. I also freelance as a production assistant on various film and television shoots. Nearly half of my friends from Wesleyan University moved to New York after graduation, so I know fun, artsy people all over the city. At night, I dress up and attend their theater debuts and gallery openings. During the day, I brush elbows with indie film stars.

But for the past couple of months I haven't been able to shake this feeling that my life in New York has become one big, glitzy distraction. I spend seventy or eighty hours a week working to bring someone else's vision to the television or movie screen, yet I still haven't finished the screenplay I started writing two years earlier. I find myself daydreaming about a new job as a journalist. This isn't entirely out of the blue—I worked on my college newspaper, as a contributing writer up through editor in chief, and I interned at the *Hartford Courant* for a summer. I know that journalism won't be as glamorous, but I'll hear interesting stories and get paid to write every day.

So the night after Christmas in 2003, I flip open my laptop and go to a job site for journalists that I browsed regularly when I was in college and envisioned a post-grad life as Lois Lane. I search for staff writer positions in New York. Forty-some jobs pop up, but each one requires more experience than I can eke out of my résumé, even with the cleverest phrasing. On a whim, I rerun the search, this time for openings in Oregon, Washington, Idaho and Montana, just because the Northwest caught my eye on a road trip one time. Voilà: features reporter in Idaho Falls, Idaho; sports reporter in Columbia Falls, Montana; news reporter in Bend, Oregon. Eleven jobs in all. As I read the descriptions, the hair on my arms perks up. Each job is at a small newspaper in a small town, the kind of modest post I might be able to land with my handful of bylines clipped from the *Courant*. The exact place doesn't matter to me at all. I love the idea of a new career in an exotic setting. I stretch out on the floor and start to tap out a catchy, one-size-fits-all cover letter:

Dear _______,

Don't let the address at the top of this letter fool you. I'm not just a city slicker looking for a Western adventure.

But the truth is, that's exactly who I am and what I'm looking for. As midnight gives way to early morning, and I polish up the letter, I also compile a mental list of reasons why moving to the rural West is not just an exciting idea, but also a smart one. I'll learn so much about myself by branching out and living on my own. I've always loved the idea of being outdoorsy; here's my chance. It sounds like a movie: spunky city gal becomes country muckraker. I already own two pairs of cowboy boots. A year or two at a small paper will provide the experience I'll need to get a better reporting job back in New York.

The next day, I walk to the post office and mail out eleven applications.

Seven weeks later, on Valentine's Day, my friend Larrison and I pack my belongings into a rental truck and head west toward Bend, Oregon, where I have accepted an offer to write news for the local daily, *The Bulletin*. Larrison has generously taken time off from her job in the writing department of the soap opera *As the World Turns* to drive with me to Bend before flying home. Somewhere in Wyoming, an honest-to-goodness tumbleweed bounces across I-80 and we squeal. The desiccated shrub looks as if it rolled right off a Western movie set. But this is real life. In the wild.

Our first stop in Oregon, just across the border from Idaho, is a gas station in a tiny farming town. I start filling the tank while Larrison heads into the store for a soda.

"What are you *doing*?" A stocky young man in a baseball cap stomps toward me.

I look down to make sure gas hasn't spilled over the side of the truck. "Uh... filling up?"

"You can't do that," he says. "Oregon is full-service only."

My heart drops. I can't pump my own gas here? Have I really left my job and all my friends and driven four days straight only to find myself in the New Jersey of the West? I suddenly realize how little I know about my new home. I wonder how much of a hassle it will be to move back to New York in a few months if coming here turns out to be a disaster. I'll have to find another apartment, not to mention face the embarrassment of telling all my friends and family that my Western Adventure was a bust.

"You must be from out of town, huh?" he asks.

"Yeah, New York City."

"New York City!" He drawls when he pronounces the name, like one of the dismayed cowboys in that old salsa commercial. "What are you doing here?"

"I'm moving to Bend."

He nods, as if this makes perfect sense. I've heard that Bend's population is booming, but now I wonder if there's a steady stream of New Yorkers driving U-Haul trucks into town.

Later that evening, Larrison and I pull into Bend, where we've booked a hotel room for the night. It's just after eight and the traffic lights are already switched off and blinking. The next morning, I check the classified ads and find a one-bedroom apartment on the first floor of a former boardinghouse downtown. It has polished wood floors, a graceful arch leading into the kitchen and built-in dressers in the bedroom and bathroom. Best of all, I have the place to myself. And for just $495 a month—$305 less than my share of the tiny Harlem apartment. Larrison helps me lug my bed, clothes and futon inside. We spend the next couple of days unpacking and taking breaks to browse secondhand stores and people-watch downtown.

At a sandwich shop, we wait in line behind a thirty-something couple wearing head-to-toe spandex. I can't help but stare past the corporate logos covering their garb to the sculpted curves of their calves and thighs. I look around the restaurant. Everyone here is thin, but not New York–smoker skinny. They're muscled.

"I've never seen so many good-looking people in one place," Larrison whispers.

"I know," I say. "Everyone is so fit."

"You're lucky." She arches an eyebrow.

"I'm intimidated."

The night before she flies back to New York, we mix Manhattans—an ode to my former home—and sip them from glasses perched on cardboard boxes.

As I hug Larrison good-bye at the airport the next morning, I know that I should feel nervous and sad to see her go. Here I am alone, in a town I barely know, three thousand miles from my family and friends, yet I'm too excited to care. Larrison's three-day stay has been like training wheels on my new life in Bend. I can't wait to get started.

The next week, I report to my new job. I was told during my interview a month earlier that the newspaper's circulation is only about thirty thousand. But *The Bulletin* is the only daily in the region, and more sophisticated than its size would suggest. It's housed in a brand-new building on the western edge of town that looks more like a modern ski chalet than a newspaper office. Windows stretch the two-story height of the lobby, with walls of stacked native stone and arched ceilings paneled in stained wood.

The sixty-five-or-so people who work in the newsroom are not locals who learned the business because it was an available job, but professional journalists—mostly city folk like me—who moved here for their careers. Editors came from the *Detroit Free Press*, *Minneapolis Star Tribune* and *St. Petersburg Times*. Reporters moved here from Denver, San Francisco and San Diego. Two news reporters even grew up in the same Maryland county that I did.

I've been hired to cover a rural area that stretches hundreds of square miles southeast of Bend. I spend the first day scoping it out from the driver's seat of a used Ford Ranger pickup truck that I found in the classified ads and purchased the day before.

Bend is almost the geographic center of Oregon. Sagebrush-studded high desert splays out to the east of the city. To the west, ponderosa pine forests creep up the volcanic Cascade Mountains. Unlike the Rockies—steep, tightly stacked peaks that form walls of granite stretching beyond the horizon—these mountains rise gradually, one at a time, like snowcapped sand castles. Together, the Cascades form a sort of sky fortress that traps clouds moving eastward off the Pacific Ocean and clutches them over rainy Portland and Eugene, freeing Bend's skies for a rumored three hundred days of sunshine a year.

In February, in the dead of central Oregon's long

winter, the landscape looks drab and dreary despite the sun. I drive past gnarled trees, scrubby shrubs and clumps of tall, native grass, dried and yellowed by the cold. Bare, reddish ground peeks between each of these plants. Unlike the wetter climates I'm used to, there is no fast-growing underbrush coating the soil here. A few dirty patches of snow cling to the shadiest spots. The sparse needles of the juniper and pine trees look dusty, more gray than green.

I get on Highway 97, and as soon as I cross Bend's southern boundary, the exit signs abruptly end, along with any other symbols of civilization. This is not like the East Coast highways I am used to, where one town peters out as another builds steam, with no discernible gap in between. Here, city ends. Country begins. I drive over a steep, craggy mound called Lava Butte. It erupted seven thousand years ago and covered nine square miles with black, porous rock. NASA actually trained astronauts for the moon landing on these desolate lava beds. As I travel south, the elevation rises, and a mat of snow blankets the ground.

I am winding down an unlined road toward a tiny regional airport when suddenly, in the middle of the asphalt in front of me, I see a gray, fluffy, dog-like animal. It's lying down but alert, with its head up, facing me. I brake and lean closer to the windshield for a better look. It's too big to be a fox. A wolf, maybe? I gasp at the possibility. As my truck rolls closer, the animal gets up and trots off the road, its tail floating perfectly straight behind it, parallel to the ground. It stares at me with pale, intense eyes as I drive past. Then I watch in my rearview mirror as it flops itself back down on the sun-soaked asphalt.

When I return to the newsroom, I run up to the environmental reporter and recount what I've seen.

"A wolf?" She laughs. "I doubt it. There are wolves in Idaho, and they may be starting to move into Oregon, but not this far west. It was probably a coyote."

Of course. A coyote, not a wolf. But I'm not disappointed. I'm in awe. A coyote is a real wild animal, infinitely more exciting than a tumbleweed. The Western Adventure has officially begun.

My life here is a distant cousin to the one I led in New York. There are no gallery openings to speak of, no theater debuts. I have to remind myself each morning to dress more casually than I'm accustomed to, so I don't stand out too much. My high-heeled shoes are getting ruined anyway, chewed up by the gravel that covers so many parking lots and paths.

Co-workers invite me to parties, where I quickly grow tired of Bend's unofficial winter greeting: "What did you ski today?" I've tried the sport a few times but don't consider myself a skier. Each time I'm offered this opening line, its speaker is so taken aback by my answer—I don't ski—that he immediately looks away and repeats the question to someone else. Someone more... Bend. I eavesdrop as other, fitter partygoers recap their mountain conquests, and I'm surprised by the level of detail in their answers.

There are two official ski areas here: Hoodoo and Mount Bachelor. People also cross-country ski on trails through nearby forests. They travel all over the West to ride different lifts for a weekend. And they backcountry ski, which involves hiking up a mountain and then skiing down it.

When a Bendite explains *what she skied today*, she doesn't simply name a location. Just as the Inuit supposedly have a hundred words for snow, so, it turns out, do ski bums and snowboard dudes in Bend. There's powder (dry, fine snow), breakable crust (an icy layer that skis sometimes fall through), dust on crust (a thin layer of fresh snow atop breakable crust), boilerplate (ridged ice), ball bearings (loose ice pellets), wind pack (crust formed by wind, not sun), mashed potatoes (wet, creamy snow), death cookies (hunks of ice hidden beneath smooth-looking snow), corn (hard, old snow that sunshine has softened into grains), Cascade cement (ultra-thick snow that grabs your skis), slush (even non-skiers know this one) and many more.

Of course, not everyone in Bend is my age. Young families are moving here, and retirees, too. Most come from California, to escape the traffic jams of Los Angeles or the skyrocketing housing prices of San Francisco. Outdoor recreation is what draws most of them—they ski, golf or mountain bike. Or they simply appreciate having so many sunny days in which to walk along the river and gaze at the mountains. Construction workers flock here, too, to help meet the growing demand for homes. I find myself in the minority not because of where I come from but because of what led me here: a job. In an office, no less.

Unlike all of these outdoorsy folks, I'm not sure what to do in my spare time.

I yearn for friends to discuss books with, lazy friends, friends who consider two o'clock a reasonable hour for brunch. Friends who want to unwind at the end of a long week with a movie and a bottle of wine, not a sixty-mile bike ride. Friends with loud political opinions. Desperate for an indoor activity, I enroll in a pottery class Tuesday evenings at the community college. Hunting is still farther from my mind than just about anything. But it is about to move a big step closer.

The morning after my twenty-fourth birthday, a Friday, I stop at a coffee shop on my way to work. You can't drive two blocks in Bend without passing a coffee shop. One intersection actually has drive-through espresso huts on three of its four corners. I don't usually drink the stuff but I've got a slight hangover from the mint juleps I downed with some co-workers last night, and I figure, what the hell, I'm trying to go local anyway. As I hand my money to the cashier, I hear my name.

"Lily! Hi!"

A tall, fit woman in her sixties waves and starts walking over from the other side of the store. She's in my pottery class, but I can't remember her name. I can't believe she knows mine.

"Hi! How are you?" I rack my brain for names. Barb? No, Barb has longer hair. Is it Ann? Or Annie?

"Oh, I'm so glad I ran into you," she says, as if we're old friends. "I need your phone number because I'm fixing you up with someone."

What? In New York, some co-workers offered to set me up on blind dates a few times, but unlike this woman they asked me first. I always declined. And how does she know I'm single, anyway?

"One of your business cards would be fine," she adds.

Still struggling for a response, I reach into my purse and pull out a card. The moment she snatches it, I realize that it's too late to say no to the setup. The card was my consent. I stare at it, in her hand, and I fumble for a polite way to ask for it back.

"Thanks. Well, his name is Scott, and he's just *so sweet*." She draws out those last two words as if she's describing a puppy. Not a good sign.

Shit. The Western Adventure has taken an awkward turn. I climb into my truck and pretty soon the whole incident slips my mind completely. Monday, I get to the office and check my voice mail.

"Hi, Lily, this is Scott. I work with Janet Windman."

Janet. I wasn't even close.

"Anyway, Janet gave me your phone number and told me that you're new to town, and so I figured, if you'd ever like to grab a cup of coffee or a beer or something, just give me a call."

The high-pressure date suddenly deflates into a casual chance for a new friendship. I write down his phone number. That night, we make plans to meet Thursday at Deschutes Brewery, a local pub.

After work on Thursday, I hop on my bicycle—still trying to go native—and pedal the five blocks to the brewery. As I crouch down to lock up my bike, I scan the front of the restaurant. A few groups of people in their thirties and forties stand in loose circles on the sidewalk, waiting for tables inside. Only one man in his twenties leans against the wall near the front door. He's wearing sunglasses, jeans and a red fleece vest over a white button-up shirt with the sleeves rolled up a few inches. He's about six feet tall, with thick brown hair and a slight suntan.

"Scott?" I try not to sound too hopeful.

"Lily?"

"Yeah. Hi, nice to meet you." We shake hands and walk inside, where Scott puts his name on the list for a table. We each order a pint and sit down near the bar.

"I liked your story in today's paper," he says, referring to an article I wrote about a local toad population that has made an unexpected rebound.

"Thanks. It was fun mucking around in ponds for a day. What did you do today?"

Scott exhales.

"Actually, I got audited."

We both laugh.

"Really?"

"Yeah, really. It's over, though."

"Well, cheers to that."

We clink glasses.

"Thanks."

It turns out Scott and Janet work for a small nonprofit that collaborates with local farmers and ranchers to restore the Deschutes River, the main branch of which runs through downtown Bend. Every summer, 97 percent of the river's flow gets diverted into canals for irrigation. Janet volunteers in the office. Scott runs programs: helping farmers switch to newer irrigation systems that use less water, piping canals to reduce the amount of water that leaks into the ground, and buying and leasing water rights to put back in-stream. The job gives Scott an interesting window into fish and wildlife populations, which I will appreciate later.

Scott grew up in the Willamette River Valley, which I immediately recognize from a computer game that I played in elementary school, The Oregon Trail. To win, you have to make it to Oregon's Willamette Valley, on the other side of the mountains from Bend. I am startled when Scott pronounces the name of the valley *will-AM-it.* It sounds harsher than the way I always pronounced it, *willa-MET.*

Both sides of Scott's family came to Oregon on the wagon trail. His parents were high school sweethearts who grew up in a small town not far from Bend and moved to Portland just after their wedding. They owned a small chain of clothing stores in the Portland area, but Scott spent his summer vacations and as many weekends as possible visiting his grandparents in eastern Oregon. This—the sunny high desert, with air that smells of juniper and sagebrush—is his home. He still spends his weekends exploring it, on skis in the winter and in waders with a fly rod in hand during the summer.

When our table is ready, we sit down and order burgers and more beer and keep talking. I tell him about my own childhood, in Takoma Park, Maryland, a city of seventeen thousand that hugs the northeastern edge of Washington, DC. Sometimes called the People's Republic of Takoma Park, or Berkeley East, its social hub is a Sunday farmers' market that straddles the DC-Maryland boundary, just two short blocks from the bungalow where I—like my brother, Nathan, before me and my sister, Gretchen, after me—was born. (Yes, we're the products of home births, midwives and all.)

Takoma Park is famous for its quirky residents, including Motor Cat, a tabby feline who wore a custom-made helmet and rode on his owner's motorcycle by digging his claws into a thick patch of Berber installed in front of the driver's seat. Years before it became hip to raise backyard chickens for eggs, a wild rooster appeared in Takoma Park. Residents named him Roscoe. He migrated between pocket parks and postage-stamp yards. Sometimes he even strutted down busy sidewalks. This went on for years before Roscoe was discovered early one morning, flattened, in the middle of Takoma Park's main drag. Angry mourners blamed the hit-and-run on a gas-guzzling SUV, probably driven by a Republican. A statue was erected in the bird's honor, and a pizza parlor adopted Roscoe's name.

But throughout my childhood, Takoma Park faded

into the background while I transported myself to Green Gables, Narnia or a secret garden. Books sustained me at least as much as food did. In fact, no week was complete without a few walks to our local library. My favorite books were Laura Ingalls Wilder's *Little House on the Prairie* series, which perhaps foretold my eventual journey west. At first I loved the details of life on the frontier: how they churned butter, cured meats and built a sod house. As I got older, I reveled in the emotional undercurrents, such as Ma's desire for a stable home pitted against Pa's wanderlust.

Scott listens to all of this and suggests a book that I might like as an adult: *Angle of Repose*, by Wallace Stegner, one of his favorite authors. As he summarizes it for me, I can't help but smile. Here we are, in Bend, discussing books.

It's nearing midnight, so we split the bill and walk outside. We unlock our bikes—Scott rode his to the brewery, too—and stroll with them toward my apartment. We arrive at my doorstep too soon, despite my efforts to walk at a fraction of my usual pace. We shake hands good-bye and make plans to meet up on Saturday, when, it turns out, we have both been invited to a barbecue at the home of one of my co-workers.

That night I toss and turn, unable to stop thinking about Scott long enough to give in to sleep. He seems both outdoorsy and indoorsy, funny and serious, smart and kind. When he spoke of his grandparents, his love for them was so real I could almost touch it. He made me laugh out loud when he admitted that he has accidentally joined several parades. (In one, he and his brother took a wrong turn and found themselves driving between floats in a gay pride parade. "What did you do?" I asked. "We smiled and waved.") Our blind date replays in my head on a continuous loop. I can't wait until Saturday. I want to learn everything about him.

Exhausted, I drag myself through work the next day. When I arrive home, a paperback book is leaning against my door. *Angle of Repose*. It is inscribed: TO LILY, HAPPY BIRTHDAY AND HAPPY READING! SCOTT.

I call to thank him but he's not home, so I leave a message. As soon as I hang up, I phone Larrison to recount the date and analyze the gift.

"I think he likes me," I conclude. "If he just wanted to be friends, he would have lent me his copy of the book, not bought me a new one."

On Saturday, Scott swings by my apartment in his twelve-year-old red Toyota pickup and drives me to the party. Afterward, we go to his house, where I meet his giant white dog, Bob. Bob barely swishes his tail at me before lunging toward his leash. He wants a walk, and we oblige. We walk down dark streets and through empty city parks. We walk over footbridges that span the black, swirling Deschutes River. We walk past lit windows framing families washing dishes and winding down for bed. We walk slowly, to let Bob sniff around and, mostly, to savor each other's questions, stories and jokes.

For the next week, I come home from work each evening and fix myself a quick dinner, then go to Scott's house. Together, we walk Bob all over town. We talk and laugh and listen. And then for some reason, when bedtime beckons and it's finally time to say good night, shyness overcomes us. Nine days pass by—countless hours spent talking about everything we can think of—before we work up the nerve for one kiss. Don't get me wrong, the kiss is slow and sexy and loaded with sweet promise. It's just not enough.

The next night, Scott has tickets to hear an author, David James Duncan, give a reading. I've never read any of his work, but Scott's a fan. As we walk to the Tower Theater, an old art deco building, Scott tells me about *The River Why*, Duncan's philosophical novel about fly-fishing. Instead of bait or lures, fly-fishermen try to attract their prey using pieces of fur and feathers tied to a fishhook to mimic a real, juicy bug. Many fly-fishermen, including Scott, catch a fish for the thrill of it, then let it go. Fishing is a passion of Scott's, and apparently many others in Bend share it, too, because the theater is packed when we arrive. We settle into our seats, and I make sure our arms are touching on the armrest.

Before one reading, Duncan explains that one of his students was recently bothered by the practice of catch-and-release. She told him that it amounts to taunting fish since it serves no practical purpose like, say, harvesting food. So Duncan responded with this humorous essay from the fish's perspective. It opens with a fish feeding on insects as usual until one particularly ferocious bug bites back, piercing the fish's lip. The fish panics as the vicious fly refuses to let go. Suddenly, a benevolent angler steps out of nowhere, finally offering relief from the evil bug. Then the angler, the hero, sends the fish on its way.

As Scott walks me home, our arms linked, I ask him to take me fly-fishing sometime. But by the time we get to my apartment, I've forgotten all about fishing.

"Want to come in?"

"Sure." Scott smiles.

I unlock the door and rack my brain for what to say next. Something witty. Something about how much I like him. Without a hint of desperation or overthinking. The door closes behind us. We look at each other.

I don't need a line. I need *him*.

We collide like two black holes. Lips. Arms. Tongues. Legs. Teeth. There's no time to build sensibly from delicate pecking. We've wasted so much time already, with our stupid talking and walking. We rush to uncover the physical facts that have been ignored during this otherwise thorough courtship.

Summer arrives, and the days get longer but never long enough. The nights are also too short. On weekends, we spend every minute together. Some days we sleep late and walk downtown for brunch at eleven—it's two o'clock in New York, I tell myself. When the snow is melted off Black Butte, per a local rule of thumb, we plant a garden in Scott's backyard. And then one weekend, he takes me fishing. Fly-fishing will turn out to be my gateway drug to hunting.

We drive north along the Deschutes River, following the water downstream for a little over an hour, until we get to the head of a desert canyon. We park and Scott pulls out two pairs of overall-style waders. The pants and bibs are made from the stuff of raincoats. The feet are neoprene, similar to the fabric used in wet suits. My waders are way too big, and I joke that I belong on a lobster boat. We hike into the canyon, each of us carrying a fishing rod so lightweight that it might as well be a dried reed.

In fly-fishing, Scott tells me, you try to give the fish every possible advantage without totally sinking things in its favor. This will be more complicated than the kind of fishing I tried a few times as a child: Spear a worm on a hook, dangle it in the water, wait. And wait.

Scott chooses a fly for me and ties it onto my line. On a trail overlooking the river, he helps me decide where to cast. A deep pool, perhaps, where fish seek refuge from the hot sun. Or a fast-moving riffle, where they gorge themselves on insects caught in the current. A submerged fallen tree makes a good hiding spot for fish, he tells me, but it can be a deadly trap for a fishing line.

To get my fly to this perfect piece of water, I will have to cast it. Picturing Brad Pitt in *A River Runs Through It*, I request a demonstration. Scott wades a few feet into the water and flings out yard after yard of graceful, looping line. It careers in front of him, then behind him, then in front again. With each toss, the flex of his fishing rod pulls out more line. When he has released enough line to reach a particular spot on the river, he lays it down in one smooth wave across the water, so the fly at the end lands without a splash.

"Now it's your turn," he says, reeling in his line.

"Uh, I'm not sure I can do that."

"You'll work your way up to it. Trust me."

I step into the water and feel the cold river swirl against my legs. It's a strange sensation, wading into a river without actually getting wet. Scott stands behind me, the length of his body pressed against mine, his arms wrapped around my arms. I feel his breath on my neck as he swings me through the motions of a perfect cast. I feel graceful, as if his fishing know-how is becoming mine through osmosis. It's easy. It's natural. I should have been fly-fishing all my life.

Then Scott steps away so I can try on my own. I let out just six feet or so of line, and flick my rod backward, then forward. The line plops in a depressing coil in front of my legs.

"That's good! Next time, try to keep your wrist straight."

When I finally do manage to launch my fly and keep my line straight, I puff with pride until I notice that the fly is so close to me, any fish who can see it can also see my legs.

I cast again and watch the fly drift past me downstream.

"Eat it!" I whisper.

"If you cast twice in one place and nothing bites," Scott pipes in, "take a step downstream before you cast again. There's no point in casting to the same spot over and over."

"But aren't they swimming back and forth down there?" I imagine that my fly lands like a neon billboard atop a fish superhighway. Sure, one picky commuter might swim by. But any second now, another will be tempted to stop and take a bite.

Scott stares at me. Apparently he has a different vision of underwater life.

"Uh, yeah," he starts slowly, "they're swimming. But not everywhere, all the time. The more water you cover, the more fish will see your fly."

On my next cast, the hook snags on a branch behind me. Scott hurries up to it and picks it off the shrub. Soon, satisfied that my casting is improving even though it still feels nothing but awkward to me, he wades downstream to start fishing himself.

"You're doing great," he says, his eyes locked on to a promising riffle. "Yell if you need me."

A few casts later, my hook grabs onto a thorny wild rose and won't let go. Scott is too far downstream, casting what looks like a quarter mile of line, to reach it for me. Branches scrape my arm as I reach for my fly. The line is not only hooked, it's actually wrapped around the tip of a branch.

After an hour or so, my right arm is tired, even though the rod I've borrowed from Scott weighs less than a pound. Something—I'm not sure what—keeps pulling back my arm for one more cast. Then one more after that.

"Fly-fishing is a sport for optimists," I remember Scott saying.

I haven't seen a single fish, but Scott has told me to trust that they're there. And if I ever stopped to think about it, I'd believe him. But I don't stop to think. I'm so caught up in my own actions—perfecting the motions of my arm, trying to let out a little more line, scanning the river for other promising spots—that I forget all about the fish.

As my fly drifts past me, I try to memorize Scott's pointers: Cast from the elbow and shoulder, not the wrist. Lift as much line off the water as you can before flinging the fly upstream. Give the line time to unfold in the air before jerking it in another direction. Keep your fly on the water as long as possible—you can't catch a fish if your fly is in the air.

I'm so wrapped up in all of this that I barely notice when the fluorescent bit of foam that is stuck to my fishing line pops below the surface of the water. I see it, and even feel a little tug on my rod, but I don't think anything of it. In fact, Scott notices before I do.

"Hey, you've got a fish!" He hurries upstream toward me.

Fish! I'd forgotten about fish. Adrenaline floods my veins.

"Keep your rod tip up," Scott says calmly. "And keep your line tight. When the fish pulls against you, let it take as much line as it wants. As soon as it stops, reel it in."

I crank the reel, slowly at first. The light tug at the end of my rod becomes a hard, twitchy pull. Something in front of me leaps into the air and flexes back and forth before it splashes back into the river. Stunned, it takes me a moment to realize that the thrashing crescent I just saw fly out of the water is a fish. My fish. A glistening, feeling, living being is on the other end of this thin thread of plastic. It's practically touching me.

"Reel it in!" Scott jolts me out of reflection. I crank again, dragging the fish closer with each rotation. Scott wades toward it. The fish is just a few yards away from him when it decides to make one more run for its life. This time, the reel spins so fast that I can't keep my left hand on it. It bruises my knuckle.

As soon as the reel slows, I crank again. Scott takes one more step and closes his hand around my line.

"Wow! Beautiful fish." He bends down and unhooks the fly underwater. He lifts the fish out of the river. It has gold speckles and shimmering green and red stripes down its sides. It *is* beautiful, though I've never before used that word to describe a fish.

"Do you want to hold it?" he asks.

For some reason, seeing it up close—this slippery alien cradled in Scott's hands—I chicken out. I'm not sure whether I'm scared for me or for the fish.

"No, that's okay." Immediately, disappointment sinks in. The fish gulps at the air. What am I so afraid of?

"Wait," I say. "I want to touch it."

"Okay, wet your hands first."

I dip one hand in the water and brush my dripping fingers along the fish's taut side. When I pull my hand away, a clear, slimy film coats my fingers. I rub them together, a souvenir.

"I should let him go," Scott says. "He's been out of the water long enough."

"Bye, fish."

He lowers it into the water and holds it there, its mouth facing upstream. It catches its breath for a second, then darts out of his hands and disappears faster than I knew any living thing could swim.

A week or two later, we return to the same stretch of river with a couple of Scott's friends and I am surprised by how well I recognize it. Not from the houses or man-made markers by which I usually orient myself; there aren't any of those here. Just a few months ago I would have described this place as barren, empty: a river and nothing else. But today I notice the lichen-covered rock that I tried to cast from. The giant alder where I lost two flies in a row. The stump where I sat and untangled my line after Scott moved downstream.

In late July, Scott flies to the East Coast with me for a long weekend, to meet my parents and some of my extended family. While we're there, we borrow my mom's station wagon and head toward the Beltway when Scott suddenly asks: "What river did we just drive over?"

I pause, wondering if he's joking.

"We drove over a river?"

"Yeah, what's it called?"

"I have no idea." I feel ridiculous. In all the years I

lived here, I never noticed that this bit of highway crosses a ravine. I never glanced at a map and learned its name. I never got to know a stretch of its banks as well as I already know one stretch near Bend.

Throughout the summer, we camp and fly-fish almost every weekend, and I start to pay more attention to rivers. When I drive over a bridge, I glance at the water and then look for a sign bearing its name. I check to see if the banks are raw and eroding or anchored by thick rows of willow. I look for a mix of fast-moving riffles and slow, deep pools, knowing that fish like variety. If we are on foot, I put on my polarized sunglasses and scan the water for dark, slender lines that hover, gently waving, over the streambed. These, I know now, are fish, though to the untrained eye they might look like weeds or shadows or like they don't exist at all. I pick out the fishiest-looking spots, where I would cast my fly if I had my rod and more time. Now a river catches my eye because I know how to decipher some of its secrets. Fly-fishing is teaching me how to imagine some of the life that takes place beneath the water's surface.

But more than two years will pass before I one day look up from the river, to the surrounding valley, and wonder if it, too, speaks in a language I don't understand. The place that I have, by dumb luck, chosen to make my home will soon embrace me and steer me toward an unlikely decision. I will wonder if perhaps hunting could teach me to read landscapes the same way that fly-fishing is teaching me to interpret rivers and streams.

CHAPTER 2

PULLING THE TRIGGER

It only takes a couple of weeks for me to realize that the town where I live and the rural area where I work are worlds apart. Bend, my new hometown, is a thriving example of the New West. Back in the 1960s and 1970s, more than 10 percent of the city's twelve thousand residents worked full-time at the town's two lumber mills. Others felled trees in the woods or transported logs. Today those mills have been replaced by a Gap, a Victoria's Secret and a movie theater—part of a shopping area called the Old Mill District.

Tourism and construction have replaced timber as Bend's economic anchors, and the once sleepy town has been wallpapered with an urban aesthetic. McMansions, groomed ski runs, mountain bike trails and Pilates studios beckon city-dwelling vacationers or retirees. Newcomers want more living space than they had in the city, but not so much space that maintenance becomes a burden. To meet this demand, owners of farms, ranches and forestlands eagerly subdivide their properties into ten-acre, two-acre, half-acre lots. The newspaper is filled each day with stories of rampant development, skyrocketing home values and a steady influx of new people. This is a good thing, economically speaking. New residents bring money and job opportunities. We also provoke a sort of identity crisis. Transplants like me, from bigger cities, are becoming the majority.

My beat has two population centers: Sunriver and La Pine, both unincorporated. The rest of the area is composed of vast ranches and federally owned forestland. Sunriver, just fifteen miles south of Bend, was a World War II military camp that is now a densely populated resort with paved bike paths linking fancy homes to golf courses, riding stables and restaurants.

La Pine, on the other hand, is a rambling, ill-defined cluster of neighborhoods that begins thirty miles south of Bend. Most of these neighborhoods were ranches until the 1950s and 1960s, when they were subdivided haphazardly. A typical homesite here is about one acre and includes a mobile home and three or four other structures. The owner is likely a former logger or mill worker, retired before he planned to because most logging in the region ended by the mid-1980s.

By my third or fourth week on the job, I am already in the habit of driving right past Sunriver to get to La Pine. La Pine, I figure, is where the real stories are. The poverty, ramshackle homes, nostalgia for bygone days of logging largesse: These are the makings of a blockbuster movie or a great novel or at the very least a readable newspaper article. Driving to La Pine each morning feels a bit like driving back in time to Bend in the 1980s. The community is still struggling to recover from the collapse of the timber industry.

Early one morning, I visit a married couple in their sixties for an article I'm writing about a neighborhood-wide property survey dispute. The man is a former mill worker who has suffered some sort of accident. He's missing one eye and an ear. Smooth, translucent skin has been grafted over the dent where his eye used to be—a no-fuss patch, I suppose. The woman wears glasses and has her short hair set in curls. She does most of the talking as they welcome me inside with handshakes and a mug of instant coffee. We sit around an old metal table in the kitchen.

As a child, the woman spent summers here visiting her aunt, who owned this very piece of land. She remembers when a water witch paced around the property holding a willow bough to sense where to drill the drinking well, which they still use today. She inherited this property as a young adult from her aunt, and the couple built their dream home here. As she says the words "dream home," she opens her hands proudly. I look around the small kitchen—a worn linoleum floor, cabinets with chipped paint, a hulking cast-iron stove that takes up one-third of the still-chilly room, a calico curtain covering the room's sole window—and marvel at the modesty of their dream.

As she speaks, her husband gets up and opens the door to the stove. Amber flames warm my face from across the kitchen. The man picks up a pail and begins shoveling its contents into the glowing cavity. An orange peel and a wrinkled piece of plastic wrap fall to the floor.

It's trash. The man in front of me is burning household waste to heat his home. I stop taking notes and watch as he bends down, picks up the peel and wrapper, and tosses them into the stove. He swings the door closed and then sits back down at the table. Soon smoke fills the kitchen. The man opens the window and sits again, but still the smoke thickens. The woman keeps talking. A few minutes go by and the man gets up to turn on an old wire fan.

"Maybe we should move into the living room," he says over his shoulder.

"Good idea," the woman says cheerfully.

Over the next few years, I will often think of this couple as I write my articles. I'll imagine that they read the newspaper every morning before feeding it, section by section, into their stove. I will try to anticipate their questions, to explain how a ballot measure or a new county ordinance could wriggle into their dream home and make the life inside it a little better or a little worse. I also imagine how they would look at my own life, at my closet stuffed with designer clothes, at my iPod and laptop computer, and scoff at all the excess.

★ ★ ★

That fall, though I continue to write about La Pine, I also take on a new beat at my newspaper, covering natural resources and the environment. I meet a particularly gruff, beefy logger in northeastern Oregon. We are sitting together on a bus, touring various logging projects and chatting about our lives, when he tells me, unprompted: "You know, the Western larch is my favorite tree." He has a favorite tree! And when he talks about it, he sounds more like a giddy child than a jaded woodsman. He loves the larch not for its value—the pine-like wood is soft but rot-resistant so its primary use is railroad ties—but for sentimental reasons. *Larix occidentalis*, sometimes called a tamarack, is that rare combination: a deciduous conifer. In the spring, it boasts feathery tufts of pale green needles that turn gold in autumn and then drop to the ground. The next spring, new needles appear.

I'm embarrassed to admit this, but I'd assumed that because he spends his days cutting down big, beautiful trees, he must not have as much affection for them as I do. Yet he has chosen a vocation that puts him in the woods all day, and requires him to distinguish species and determine the health and value of each specimen. He manages to be realistic about our need for lumber and, at the same time, romantic about the trees themselves.

I begin to question my views about environmentalism in general and logging in particular. By the 1980s, most of the big trees in central Oregon had been cut down.

Environmentalists filed lawsuits to slow the logging of remaining forests, arguing that logging displaces wildlife, erodes hillsides and pollutes streams. Old-timers blamed these conservationists for killing the local industry. When I first moved here, I had little sympathy. What kind of people would we be if we willingly traded the splendor of these forests and the life that inhabits them for toilet paper and two-by-fours?

I have come to see it differently. Mill companies in central Oregon have shut down local operations and moved to Lithuania and Bolivia, where environmental regulations are lax and labor is cheap. Meanwhile, communities like La Pine have paid the price. I think of the disfigured mill worker and his wife, who live on so little. More logging and milling here would not make them obscenely wealthy, but every little bit would help.

Besides, no matter how much we love trees, every single one of us needs some of them to be chopped down so we can go about our lives. Fallen trees make up our floors, walls and furniture, fuel our fireplaces and feed our printers. My newspaper career is dependent on paper. When some of our wood comes from local sources, we can carefully plan and monitor harvests with sustainability in mind. Otherwise, we're simply outsourcing our environmental damage.

As it turns out, loggers are not the only population with whom I'm surprised to find myself sympathizing.

★ ★ ★

Jim Court, the community's fire chief at the time, becomes the first hunter I meet in real life. During my first interview with him, when business is done, I ask about his family. Then what he likes to do for fun.

"Oh, I just love to get outdoors," he says in an aw-shucks sort of way. "You know, fishing, bow hunting..."

"Bow hunting?" I must have misheard him. "Like, hunting with a bow and arrow?"

"Yeah." He pauses. "What did you think I meant?"

I shrug. What I don't say is: People still *do* that?

Here we are, the fire chief and I, sitting in his office. And suddenly, much more than a desk is dividing us. Fear is part of it. Court is a hunter, which means he is capable of killing a living thing. Not to mention, he owns weapons. (The mightiness of the pen offers little reassurance next to a high-powered bow.)

There's another difference between us, too. And this one surprises me.

Court goes on to explain that bow hunters, unlike rifle hunters, must sneak within about thirty yards of their prey to have any chance of making a kill. This requires intimate knowledge of both the animal and the forest it lives in, not to mention a little luck. I will be grateful that my first conversation about hunting happens to be with a bow hunter. With guns removed from the equation, it's harder to fall back on the negative stereotypes that have, for my whole life, defined a modern hunter.

Growing up, I never knew any real-life hunters. And I never talked or heard much about them. So without even realizing it, I based my opinion on the anti-hunting propaganda that had bombarded me from a very early age. Disney's 1942 film *Bambi*, the first movie I ever saw in a theater, gave me the evil poacher who shoots Bambi's mom and the unethical hunters who ignite the forest to drive Bambi and other deer out of the woods. Looney Tunes offered me Elmer Fudd, who is foolish, inept and clearly no match for the wily Bugs Bunny. In more-serious depictions, such as the novel *Lord of the Flies*, civility implodes when the characters start hunting wild pigs. In each of these cases, the hunter and the prey are enemies. So it wasn't a stretch for me to assume that hunters, as a rule, hate their prey.

This portrayal of hunting couldn't have been more different from my own upbringing, in which animals were cherished companions. Like most modern-day Americans, my closest animal relationships involved pets. Throughout childhood, my sister, Gretchen, and I adopted fleets of gerbils and guinea pigs. We celebrated their birthdays. We bought them Christmas presents. And those were just the rodents. The center of our universe was Daisy, a tall, exotic supermodel of a mutt whom my brother found loping down the street when I was just five years old. We included her in our make-believe games. We shared our Popsicles with her.

During high school, I worked at a veterinary clinic and then a veterinary research laboratory. Yet outside of work and the home, I gave only superficial thought to the animals who contributed to my daily life. I tried not to buy cosmetics or toiletries tested on animals, though I approved of animals' use in medical research. I've long respected vegetarians because they seem to be animal lovers who are consistent in their beliefs. I've noticed, however, that when I don't eat meat for a few weeks, I start to crave a hamburger as a vampire does blood.

Court's description—of rolling in pine boughs to cover up his scent, then creeping through the woods wearing camouflage face paint—reminds me of stories I've heard about American Indian hunters blending into their surroundings, almost becoming more animal than human.

Despite my disapproval of hunting, I have always granted a special immunity to American Indian hunters. Perhaps that's because Indians are, as I learned in grade school, careful to use every part of the animal. Or because they didn't traditionally use guns. Or maybe I simply recognize the cruelty of condemning a lifestyle that has all but vanished at the hands of my own non-native ancestors. In any case, bow hunting sounds more like Indian hunting—the good kind—than the gun-fueled Disney kind I am used to.

After that conversation with the fire chief, I go on to meet more hunters nearly every day. Sure, some of them perpetuate the gun-crazed stereotype that I grew up with. But more often than not, I find myself talking to hunters who are surprisingly thoughtful about their prey. They know all of these tiny, fascinating details about the animals they pursue, and they manage to poke this knowledge into unrelated conversations.

When I act curious—which I am—they speak a little louder. Soon they forget to pause politely to let me respond or ask a question. Their eyes widen. They gesture wildly, absorbed in this world of non-human life. They remind me of enthusiastic volunteer docents at a zoo, not testosterone-fueled gun nuts. In fact, these hunters will do something that I never could have expected: teach me to rethink a sport that I once wrote off as the pastime of rednecks. Eventually, this will change the way I feel about my new home. It, along with falling in love with Scott, will compel me to stay beyond the one or two years I originally envisioned.

That fall, a handful of small businesses in La Pine lock their doors and hang CLOSED UNTIL NOV. 8 signs in their windows. The flooring store. The lunch truck that is painted like a cow and drives around selling sandwiches and espresso drinks. The house painters. Even a mechanic's shop.

"What's going on?" I ask a local know-it-all.

She looks at me as if I've just asked what year it is. "It's elk-hunting season."

"Oh. Right, of course." Though I've met many hunters, we haven't discussed details such as the time frame for hunting a particular species.

As I drive back to Bend, where hunting season is about as noticeable as National Lead Poisoning Prevention Week (also the last week in October, apparently), I think about all the hunters I've met on my beat.

I think about the twenty-something man who tracked a group of antelope in late summer so that on opening day, he not only knew where to find the herd, but also knew which individual animal he was going to kill and how to identify it. I think about the middle-aged elk hunter who tells me: "Most years, it doesn't even matter if I get an elk. I love the meat, and it's great to fill the freezer when I get the chance. But mostly, I just love to be outside." It surprises me to hear that the slaughter is not necessarily the object of hunting. These people are not heartless killers as much as amateur biologists, real-life experts in the natural environment—something that I care very much about.

I have always considered myself an environmentalist. But now that I stop and think about it, my résumé is pretty weak. I speak fondly of big trees, buy organic lettuce and chide SUV owners. Until now, that has felt like enough. Lately I've noticed some disturbing habits, though: I talk about the destruction of nature as if I play no part in it myself. The *timber companies* got greedy and cut down every tree. *Oil companies* are stabbing at the bottom of the ocean. *Gas companies* are scraping bare the Rocky Mountains. Meanwhile, I'm printing on paper, working and living in buildings made of wood, driving a car and heating my apartment with gas. It's me. I'm fueling all of this destruction, even if I've arranged my life so I don't have to see it or admit it. My hands may be clean and free of calluses, but there is blood on them just as there is on the hunter's or the logger's.

Perhaps the hunters I've met are, in a way, responsible

environmentalists like I aspire to be. They come face-to-face with what they take from the earth. They have a vested interest in making sure that wildlife populations are sustained in the long term. They have a better understanding than I do about how those populations fit into the balance of the world. I begin to grow jealous of the conservative hunters whom I interview and write about. They work on the land, while I merely play on it.

It's funny to find myself looking to hunters as environmentalist role models. These days, celebrity hunters such as Ted Nugent and Sarah Palin tend to be spokespeople for gun rights and right-wing politics, never for the environment. But it wasn't always this way. In fact, modern conservationism was first promoted by hunters, including our Hunter in Chief Theodore Roosevelt. In his biography *The Wilderness Warrior*, Douglas Brinkley points out that, in the 1870s, while the poet and naturalist Henry David Thoreau "contemplated nature preserves in the *Atlantic Monthly*," hunters' organizations such as the Adirondack Club and Bisby Club were taking real action, setting aside private wildlife preserves.

In 1887, more than a decade before taking over the Oval Office, Roosevelt co-founded the Boone and Crockett Club to create bison, elk and antelope preserves for future generations of hunters. Two years into his presidency, he declared Florida's Pelican Island the nation's first federally protected preserve and breeding ground for native birds. By the time he left office in 1909, Roosevelt had created or enlarged

150 national forests, fifty-one bird reservations, four game preserves, six national parks and eighteen national monuments. In total, he spared some 234 million acres of our country's most valuable land from development. Roosevelt believed in a goal of conservation that his friend and colleague Gifford Pinchot defined as "the greatest good to the greatest number for the longest time."

Roosevelt was condemned by some (including Mark Twain, one of his most vocal critics) for his frequent hunting excursions. Other Americans didn't bat an eye at photos of their president standing over a dead animal. Today, of course, hunting is nowhere near as widely accepted. Urbanization is a big reason why fewer Americans hunt or know someone who hunts. In 1950, two-thirds of Americans lived in cities or suburbs. Now more than four out of five Americans are packed into 366 metropolitan areas.

Our image of environmentalism began to shift in the 1960s, when Rachel Carson's *Silent Spring* warned of the dangers of industrial pollution and Paul Ehrlich's *The Population Bomb* foretold the environmental devastation caused by rampant population growth. Humans—including hunters—were pegged as a growing threat to the natural world. "Nature seems safest," writes historian Richard White, "when shielded from human labor."

Popular culture has also changed how we think about hunting. I learned only as an adult that the film *Bambi* is loosely based on a 1923 Austrian novel, *Bambi: A Life in the Woods*, by Felix Salten. Like the movie, the book tells the story of a forest from the point of view of an anthropomorphized deer. But the book is dark, with death as a central theme. Humans shoot birds and deer, including Bambi's mother. Bambi must learn to avoid snare traps. A fox kills a pheasant in broad daylight as Bambi and others look on. A crow tears apart a young hare and it dies slowly, moaning. A ferret wounds a squirrel, who bleeds to death before magpies land and feast on him. In this *Bambi*, life is beautiful but also harsh and full of danger.

The Disney film, however, portrays a utopia of frolic and fun. The animals never kill or procreate or even poop. They have just one enemy: man. Though humans never appear on screen, they violate hunting ethics and, in all likelihood, state law. In the film's most famous scene, Bambi's mother is shot during spring, with young Bambi still by her side. Later, humans use dogs to track deer and ignite a forest fire to chase the animals out of the woods. In other words, these people aren't hunters, they're poachers. There are poachers in real life, too, but they don't represent hunting any more than counterfeiters represent artists. Yet as anthropologist Matt Cartmill writes, the name "Bambi" has become "virtually synonymous with 'deer,'" and has shaped many Americans' views of hunting.

I've been living in Bend for a year when I take a week off work and meet my family in Recife, Brazil, where my brother now lives and teaches English. Nathan moved from Washington, DC, to this beachfront metropolis in Brazil's Afro-Caribbean northeast just a few months before I moved to Bend. The strangeness of his new home puts my life in Bend into perspective. In just a little over one year, he has become fluent in Portuguese, adopted the local diet of grilled meats, bean stews and exotic fruits, and embraced the natives' casual attitude about time (goodbye, wristwatch!). Meanwhile, in Oregon, all I've really had to do is learn to live without Vietnamese food and make small talk with cowboys and loggers.

Seeing Nathan so engaged in his new setting makes me notice something else. I've always considered myself to be a good, adventurous traveler. I approach a mysterious place with my senses wide open, eager to inhale its smells, tastes and history. But my curiosity fades as soon as I go home. Or, at least, it always used to. I think about how much I took for granted in Takoma Park, including rivers that I never knew existed. I think about the efforts I've made, since moving to Bend, to assimilate—drinking lattes, riding a bike, learning to ski. All of that has made me more like the people who migrated here in the past decade, not the ones who have lived here their whole lives. In La Pine, which has become my reference point for the Old West, I still have very little in common with the people I meet.

I frequently hear old-timers complain about Bend's ongoing suburbanization. "Why do people come here if all they want to do is turn it into the place they left?" one asks me. At first, I'm defensive. I'm not trying to change central Oregon into something it's not. Am I?

I wonder how I would feel if my own hometown,

Takoma Park, were suddenly overrun with people who wanted to shut down the hippie food co-op, cancel the raucous Fourth of July parade or otherwise disrupt the local traditions that I grew up with.

Bend is attracting an athletic crowd that favors sports like running, yoga, cycling, golf and, of course, skiing—things that can be done in a well-groomed park or even indoors. Hunting and fishing, on the other hand, require a living, breathing ecosystem. They're messier, more complicated activities. Some depth of knowledge is required to excel—knowledge about a specific place and an animal's life cycle or, at the very least, its preference for food or cover. In a sense, hunting is the embodiment of rural America. Not the romanticized version that appeals to city slickers, necessarily, but the grisly, utilitarian truth of it.

In central Oregon, public opinion about hunting has soured. We newcomers have fueled thousands of acres of development and passed anti-shooting ordinances in surrounding areas, too. It's simple geometry that as neighborhoods and trails expand, there is less room in the woods for hunting. So perhaps it's no surprise that nationwide, hunting has been on a steady decline. The age of the average hunter is rising, too, as older hunters retire to less-strenuous activities such as hiking and bird-watching, and fewer of their children take up the sport. Surveys have found that just one in four children raised by hunting parents will learn to hunt.

When I return to Bend from Brazil, Scott picks me up at the airport. As we drive home, I scan the now familiar scenery—dusty juniper trees and slack barbed-wire fences corralling swaybacked quarter horses. I close my eyes and make a promise to myself: I will not take my new home for granted. I will scour the high desert like a hungry tourist. Or better yet, like a true native.

The following summer, I sit down to dinner one evening and cut into a chicken breast that Scott just grilled. I think back to a conversation that I, a lifelong meat eater, had with my aunt Nina, more than a decade ago.

Nina has been a vegetarian for as long as I can remember. I was a teenager when I finally asked her why she didn't eat meat. She thought about it for a moment.

"If it ever came down to it," she said, "I don't think I could kill an animal. And so it seems hypocritical for me to pay someone else to do it."

Though I'd never thought about it before, buying a package of chicken thighs at the grocery store suddenly seemed a bit like hiring a hit man: not only vicious but cowardly, too.

Hunting—this theme that keeps showing up in my life, this intersection of culture and politics and history—offers a chance, as Michael Pollan writes, "to prepare and eat a meal in full consciousness of what was involved." I spear a chunk of meat with my fork and stare at it. A question bursts into my head like a firework: *Do I have what it takes to find and kill my own dinner?*

This same reasoning that led my aunt to a life of

vegetarianism is what convinces me to pick up a gun and hunt my own meal. After more than twenty-six years of eating meat, I want to know if I have what it takes to cut out the hit man and kill my own dinner.

When I first moved here, my plan was to get some experience under my belt, have some outdoors adventures and then move back to New York. But now I have fallen in love with a man, and perhaps I'm being seduced by a place, as well. My once flippant goal of "becoming outdoorsy" has morphed into deeper questioning of what kind of relationship I want with nature. My glimpses of La Pine have convinced me that rural life—perhaps even its uncomfortable traditions such as hunting—is worth protecting. But I can't help protect it if I don't first understand it.

Here in Bend, where the dominant culture more closely reflects my own, it's easy to marginalize the old ways. An element that is being dismissed is hunting for one's own food and the vast knowledge necessary to do it. We risk losing something essential if we deign to define the New West without first understanding its old cultures and traditions. Besides, my liberal heart bleeds for any species on the decline, even if that species happens to wear camouflage and carry guns. It's time for me to stop talking about giving back to the earth and start understanding how much I really take from it.

I decide to do the unthinkable: I will learn to hunt.

I break the news to Scott. At first, he's confused.

"If you want to understand where your food comes

from, why not get a job at a feedlot?" he asks. "Or a slaughterhouse?"

But it's not just about the food. Hunting could teach me to read landscapes the way fly-fishing has taught me to read rivers. It could connect me to the rural culture that has shaped my new home. Hunting could also help me delve into the disjointed motifs of urbanization and environmentalism that have haunted my work and my thoughts for more than two years now.

"I'll help however I can," Scott says. "But I don't want to hunt."

He tells me about shooting and skinning a rock chuck (local slang for a marmot) when he was a kid. At the time, he says, he felt like a pioneer, and killing the animal was an adventure. Now he feels guilty about it.

The way I imagine it, learning to hunt will be a similar adventure. But I wonder if I, too, will kill an animal and then regret it for decades.

CHAPTER 3

GUN-SHY

As a reporter, research is my bread and butter, so I start by going straight to a source: Tony, one of the elk hunters I've met on my beat. Tony and his wife own a truck, painted black and white to look like a cow, which they drive around La Pine, selling sandwiches and coffee.

"What would you do if you were in my position?" I ask him. "You're an adult and you've never hunted in your life but you want to try. How do you get started?"

He rubs his beard and purses his lips for a minute.

"I have no idea. Just get out there and do it."

It may be true that the only way to learn to hunt is by hunting. But I don't know where to go, what kind of gun to carry or how to shoot it. I visit a couple of local bookstores but strike out, because they don't carry anything related to hunting. (It's Bend, after all; there's not much shelf space between the Pilates and hiking books.) Next, I visit the library, where the relevant books are far too specific (say, *Mule Deer Hunting in Eastern Oregon*) and assume a base of knowledge that I don't yet have. On Google, I slam into the same problem: All of the sites about hunting are for people who already hunt. And then it dawns on me: Hunting is a twenty-first-century rarity—something you can't learn online or in a book. There's no *Hunting for Dummies*. There are no intro classes at the local community college.

Most hunters would have you think that theirs is the sport of the everyman. But I'm finding it to be oddly exclusive. Hunting isn't so much a hobby as an inheritance, passed from one generation to the next. You have to learn from someone, and that someone is usually your dad. But where does that leave me—an adult whose parents are openly disgusted by the idea of killing an animal in the wild? Maybe this decision to hunt was a stupid one. Maybe you're born a hunter or you're not. I'm not.

Then one afternoon, Scott invites me to have lunch with him and one of his co-workers, Andy Fischer, a former professional bike racer with red hair and a wiry build. Andy grew up hunting and fishing in Montana. He and his younger brother, Kit, observed this household rule: If you shoot it, you eat it. Once, their parents made them choke down a couple of squirrels the boys had shot for fun. The meat didn't taste too bad, Andy said, and so later, they tricked their parents' friends into eating teriyaki-grilled squirrel during a dinner party. The more I hear about Andy's upbringing, the more it sounds wild, adventurous and, dare I say, idyllic.

At first, Andy doesn't have any ideas for how I could get started hunting, either. But when I press him for details of his own hunting education, he says something that catches my attention:

"I was probably in sixth or seventh grade when a lot of my friends started taking Hunter Safety, so I took it, too."

Hunter Safety. It turns out that every state in the Union offers a similar course, usually with a long proper name but known informally as Hunter Education or Hunter Safety. Most are geared toward children because, at least in Oregon, all minors must pass a Hunter Safety course before purchasing a hunting license. Adults don't need any such qualification.

The next day, I call the Oregon Department of Fish and Wildlife to sign up.

The woman who answers says that although a class for adults is offered every year, I've missed the 2006 class by a few weeks. It's late summer; I can wait eleven months to take an adults-only class or enroll in an all-ages class starting next week in Culver, a small farming town about forty-five minutes away. Culver has more in common with La Pine than Bend, but it's a closer-knit community than either of the areas where I spend my days. This class meets two evenings a week.

Eager to get started, I give the woman my name and

phone number and promise to bring five dollars—tuition for the entire four-week course—to the first class. Then I admit that I feel silly enrolling in a class for children.

"Don't worry," she says. "Adults take this class all the time."

The following week, I walk into the Culver City Hall, which doubles as a fire station, and am greeted by a gaggle of about twenty kids much younger than I was expecting. A few grown men are scattered around the room, too. Two of them are obviously instructors, so I home in on a third, a man who looks about ten years older than me. I slide into a plastic chair beside him. He smiles at me, and I smile back and relax a little, taking his friendliness as a sign that he's a student, too.

One of the instructors sidles between rows of chairs, handing out forms. When he gets to the man next to me, the man shakes his head.

"I'm just here with my son." He puts his arm around a small boy—maybe ten or eleven—sitting on his other side. I look around the room and sure enough, I'm the only adult holding a form.

Our instructors introduce themselves. The first is E.V., Culver's one-man public works department. He pronounces his initials like the girl's name Evie. He stands tall, a stern, no-nonsense military veteran with a trim gray beard and glasses. The second instructor is Jack, a redhead whose seemingly endless patience with children is matched only by his seemingly endless girth, and who also happens to be the county sheriff.

I quickly glean that the primary purpose of Hunter Safety is to instill a healthy dose of fear in these soon-to-be-gun-toting tots. But for me—an adult who arrived already afraid of guns—the fear is paralyzing. We begin by watching a scripted movie called *The Last Shot*, in which a thirteen-year-old boy accidentally shoots his best friend. At the end, Jack offers one more warning, in case we somehow missed the seriousness of the film.

"That boy?" he says, pointing to the screen where the final frame shows police officers approaching the shooter in the hospital, just after his friend is declared dead. "He is never going to be allowed to hunt again. Think about that."

The room is silent, but I'm pretty sure no one is focused on the kid's lost hunting career. That boy is never going to be able to sleep again. Or look at himself in the mirror. Or enjoy one ice-cream cone without guilt beating his insides to a pulp. For the next few days, I question my decision to hunt based on the human safety risks alone. It all comes down to the guns. What could possibly be worth the danger of carrying a loaded rifle through the woods?

E.V. starts the second class with an order: "If you are twelve years of age or older, raise your hand."

I raise my hand along with about half of the class. The Under-Twelves are sent with E.V. to practice safe gun handling, while we wizened Twelve-and-Overs sit with the sheriff to talk through a chapter of our textbook.

The book, which is more like a thick pamphlet, includes sections about animal identification, wilderness survival and basic gun information. Today's lesson is about guns—different types, gauges and calibers and their various parts. Earlier this year, I happened to learn the difference between a shotgun and a rifle during the non-stop news coverage of Dick Cheney's hunting accident. The vice president was hunting for quail when he accidentally blasted his friend in the face. (Yet he managed to avoid the horrific consequences we were warned of in the video.)

A piece of shotgun ammunition is called a shell, and is basically a plastic tube filled with metal pellets. Because a shotgun shoots a spray of metal, it doesn't need to be aimed as carefully as a rifle or handgun, which usually shoot one bullet at a time. This makes a shotgun useful for fast-moving, airborne targets such as birds. A single pellet, as long as it's heavy enough and moving fast enough, will kill a bird by intersecting its neck. A gun's gauge inversely corresponds with the diameter of its barrel, so the larger the number, the smaller the gun. For example, Cheney was carrying a diminutive 28-gauge shotgun during his quail-hunting accident. But if he had instead used a burly 10-gauge (typically reserved for downing geese), his friend probably would not have survived.

It makes sense, then, that hunters tend to own so many guns. There is no one-size-fits-all. You start with a rifle to shoot deer or antelope. Then you decide to buy a bigger rifle to shoot elk or moose. You have a 12-gauge shotgun for rabbits and turkeys. And you don't want to pulverize small game like quail, so you probably need a smaller shotgun for those.

In class, we will be handling rifles only. According to Jack there are many parts to a rifle, but to me, there are only four that you really need to know:

1. The trigger: what you yank to shoot. Avoid this.
2. The muzzle: the hole through which a bullet exits the gun. Keep away from this *at all times.*
3. The action: a movable piece of metal that holds the bullet in place in the chamber so that the gun will fire properly. When this part is open, the gun should not be able to shoot. People who know guns can see that the action is open and feel reassured that you're not about to open fire.
4. The safety: a switch or lever that should be left on—to prevent the gun from firing—until your target is in sight and you're ready to shoot. Jack frequently reminds us that, like any mechanical part, a safety can fail at any time.

Jack never misses an opportunity to wedge a safety warning into the lessons. In the video we watched on the first day, for example, the boys made the repeated mistake of dismissing their .22-caliber rifle as *just a .22.* Hunters call these small rifles "varmint guns" because they're not powerful enough to take down an animal much bigger than a squirrel or rabbit. However, Jack reminds us again and again that no gun is too small to be taken seriously.

"A .22-caliber rifle is never *just a .22,*" he tells us. "Don't ever call it that. It can kill you."

At one point, Jack mentions that in all his years of teaching the class, only two people have ever failed Hunter Safety. Both failed, he tells us, because they could not demonstrate a level of maturity necessary to hunt safely. Ten years older than my oldest classmate and fourteen years older than the average one, I quietly hope that we'll be graded on a curve.

Soon, however, my classmates' youth will begin to intimidate me. People like to say that you should never stop learning. The truth is, it's hard to learn something new—really new—as an adult. As we age, our obligations pile up and free time becomes scarce. But that's only part of the problem. We also face a relatively short list of acceptable activities to try for the first time. Tell your friends you're learning to speak French and they'll likely congratulate your gumption. Mention your beginner gymnastics lessons and they'll probably laugh. Even activities that are considered lifelong sports, like tennis, are things that people usually learn as kids and then hone as adults. In Hunter Safety, it's easy to see why. As I've aged, I've lost my willingness to be bad at something and stick with it anyway. Kids have no choice in the matter—they're pros at being novices. When the teacher mocks or scolds them, they shrug it off. They're kids—being publicly humiliated is practically their job. For me, it's . . . humiliating.

Before we move to E.V.'s portion of the class, we take a short break and I head to the ladies' room, along with one of two other females in the class. I'm in the stall when I hear her voice from the next one.

"So," she starts quietly, "how old *are* you?"

"Twenty-six."

"Oh my God." She practically gasps, I'm so much older than she'd imagined. Then, after a pause, she adds: "I'm thirteen."

We walk out of our respective stalls and introduce ourselves. Her name is Jade. She wears thick black mascara and clear lip gloss. As we leave the bathroom, she asks, "So, what are you doing here?" I search for a concise explanation of my heady ideas about environmentalism, food and my relationship with nature. Leave it to a thirteen-year-old to tongue-tie me.

"You know," I say, sighing, "I'm not really sure."

We sit down in front of E.V., whom we all know is about to hand out the guns. The room is silent. E.V. wastes no time; he turns and addresses the elephant in the room by name.

"Who here is afraid of guns? Raise your hand."

I fling up my arm before looking around to notice that everyone else is comfortable around weapons or at least has the good sense not to openly admit their fear. My admission bothers E.V., and he will not let up on me until the course is almost over. Each evening, he will ask if I am still afraid of guns. Each evening I will make the mistake of answering truthfully.

"But you drive a car," he will plead on the fourth or fifth class, obviously irked. "Do you get scared every time you get into your car?"

"No."

"Yet cars kill way more people every year than guns do."

"But I'm used to cars." My whole life, since before I could walk or speak, I've tacitly accepted the risk inherent in cars. Not in guns. Guns are a big, sudden leap.

Eventually, E.V. accepts that my fear will take more than a few night classes to overcome. He works on my vocabulary, instead.

"The next time someone asks you if you're afraid of guns," he says, "tell them, 'I respect guns.' Because there's nothing wrong with that; you should respect guns."

I am 99 percent sure that the next time someone asks me if I'm afraid of guns, it will be E.V.

"Okay," I say, "I will."

Tonight, E.V. marches over to a wardrobe to hand out the guns: .22-caliber rifles with bulky padlocks attached, so that they can't actually be loaded. According to E.V., the padlock keys were lost, sentencing these guns to a lifetime of nervous handling by Hunter Safety students. The locks do nothing to allay my fears, however. I've been reminded too many times today that guns misfire, that accidents happen no matter what. Perhaps my gun will be the one with an ancient piece of shrapnel wedged inside, waiting to be freed by my careless brush against the trigger.

When I get to the front of the line, E.V. holds a long metal rifle out to me. My palms are sweating so much that I worry it might slip through my hands.

"Action open, safety on," he says, glancing first to a hole where—God forbid—the ammunition would go, then to a small switch. He looks me in the eye. "Got it?"

I grasp the gun with both hands.

"Got it."

He lets go of the gun. I hold it away from my body, like a stick of lit dynamite, and walk slowly to join the row of already armed students.

We practice picking the guns up and setting them down to climb over an imaginary fence. We practice handing them to one another across an imaginary stream. We practice carrying them as we walk alongside other hunters. I am nervous the whole time. I only relax when I hand my gun back to E.V.—*Got it? Got it.*—and he stows it back in the cabinet.

Other students don't admit it, but many of them are afraid of guns, too. Much more than age, fear is what divides our class. I recognize my fellow scaredy-cats by the way they nervously chew their lips while their rifles lean against their shoulders. They turn the guns a little too often, searching for a comfortable position but never quite finding one. Some of the kids are younger versions of me, with no significant firsthand experience with guns. Of course, they didn't grow up near Washington, DC, like I did, during the city's reign as murder capital of the world. Police officers visited my elementary school and said things like, "If you see a gun lying on the ground, do not touch it. Back away and call the police." Whenever a gun appeared in the newspaper, it was usually illegal and it was, without exception, bad news. Mostly the news was grisly and tragic. In rare, lucky cases it was a reminder of all that could have gone wrong.

To us in the fearful half, this is what we know: Gun = death. QED. Even if the gun is shooting innocuous holes in paper, those shots are preparation for death. Even if the gun is shooting a rodent—even a small, pesky rodent that threatens something we value such as crops or songbirds—the finality of its death is sobering and terrifying. Our teachers hammer that into our heads by saying things like, "Once you pull the trigger, you can never take that shot back."

They repeat, again and again, how much is on the line when you're carrying a loaded gun. "You can't afford to make a mistake," E.V. tells us. This is what scares me the most about hunting: I can't think of a single thing I've done *without* making a mistake. In a way, I lied to E.V. when I told him that I'm not afraid of cars. I drive almost every day, but when I stop and think about it, I'm still scared of driving. I've had my share of fender-benders. I've had close calls. The reason I've never hurt anyone behind the wheel, or been hurt myself, is probably as thin and flimsy as *I've been lucky*. I might not be so lucky with guns.

But then there's the other half of the class, the fearless half. To them, guns are not confined to our sad, simple equation. Guns are a means to death, of course. But they are also a means to adventure, to food, to family bonding. This half includes kids like Grayson, a skinny fourth grader with dimples, a crew cut and a wide grin, who is barely as tall as my waist. Grayson clearly knows guns and is only here to meet the state requirement for buying a hunting license. When he picks up a rifle, checks to make sure it's unloaded and hands it to a classmate, for example, he automatically points the muzzle in a safe direction. There is no moment of shock as he realizes, woops, the muzzle is aimed straight at his buddy, then fumbles guiltily to readjust it. Safety isn't a panicked afterthought but a steady instinct. His face and body remain relaxed. To Grayson, a gun is an object or a tool. When he thinks of a gun, he thinks of hunting with his dad, and all the richness that comes with it. He knows how to take his own gun apart and clean it. He even knows, he tells me modestly, how to skin and gut the animals he has shot—a few birds, some rodents.

If you think guns are powerful, you should see one in the hands of a competent child. Watching Grayson handle a gun for thirty seconds makes me think that even if I quit my job and moved into the woods and dedicated my life to it, I would never truly be a hunter. Not like Grayson is. Sometimes I look around the room, filled with people big and small, young and old (parents and grandparents of my classmates occasionally show up to observe), and think about whom I would trust as my hunting partner. Together, we would head out into the forest with loaded guns in our arms and feel safe, trusting each other. Most of the time, I'd pick Grayson. Sure, our instructor knows a lot about first aid and wilderness survival. And one of the kids has a dad who's a cop. But Grayson is quiet and pleasant and curious and looks, at least to me, like he handles a gun as safely as anyone.

In fact, seeing that rifle in Grayson's hands is, oddly

enough, reassuring. He is proof that children can be raised to handle guns safely and with respect. Of course, plenty of households do contain both guns and children and manage to avert disaster. In 2009, for example, 138 children, ages eighteen and younger, were killed by accidental gunshots in the United States. This figure—any number over zero, really—is undeniably tragic. But consider that more than seven times as many children—1,056—accidentally drowned that year. Or that a whopping 6,683 children died in motor vehicle accidents.

The rate of gun-related accidents seems startlingly low given the astonishing number of guns that we, as a nation, own—roughly 250 million. Our response to the risk of drowning, for example, is not to discourage children from being around water, but rather to educate them, to teach them to swim and be safe. Our preferred response to the risk of guns is the exact opposite—total avoidance.

Accidents represent just a fraction of gun-related deaths—about 2 percent in 2007. Nearly three out of five gun-caused deaths are suicides. This is all to say: Chances are slim that you or I will die by gunshot. The lifetime odds that you will be murdered by a firearm are 1 in 306. Your chances of dying by accidental gunshot are 1 in 6,309. Your odds of death by cancer, on the other hand, are 1 in 7. Heart disease, 1 in 6. The odds that you will die of any cause? One in one.

Why we own all of these guns is another story altogether, and one that has little to do with hunting. Forty percent of U.S. households contain at least one gun. One in four adults owns one or more guns. Yet only 11 percent of firearm-owning households say they hunt. The rest keep their weapons primarily for self-defense.

The idea of keeping a gun for self-defense sounds crazy to me. And the more I handle a gun, the crazier this idea gets. The only way for a gun to be useful in, say, defending one's family from a burglary would require that the gun be stored unlocked and with ammunition either inside the chamber or close by—a dangerous situation in any household, with or without kids.

Despite my fear of guns, I enjoy the class. More than anything, I enjoy my classmates. At first, they're amused to have an adult in their ranks. But eventually, they treat me like I'm one of them. One evening, a boy comes in all puffed up because he learned to whistle earlier that day. He demonstrates for me—sucking in one long, sustained note until finally he has to stop to exhale. I congratulate him.

"Can you whistle?" He's not issuing a challenge, he sounds genuinely curious.

"Yes." I toot out a simple tune.

His eyes widen. "Wow, you're really good."

"Thank you. I've had a lot of time to practice."

Later, when E.V. scolds me for picking up a rifle without checking to make sure it's unloaded first, my new buddy reassures me.

"At least you're good at whistling," he says cheerfully.

One of the students lives on a remote farm with his large family. He is homeschooled, which makes our crowded classroom exotic to him. He tells me that his family raises pygmy fainting goats, which freeze and topple over when they get scared.

"Really?" I can't tell if this kid is pulling my leg.

"Yep. You go like this"—he smacks his hands together—"and they just fall right over."

I rush home from class, excited to tell Scott about my latest friend from Hunter Safety.

On the last two days of class, we meet at an indoor shooting range. Technically, this will not be the first time I have shot a gun. In high school, my friend Nick and I went with his father to a shooting range and gun store next to a prison in Jessup, Maryland. Nick's dad rented the gun for us—a 9mm or maybe *just* a .22, I can't remember. Either way, it was a semi-automatic pistol. The store clerk taught us how to shoot. First, he showed us how to snap the magazine into the bottom of the gun. Then he explained what he called "the push-pull technique": Hold the gun with two hands, fingers interlaced. Push against the gun with the heel of one palm, pull toward your chest with the other. Apparently this steadies the gun. He reminded us to stand with our feet shoulder-width apart and our arms straight in front of us.

We took turns. Nick walked into the range and shot at a paper target with a silhouette of a human bust until the magazine was empty. Then he came out and handed me the gun, and I shot through the magazine. I don't remember much about the whole experience except that I'd anticipated it would be a thrill. I'd seen *Pulp Fiction*; I knew how much swagger came with shooting a gun. But in the moment, I was terrified. Even just holding the gun, unloaded, made me sweat.

I am even more nervous now. Today feels more momentous than my onetime target practice in high school. Back then, I was casually trying out a new experience. This time, I am preparing for something in which I have already invested a great deal of time and energy.

Before the rifles are handed out, E.V. gives a stern recap of gun safety principles. Then the kids and I line up to receive our guns, and it occurs to me that if one of them is an undiagnosed psychotic, he could turn his back to the paper targets and open fire on all of us right here, right now. I shudder. But I remind myself again that this situation—blindly trusting the people around me to do the right thing and keep us all safe—is not remarkable. We all do it every day. When I lived in New York City, for example, a fellow subway passenger could have shoved me into an oncoming train, on purpose or by accident, at any moment. But for the most part I choose to believe, as I always have, that it won't happen today.

Even so, as an adult hands out rifles to a group of eleven- and twelve-year-olds, this social contract into which I've placed my well-being is looking awfully fragile. Then I look at Grayson, who is somber and calm. I take a deep breath. These kids have shown me, for four weeks now, that they take guns more seriously than a lot of adults I've read about in the news. I heave away my doubts. It's time to shoot.

I pull on a pair of clear plastic glasses, to protect my

eyes in case the gun backfires, and what look like giant headphones, to protect my ears from the noise. The voices around me are muffled, and my heartbeat becomes the dominant sound. I lay my gun on a small desk and walk deep into the shooting range to pin up my paper target, a round bull's-eye. My heart thumps ominously as I walk back to my gun. Da-dum. Da-dum. Da-dum. E.V. walks up and down the line of desks, helping us drape ourselves from the chair onto the desk and gun. The idea is to keep the rifle perfectly still by anchoring as much of my quivering body as possible—feet, butt, elbows, forearms—to the ground, the chair, the desk. I press my right cheek into the cold metal gun, peering down its black neck to line up the metal sights with my paper target. On the desk next to me is a wooden block with five little copper bullets poking out.

"When you are ready, go ahead and shoot all of your rounds," E.V. barks to the class.

Immediately, guns start blasting. The booms are muffled by my headgear, and sound more like a heavy book dropping on the ground than an explosion. I take a deep breath and load my first round. I take another deep breath and pull the trigger. Bang! My face, arms, torso—everything that is connected to this gun—flinches for a moment. The gun is small by most standards, and according to E.V. it doesn't kick, but this burst of motion still startles me. I smell gunpowder. I sit up and look around before reassuming my shooting position and reloading the gun. This time, I tense up, bracing myself for the jump when I shoot. I pull the trigger. Bang!

I reload the gun and before I have time to get settled, E.V. is crouched down beside me.

"You're closing your eyes, Lily."

No kidding.

"Concentrate on keeping them open when you pull the trigger. You'll shoot straighter." He stands up and heads to someone else in need of coaching.

Before each of the next three rounds, I try to take a few deep breaths, to calm my nerves and will myself to hold my eyes open. But with each shot, I flinch and shut my eyes.

When everyone has shot all five rounds, E.V. tells us to go get our targets and see how we did. I take off my earphones and relax when the sound of my heartbeat is eclipsed by the familiar squeaking of sneakers and whispering of kids. My target has five shots on it, one of which is actually touching the bull's-eye. I'm proud of myself, but worry that my best shot was the first one, before my body knew to flinch with the trigger.

On the last day, after another session at the shooting range, we go into a classroom and take a multiple-choice test. As we are finishing, the kid next to me turns to his father, who is sitting behind us, and whispers, "Dad? How do we spell our last name?"

After a few minutes of grading tests, our instructors announce that each one of us has graduated from Hunter Safety. We cheer and give each other high fives. E.V. calls our names, one by one, and we walk up to shake his hand and receive a small cardstock certificate.

When I get home, I show off my certificate to Scott. Attached is a letter addressed "to the parents of the Hunter Safety graduate," reminding them that safety should be a lifelong practice. By now, I have described so many of my classmates and recounted my funny interactions with them that Scott can't help but find the whole Hunter Safety experience charming. By extension, he is starting to get excited about the idea of me hunting in the near future.

Scott's growing enthusiasm has encouraged me to do something I've already put off for way too long: tell my parents about it. The next month, in October, they arrive in Bend for a weeklong visit. The first morning of their trip, we go to a small, empty coffee shop and settle into a pair of sofas. I can think of no way to ease the shock of this news, so I just spit it out.

"Mom and Dad?" I take a deep breath. "I'm learning to hunt."

They both stare at me, expressionless. I think of one of my gay friends, and how he must have felt when he came out to his ultra-conservative parents. In fact, for one brief moment, this feels like a bizarre version of coming out: I'm shocking my hippie, blue-state parents with a revelation that they reserve for heartless conservatives.

My dad breaks the silence.

"Won't you be the darling of the right wing," he says, his voice dripping with disgust.

"Dad, it's not what you think."

We spend the next hour or so talking about my reasons for wanting to learn. I explain that the way I see it, hunting will make me a better environmentalist. To my parents' credit, they listen and eventually soften up to the idea. My mother frets about the guns, but is somewhat reassured when I tell her about the Hunter Safety course. Both are amused by my stories about the kids. Over the next few years, they will become curious, enthusiastic boosters of my hunting expeditions. For now, though, the conversation stalls when they realize that I haven't actually hunted anything yet.

"How do you know you'll like it?" my mom asks.

"I don't think I can know until I get out and try it."

This is what Hunter Safety couldn't teach me. Although I am slightly more comfortable handling a gun and I now have a basic familiarity with hunting rules, I still know nothing about how to actually hunt. It feels like I've earned my driver's license without ever sitting behind the wheel and turning the key in the ignition.

CHAPTER 4

PULL

Later that fall, Scott and I decide to get married. I have no doubts whatsoever about Scott. But I am startled to find myself marrying someone so deeply rooted in central Oregon, a place that still feels foreign at times. By marrying Scott, I am setting down roots of my own, gripping this lean, rocky soil.

Friends from college ask me, "Do you think you'll stay in Bend for the rest of your life?" I hyperventilate at the thought of staying anywhere for the rest of my life. "Who knows?" I answer. "Forever's a long time."

Occasionally, I worry about what will happen if my family remains far apart. My older brother lives in Brazil, my younger sister in Los Angeles. I fret about what will happen when my parents, who still live in Maryland, grow old. But most of the time, I shoo away these concerns. I have plenty of time, and so much will happen between now and then.

We rush to arrange a small wedding during the last week in December, when my parents and sister are already planning to be in Bend for Christmas. Nathan won't be able to come from Brazil on such short notice. We make plans for a reception on the East Coast in May, and he vows to come to that.

"Maybe I'll come visit you in Oregon, too," he adds.

My family is all city folk, but none more so than Nathan. Even in Brazil, it took him years to stop dressing like an inner-city American, with baggy jeans and tan Timberland boots (he has since replaced them with baggy shorts and flip-flops). Yet for the last year or so, he has grown increasingly curious about my country life in Oregon. A voracious reader, he randomly discovered *The River Why*, by the author Scott and I heard speak shortly after we met. This novel has spurred his growing interest in fly-fishing—a theoretical one; he has never touched a fly rod. Scott accompanied me on a trip to visit Nathan earlier this year, so he knows what an odd figure my brother would cut in a pair of waders, with a fly rod in his hand. We joke about testing his touted interest in this technical and sometimes tedious sport by taking him on an overnight float trip. Nathan has little experience outdoors, amid bugs and mud. His fair skin would burn in the sun. His patience would fizzle against hours of fruitless casts and tangled line. Then again, Nathan has surprised me before.

In March, he calls with an announcement that eclipses

my nuptials: His girlfriend, Luciana, is pregnant and due in September. Nathan loves children, and he has always wanted kids of his own. But this pregnancy wasn't planned, and he stammers as he spills the news. At the end of our conversation, he apologizes because he won't be able to make it to our reception.

"It's okay," I tell him, too drunk on newlywed love to take offense. "It's not that big a deal."

"No." He sounds far more disappointed than I am. "It is a big deal."

Nathan and I have a relationship that swings like a pendulum. For a time, while he was in high school and I was in junior high, we would stay up late together, watching reruns of *The World's Strongest Man* on cable. We added our own commentary—trying to mimic the contestants' accents—and made each other laugh so hard that we struggled to breathe. Then there were years like his first year of college (and mine of high school) when we could barely look at each other without resorting to yelling. Since he moved to Brazil and I to Bend, we've slowly grown apart, and our relationship has settled into an unfamiliar coolness.

The news of Nathan's impending fatherhood highlights the distance between us. Not knowing what to say to each other when we talk on the phone, we resort to mostly small talk. It's hard to remain close—even harder to close a growing gap—when we live so far apart. But I don't worry too much about this. It's part of the ebb and flow of our relationship, I tell myself. Nathan is family—he'll always be there for me, I'll always be there for him. One of these days, something will spark new feelings of closeness and we'll swing back together.

Before I know it, 2007 has arrived and halfway passed, and if I'm ever going to hunt I had better get started. As the summer days shrink, I hatch out a plan. I will try hunting for birds, then move on—if I'm still interested—to big game. My reasons are simple: A smaller animal is less overwhelming and scary. And bird hunting has the added appeal of using dogs as helpers. But one fact is unavoidable: I will need a gun.

When Scott's family first heard that I was interested in hunting, several of them offered to loan me their heirloom guns—if I'd keep them at my house. I declined. I hated the idea of storing a firearm in my house. For weeks, I brainstormed ways to learn to hunt that did not involve a gun in my home. I considered borrowing one. Or renting a storage unit. Or just keeping my gun in the trunk of my car. Eventually, it became clear that to hunt with any seriousness I would need to own at least one gun, and the responsible thing was to keep it in my house. Locked up, of course, and unloaded, with the ammunition stored someplace else.

There are a few different gun stores in town. I decide to try the biggest outdoors store first, figuring it will have the greatest selection.

I walk in intimidated. The gun department is a long

counter with racks of guns hanging on the wall behind it. The area is crowded with men and teenage boys holding different guns and taking them apart on the counter. I stand back and wait for fifteen minutes or so until a salesman is available.

"What can I get you?" he asks brusquely.

"Uh, I'm not sure. I'm looking for a 20-gauge shotgun."

"Okay, shotguns are over here." We walk down the counter. "Is it for you?"

"Yes. It's my first gun."

He nods. It's obvious from the guy's deflated facial expression that I'm the worst possible client. I will require time, hand-holding and detailed explanations and then spend little, if any, money. I won't be able to gab excitedly about the new semi-automatic Beretta, like the man standing to my right is doing with the other, luckier salesman.

"What will you use it for?"

I don't understand what he means. What does anyone use a gun for?

"Um, hunting? Birds?"

He sighs. "What kind of birds?"

"I'm not sure. All different kinds?"

"You should get a 12-gauge." He turns around, pulls a gun off the rack behind him and places it on the counter in front of me. "This is a good entry-level gun. It's a pump-action. It comes in a wood stock, black or camo."

"Well, I think I'd rather have a 20-gauge."

"Why? A bigger gun is more versatile."

"Uh, well..." He's getting impatient. "I don't want to

have to carry such a big, heavy gun. And I want as little kick as possible."

"A 12-gauge kicks less than a 20-gauge," he says matter-of-factly.

This will turn out to be a recurring theme in my gun-related education: Hunters disagree about what makes a gun kick against your shoulder when it fires. Some stick to the intuitive rule that the bigger the gun, the harder it kicks. Others have elaborate theories about how the energy from a wider shotgun shell somehow dissipates more quickly, sending a lighter thud into the shooter's shoulder. I'm sure it's true that an expensive, well-designed gun shoots with less of a jolt than a cheap one. But later, when I have shot several different guns, I will reach the same general conclusion I'd suspected all along: To spare your shoulder, shoot a smaller gun. I will be glad that I held my ground with this salesman.

"Let's just pretend I was going to buy a 20-gauge. Which would you recommend?"

He caves, and plunks down a couple of guns with the muzzles pointing behind the counter, toward him.

Nervously, I pick one up, turn it over and put it back down. I'm not sure what to do with it. The salesman smiles. Finally, I've offered him something: amusement.

"Well? Whaddya think?"

I don't know what to think. Other than price, I have no idea how to compare these guns. I'm not even sure how to hold them. I thank the salesman and leave the store. At another large outdoors store, the salesman is more helpful.

Instead of trying to talk me out of a 20-gauge, he picks one out. As he places it on the counter, he points out the pump action and other features.

"You can't really go wrong with this one," he concludes. "It's well made; it'll last a lifetime."

"May I pick it up?"

"Sure. You've never shot a shotgun before, have you?"

"No."

"It's real easy," he says, not missing a beat.

He plants the butt of the gun in my right shoulder and softly presses my head down until my right cheek touches cold metal.

"There you go."

He points out two small metal bumps, one from the end of the barrel closest to my face, the other rising off the tip of the muzzle. To aim the gun, he tells me, I need to line up those two points and use them to underline my target.

He stands up straight.

"Now point it at me," he says.

"What?" I lift my head off the gun. "No, I'm not comfortable with that."

"Well, how else am I gonna check if you're holding it right?"

This feels like an important moment, one that could someday be reenacted and videotaped to scare future Hunter Safety students. Rule number one in Hunter Safety class: *Always treat a gun as if it were loaded.* Rule number two: *Keep the muzzle pointed in a safe direction at all times.*

That class, those rules, they are my defining experiences as a hunter so far. They are the only principles I have. If I breach them now, simply to avoid looking like the hand-wringing ninny that I am, what do I have left?

"I'm sorry, I can't."

He holds his hands up, as if to say, *I give up.*

The salesman gets out a few other weapons, too. He determines that the "youth" models fit me better than the adult-size guns.

"Lots of ladies buy youth models, because they fit the smaller frame better." He seems concerned that I'll be too self-conscious to seriously consider the smaller size. But I'm actually reassured by the idea of buying a gun meant for a child. It feels safer, somehow. Again, I thank the salesman and leave empty-handed.

A few days later, I try a small shop closer to my house that sells shotguns and fly-fishing equipment. There is just one employee, a tall, round man with a trim white beard. His shirt is missing one button. I'm the only customer, and I tell him that I'm looking to buy my first gun, a 20-gauge shotgun. Nothing fancy, just something for a beginning hunter.

"Sure," he says. He turns around and selects a gun from the rack behind the counter. "I've got a Benelli here that might be just the thing."

In an instant, he shows me that this time will be different: "Here, I'll open the action," he says, "so you can see that it's not loaded. And then we'll just leave it open the whole time we're handling the gun."

The salesman's name is Russ, and he's a retired music teacher who is so by-the-book that he might as well be a Hunter Safety instructor in his spare time. This puts me at ease, and I immediately vow to buy my gun from him.

He explains that the reason he likes this gun is because it's so simple. It's a pump-action, which means that you push the sheath of the gun forward to move the shell into the chamber. There are no semi-automatic parts to turn finicky after years of use or during poor weather. Next, Russ shows me how to take the gun apart—yes, it actually breaks into pieces—and then reassemble it. Apparently this comes in handy during cleaning.

He also shows me how to hold the gun correctly, and he does so without making me aim the muzzle at his face.

"Are you right-handed?" When I nod, he asks, "Do you know which eye is dominant?"

"My right eye."

"Are you sure?"

"I'm positive." During Hunter Safety, we learned how to determine this. With both arms outstretched and the backs of your hands facing you, make a triangle with your thumbs and pointer fingers and center the triangle on some fixed object at least fifteen feet away. With your eyes focused on that object through the frame of your hands, slowly draw your hands all the way back to your face. Your hands will end up circling one eye: This is the dominant one.

Russ tells me that women are more likely than men to be right-handed but left-eye-dominant, or vice versa. Men almost always have their dominant eye and hand on the same side.

I notice that I'm relaxed enough to remember all the questions I couldn't think of when I was nervously trying to keep up with the last two salesmen.

Although he doesn't have a youth gun in stock, Russ agrees that a smaller gun would probably fit me better. He doesn't pressure me to buy anything. But he does suggest that whenever I do get a gun, I go to a nearby gun club to practice shooting it. It turns out Russ is a competitive trap-shooter. He even offers to lend me some videos on shotgun shooting technique.

While I'm there, I notice the other, fancier shotguns. They have stocks of smooth chestnut, unlike the matte-black metal one I'm considering buying. One has a metal inlay of two dogs—one gold and one black—flushing a bird. Russ opens the action and hands it to me so I can examine it. I rub my fingers over the smooth wood and finely etched metal. It has never occurred to me before that a gun could be a work of art. I flip over the price tag—more than twenty-five hundred dollars—and quickly hand it back.

When I leave the shop, I feel upbeat for the first time in a while. With someone like Russ helping to escort me into the world of hunting, perhaps I will gain admission after all. The next day, I call Russ and place an order for the most basic gun I looked at, a Benelli Nova pump-action shotgun in the youth size. Later, I drop off a deposit, half of the gun's $419 cost. This is not a small purchase for me.

The gun costs about what I bring home in a week of work as a reporter.

A week later, when Russ calls to say the gun has arrived, my first thought is: Shit. Am I really ready to be a gun owner? But Russ sounds genuinely giddy, and a little of his excitement can't help but rub off on me, even over the phone.

I swing by the shop after work. Russ waves to me as I walk in, then turns and pulls a long white box—like an extra-long board game container—out from behind the counter. I had no idea guns came in boxes.

"You are gonna love this little gu-un," he sings. I do feel excited to get my hands on it, to see if it really fits and feels as light as I hope.

But first, the paperwork. Name, address, date of birth, driver's license number, Social Security number. A long list of yes-or-no questions about whether I've ever been convicted of various crimes. Easy stuff. Next, Russ dips the fingers of my right hand in ink before rolling each one, sideways, onto the page. He looks over the sheet to make sure I haven't left any questions blank. He asks to see my driver's license, which he compares with my answers on the page. Then he picks up the phone and dials.

"This'll just take a sec..."

As Russ reads my answers out loud, my palms sweat. For a moment, I let myself imagine: What if I fail the background check? But before I can muster a believable image of prison, Russ hangs up the phone.

"You're all set."

He sends me home with my brand-new gun and three borrowed DVDs about how to shoot a shotgun. In them, champion shooter Todd Bender demonstrates proper shooting technique. The first thing I notice in the videos is that Todd really smashes his face against the gun when he shoots. His face is also highly asymmetrical, and I wonder if these two things are related.

A couple of weeks later, I meet Russ at the Redmond Rod and Gun Club to shoot clays. I park my car in the gravel lot, at the end of a long row of American-made pickup trucks. I recently sold my Ford pickup and bought a used Toyota Echo, and it occurs to me that I might fit in better if I hadn't placed so much value on fuel economy. As I walk around to the trunk to get my gun, Russ sees me and waves. He saunters over wearing an old, broken-in hunting vest. The pockets nearly burst, they're stuffed so full of lead shotgun shells.

Seeing Russ's vest reminds me that I don't have a good way to tote around the shells I purchased a few days earlier. I grab my down coat from the passenger's seat and unzip the pockets. I throw it on, pick up my gun and shells, and trot after Russ, who is joining a group of older men lounging around a picnic table. The youngest ones are about Russ's age, probably late fifties or early sixties. The oldest are in their eighties. Most of them are wearing baseball caps that announce membership in a veterans' group or other military affiliation. Russ introduces me. I shake their hands and some—the hardest of hearing—ask me to repeat my name.

The club itself is just a flat expanse of sagebrush that has been transformed, shot by shot, into a barren patch of gravel and lead. A trailer makes up the office, where I sign in and pay three dollars per round. Several rickety buildings that look like outhouses are sprinkled around the property. These shacks house the machines that fling clay pigeons into the air to replicate flying birds.

There are several games of clay-pigeon shooting, including trap, sporting clays, five-stand and skeet. In each one, shooters move between stations and shoot "birds"—clay disks—that are flung from various positions. Trap is the simplest: A shooter takes two shots from each of five stations, taking aim at two clays from each. Trapshooting first developed in England in the 1700s, and involved shooting live pigeons as they were released from cages called traps. Today trapshooters take aim at standardized four- and five-sixteenth-inch disks, painted fluorescent colors and hurled at about forty-two miles per hour. Inside the trap house, the throwing mechanism oscillates back and forth, so although the shooter knows the general flight path, he doesn't know exactly where the pigeon will go.

There's a particular etiquette to a shooting range. When you aren't shooting, for example, you rest your gun in a wooden gun stand, unloaded. At first, this bothers me. I'm not comfortable leaving my new, expensive weapon lying in a rack where anyone could take it or tamper with it. But the shooting range is a little bit like a fraternity. Trust is important. And it's impolite to pace around carrying a gun while everyone else is sitting unarmed. Besides,

I quickly learn that my 20-gauge youth model is downright laughable to these über-shooters, many of whom own custom-made shotguns designed specifically for trap. In the rack it goes.

As a group of people shoot, Russ narrates the game of trap: Five people at a time participate—one at each station, arranged in a semicircle. Each person loads two shells at a time. When everyone is ready, the person on the far left shoulders his gun and yells, "Pull!" A sensor detects the noise and releases one pigeon from one trap. The shooter shoots it. The shooter yells, "Pull!" again and shoots a second pigeon. Now he steps back a bit and the person to his right takes a turn. Everyone gets to shoot twice from each station.

Soon my name is called. I run to get my gun. I step up to the station and shoot an entire game of trap without hitting a single disk. Even my voice when I yell "Pull!" sounds inadequate—higher-pitched and more tentative than the others, as if I'm asking a question instead of barking a command. When it's over, I slink back to the picnic table. Russ pats me on the shoulder.

"Lily, there's someone I'd like you to meet." An older, shorter man is standing next to him. He holds out his hand. "This is Del, and he's one of the best trapshooters in the whole country."

Del walks me away from the group, asks me to shoulder my gun, and gives me a few pointers. Some of it contradicts what I learned from Russ's videos. The videos spent a lot of time explaining how to lead a moving target—in other words, how far ahead of the pigeon you must aim to hit the disk where it is instead of hitting thin air where the disk recently was. Del waves off this advice.

"Don't lead the bird," he says. "You're standing close enough to it, just lock on to it—move with it for a second to make sure you're locked on—and pull the trigger."

Next time when my name is called, Del walks to the trap station with me. He stands a few feet behind me. When my turn comes, I take a deep breath. "Pull!"

The bird flies out of the house. I lock in on it and pull the trigger. Pow! The clay sprays apart like a firework.

"Good girl," Del says.

"Pull!" I lock in on the pigeon and pull the trigger. Bang! This time, I miss. The clay sails into the sagebrush, unharmed. Del steps toward me and whispers some advice.

"Keep your weight forward; you want to stand right over the gun, not behind it."

I shoot a few more rounds of trap and, with Del's help, hit about half of the pigeons. He congratulates me and I thank him for his patience. Over the next month, as I go back to the range to practice, Del waves and occasionally offers a few words of advice.

Surprisingly, none of the men at the range hunt. Their sport is shooting, not hunting. Every year, a few weeks before hunting season, hunters like me flood the range. The rest of the time, these men—and a handful of women—have the place to themselves. They arrange car pools to shooting competitions. They rib one another. They drool over one another's guns.

It's not uncommon to meet hunters and non-hunters who fetishize their guns. They name their weapons and own more than they can possibly use. I write about a burglary in which more than fifty guns are stolen from the La Pine home of one couple. Turns out they had an entire room devoted to firearm storage. Others, like Scott's family, end up with heirloom guns that they don't use but can't bear to sell or give away.

Andy starts calling my shotgun "The Peacemaker," poking fun at its diminutive size in all-business black. Eventually, I will call it that at times, too, though I never feel emotionally attached to the gun itself. Most people I hunt with will carry more elegant-looking guns and ask when I'm going to upgrade mine. Yet I will remain satisfied with my plain-Jane shotgun. It does its job. I see no need to glorify it.

Soon I agree to take in an heirloom shotgun from Scott's cousin. It's a 12-gauge, which will allow me to hunt bigger birds. I try not to think about the fact that I now live in a house with two guns.

CHAPTER 5

GUTS

I awake in a panic. It's the first day of September and I'm about to go on my first hunt for doves. As I wait for Andy to pick me up, I joke to Scott that it's symbolically appropriate for my first hunting experience to involve shooting at the international icon for peace. (This image was appropriated back in 1949, when a Pablo Picasso lithograph of a dove was selected as the emblem for that year's World Peace Council meeting in Paris.)

I won't be downing any birds, however. I've packed a water bottle and lots of neutral-colored clothing but not my gun. Leaving it at home is a strategic excuse to merely observe the other hunters. Despite all of my preparations, the idea of killing something still terrifies me. I don't want to be pressured by a group of seasoned hunters into doing something I'm not ready to do. Or, perhaps worse, chickening out in front of them.

Andy picks me up this morning and drives us to a farmhouse, where about a dozen people are going to hunt doves on a privately owned farm. Later, I learn that being invited to join this troop is a bit like drawing Willy Wonka's golden ticket. If you know someone who owns a large piece of land and lets you hunt on it, you automatically gain two advantages. One, you're not competing with other hunters for space, like you do on public land. And two, the owner of the land can act as a guide, sharing hard-earned secrets, such as where the animals tend to bed down and along which paths they travel and when.

It's chilly but clear when we pull up, and Marc, the owner, invites us onto his kitchen deck for a cup of hot coffee and a slice of store-bought coffee cake. Andy chats with the other hunters. I take a few bites of cake but I'm too nervous to eat more. I pace across the deck, peering into the sky for a glimpse of dove.

Mourning doves (*Zenaida macroura*) are considered migratory birds, though some of them stay put year-round. In Oregon, dove hunting is allowed for about one month in early fall, just as the birds are beginning to fly south to spend the colder part of the year in Mexico, Arizona or California.

Across the country, doves are a particularly popular hunting bird. In the 1970s and 1980s, biologists estimated that hunters killed fifty million doves a year—more than all other game birds combined. Despite the high rate of "harvest"—that's biospeak for death by hunters—doves remain abundant throughout the country. Car accidents, animal predators, disease, weather and miscellaneous causes account for four to five times as many dove deaths as hunters do. One 1993 study, for example, estimated that domestic cats were responsible for about 70 percent of them.

Last week, in preparation for this day, I purchased my first hunting license. Across the country, wildlife is the property of the state in which it resides. That means each state gets to set its own rules for hunting and fishing. (The one exception is for federally listed threatened or endangered species, which fall under the purview of the U.S. Fish and Wildlife Service or the National Oceanographic and Atmospheric Administration.) It also means that even if a dove lands in my yard, it's not *my* dove. It belongs to Oregon, and if I kill it without a license or out of season, I'm poaching, or stealing that dove from the state.

The U.S. Fish and Wildlife Service began tracking hunting licenses in the 1950s. License sales peaked in 1982, when about 16.7 million Americans paid for the right to hunt. That number has slipped steadily since. By 2006, about 12.5 million people hunted, a 25 percent drop even as the population grew more than 30 percent.

Buying a license was easy. I walked up to the counter at a sporting goods store and gave them my driver's license. Because I've bought fishing licenses each year I've lived here, my information was already in the state's database. The clerk printed out a license, and I signed it, paid for it, and headed home.

As the morning warms up, Marc walks us down to a flat gravel road that divides an irrigation pond from a grove of juniper trees. He explains that he's been watching the birds for weeks now, memorizing their routine. The doves are currently eating grains off the floor of nearby fields. Doves are adaptable eaters, munching on a wide variety of seeds, grasses and forbs. At about nine thirty each morning, they fly through the junipers, parallel to this gravel lane, he says. The hunters spread out across the road. I stay close to Andy but make sure I keep a few feet behind him. Everyone is fumbling with shells and loading their guns. I'm worried about the guns. There are too many for me to monitor the direction of each muzzle. To calm down, I look instead to the horizon: a small grassy hill to our right. It doesn't help, though, because I'm also worried about the doves. I hope some of them—how will I know how many are enough?—survive the shootings. Then I worry that this means I'm not a hunter, and never will be. After all, there can't be too many people who wake up early to go on a hunt and then spend the morning fearing for their prey.

My thoughts are interrupted by a stage whisper from Marc: *They're coming.* I squint. Yes, tiny spots are soaring over the hill. Doves. I roll my foam earplugs between my fingers and squish them into my ears. Andy shoulders his gun.

A flock of doves is sometimes called a dule or a dole. There are places in South America where dules can reach tens of thousands of birds. Here, there aren't that many—a hundred, maybe. As the birds reach the trees in front of us, I notice that when they stop flapping their wings and glide through the air, they strike a pose immediately recognizable from any peace flag or greeting card (but without olive branches clutched in their beaks). Andy and the other hunters begin blasting. I try to control my shoulders, to avoid shuddering with each shot.

The doves are fast, fancy fliers. They glide, then accelerate without warning. It's hard to tell from my angle, but they seem to weave left and right as they fly. This is why ammunition manufacturers love dove season: It takes a lot of shots to bag one bird. After a minute or two, when a few of the birds have fallen from the sky and the rest have flown to safety, everyone lowers their guns. Andy heads into the trees to find the birds he has downed. I follow him. I'm having trouble hearing him, so I remove my earplugs. We hike down among the trees, and I relax a little. Andy thinks one of his birds has fallen in this general vicinity.

Suddenly the shooting resumes. Another dule has arrived along the same flight path. Andy waves, then yells, "Hey! We're down here!

"I just want to make sure they see us," he explains to me. He doesn't look alarmed, as I am.

"Here it is," he says as he reaches down and picks up a bird. "I'm pretty sure I got another one, too."

We pace around, but instead of looking at the forest

floor, I keep looking at the bird dangling from Andy's clutch. He holds it casually, upside down, and the bird's head flops like a ball on a limp string.

Andy finds the second bird, and as he bends down to grab it, I hear what sound like heavy raindrops. Andy raises his eyebrows.

"Let's get outta here," he says.

The dripping sound was shotgun pellets hitting juniper branches as they fell from the sky. Some hunter is aiming too close to our vicinity for Andy's comfort. We walk briskly back to the road. I would sprint but I don't want Andy to know how scared I really am. When we arrive, Andy hands me the birds so he can shoot at the next dule.

I hold each bird by its neck, which is no thicker than a Sharpie and surprisingly warm. One at a time, I lift them up and examine them. The birds are small, with blue eyelids. Other than their tails and wings, which are mottled black and white, their bodies are covered in the tiniest grayish brown feathers I have ever seen. Their fanned tails are outlined in white, as if the edges were dipped in paint.

Later, Andy shoots another dove and finds it still alive, twitching slightly. He calmly clutches the animal with two fists and twists, like he's opening a jar. Just like that, its neck breaks and the animal goes limp. I know it's the humane thing to do, but his calm shocks me. I try to imagine what it would feel like to wring a bird's neck with my own hands. By comparison, shooting it from twenty yards away seems so detached it might be easy.

In the early afternoon, we gather around a trash can to clean the birds, which, it turns out, we'll be eating for lunch. A short, pudgy older hunter shows me how to do it. He grabs one of the birds from a nearby pile and holds it in one hand, on its back.

"These birds are so small, you don't bother cleaning the whole thing," he tells me.

I stare at the tiny, perfect bird cradled in his hand. It has no blood on it, no sign of the metal pellet that caused its death. Its eyelids are closed, which makes the animal look cold but peaceful.

"So you just breast it out, taking these two parts." He touches his thumb and pointer finger to either side of the bird's breastbone. He clutches the two areas he just touched, one with each hand, and rips, violently. I flinch. The skin tears with a little zip, like the sound of splitting pants. Inside, there is no blood, just taut breast muscle. It's purple, much darker than the thighs of a chicken.

"Then you fillet the breasts," he says, cradling the bird on its back again, now with its loose skin dangling on either side of his hand. He takes a small knife in his other hand and draws it along the breastbone, until he has separated one breast from the bird's body. He drops the dark purple slab—only slightly larger than a chicken nugget—into a stainless-steel bowl that the other hunters have already filled halfway with their boneless, skinless dove breasts. Some of these fillets have a small, dark spot on the muscle, announcing the entry point of a pellet.

Once the piece of steel is removed, the meat will be as good as new.

"And that's all there is to it," he says, starting in on the second cutlet, which he then plops into the bowl. He drops the rest of the animal into the trash can and wipes his hands and the knife on a small towel.

"You're not going to do anything with the rest of it?" I ask, peering into the can at the pile of half-emptied birds.

"Nah, there's no meat left," someone else pipes in.

"Here ya go," my instructor says as he hands me a bird.

I pinch the skin covering the breast with both hands, then look back at the man.

"Just tear it?"

"Yep."

I jerk my hands apart and am surprised at how easily the skin rips. It's no thicker or stronger than a green oak leaf. How does an animal this fragile survive the bumps and scrapes of daily life?

The bearded man nods approvingly and hands me his knife. I pierce the muscle with the tip and start feeling around for the bone as a guide. Satisfied that I'm following orders, he picks up another bird and rips into it.

By the time I'm halfway done with my second bird, the two or three other hunters have cleaned the entire pile—thirty or so doves in all.

Marc takes the bowls into the kitchen. There, he chops the meat finely, mixes it with ground chicken thighs and stir-fries it in a wok. Half an hour later, we are sitting on his sunny deck, spooning the meat, which is drenched in sweet soy sauce, onto crispy rice noodles and eating it, wrapped in iceberg lettuce leaves, with our hands. Despite my hesitance about eating it, the meal is delicious.

On the car ride home, Andy tells me he's disappointed that, beneath the sauce and the ground chicken, we didn't really get to find out what dove meat tastes like. I don't admit that after watching the slaughter, I didn't have much appetite and felt grateful that the resulting meal didn't taste any more exotic than chicken, my lifelong comfort food.

I stare out the window and wonder, yet again, if I actually have what it takes to kill an animal. Just two generations ago, almost any American would have answered this question about herself long before turning twenty-seven. Even those who didn't farm or ranch went to butcher shops where they saw their meat before it was hacked into roasts or ground into burger. They knew what went on behind the scenes. In a very short amount of time, we have become completely detached from the gory, grisly truth about what we eat.

More than 96 percent of Americans—298 million of us, give or take a few hundred thousand—ate at least one piece of meat last week. Most of us ate meat more than once a day. Strips of bacon with breakfast. A sliced turkey sandwich at lunch. A quick snack of beef jerky. Grilled chicken breast for dinner.

As a nation, we raise and slaughter nearly ten billion land animals each year—more than one million each hour—for food. That portions out to about two-thirds of a pound of meat per person per day, or 241 pounds a year, which happens to be more than twice my body weight. It's also more than twice the international average. The average American today eats eighty pounds more meat per year than in 1942.

On a global scale, our appetite for meat has literally changed the face of the earth. Add up the weight of all land animals on the planet—monkeys, mice, elephants—and domestic livestock accounts for one out of every five pounds. Thirty percent of the earth's land surface is now used to raise meat, either for grazing or growing grain for feed. All of that land was once habitat for wildlife and native plants, making livestock one of the main reasons why, every hour, an average of three species go extinct from our planet. The meat industry is a leading cause of deforestation, erosion and water pollution. It's responsible for 18 percent of all greenhouse gas emissions, which is more than transportation. The production of merely 3.8 ounces of beef—enough for just one McDonald's Happy Meal hamburger—releases about as much carbon dioxide into the atmosphere as a sedan emits by driving eighteen miles. And that's not including the gas it takes to *get* to the drive-through window.

Unless you eat wild game or raise your own meat, you can bet that the meat on your plate lived a miserable, confined existence. As Jonathan Safran Foer writes in his vegetarian treatise, *Eating Animals*, "We know that if someone offers to show us a film on how our meat is produced, it will be a horror film." Still, it's easy to live a well-fed life in the U.S. without thinking about where any of that meat came from. As a lifelong meat eater, I feel a responsibility to see for myself that uncomfortable thing that has always been at the heart of a human diet, since long before meat animals were domesticated and their upbringing industrialized: death.

CHAPTER 6

FIRST KILL

When I read about a hunting workshop offered by the state, I jump at the chance to enroll. It's part of a series of workshops—on fishing, camping, paddling and the like—bearing the unfortunate name "Becoming an Outdoors Woman" and aimed at females with little to no outdoors experience. For forty dollars, a woman may hunt for real, live pheasants with the help and advice of experienced volunteers, all of whom are male and own well-trained bird dogs. As with the dove hunt, my main goal is to leave with a better idea of what it takes to find and kill an animal in the wild. Actually shooting a pheasant would be a bonus. Or perhaps, if I feel as guilty as I expect to, a curse.

The workshop starts at eight in the morning in a cement barracks-style building where we eat donuts, sip coffee and listen to a state wildlife employee lecture us on hunting safety. Next he shows us photos and explains how to identify the gender of a pheasant. This is important because we are only supposed to shoot males, sparing the females to lay eggs and raise babies. As with a lot of birds, the two sexes hardly look like the same species. Roosters are gaudy, with a shimmering teal head with a scarlet patch around each eye, a white collar and bodies speckled with copper, blue, purple, brown and white. Hens are plain by comparison, with mottled brown-and-white feathers covering the whole body. Both sexes have long, pointed tail feathers with horizontal brown stripes.

The state employee gives us no other background on the animal, so it isn't until I return home that I learn ring-necked pheasants (*Phasianus colchicus*) are native only to Asia. Today the birds are found in farmlands all over the country, although populations in many places have declined since the 1960s. Wildlife biologists say that small American farms used to contain rugged, overgrown vegetation along the fence lines. Now, as many of those fields have been fully tilled and consolidated to make operations more efficient, there is little nearby cover left for the birds to hide and nest in.

We're in the Klamath Wildlife Area, a state-owned piece of land in southern Oregon whose primary goal is providing habitat for wildlife, especially birds. There are a few wild pheasants here, but mostly we'll be shooting the pen-raised birds that were released last week for a similar workshop.

We head to a nearby field to practice shooting some

clay pigeons, then divide into small groups and head out to hunt for real, live birds. At first, the morning feels similar to that of the dove hunt, full of nervous excitement. This time, of course, there's more gravity because I'm carrying a weapon. Another, smaller, difference is that all of the other gun toters are women. But that doesn't help me feel like I belong. All morning, I keep thinking: I'm not one of these people. These other women may not know what they're doing, but they seem confident that it's a good thing to do. Their faces strain with concentration. Each of them wants, without question, to shoot a bird. Me, I'm still not sure.

We hike along a dirt road, swing ourselves over a barbed-wire fence and start picking our way up a small, treed hill that Gerry, our volunteer guide, finds promising. We spread out and walk side by side in a straight line. This way, if a bird flushes in front of us, nobody is caught in the line of fire. This does little to allay my fears of being shot, however. I'm on the end of the row, with three other women to my right, and it's easy to imagine a scenario that ends with an accidental shooting. What if a bird flushes to my left, and an eager hunter swings and shoots before noticing me? What if a bird flushes between me and the hunter to the right, and we shoot each other? My imagination gets plenty of time to run wild because despite hiking for over an hour, we still haven't seen any birds.

Until we're halfway up the hill, that is, and my doomsday-dreaming is interrupted by some commotion to my right. Tessa's tail is wagging in double time, and Gerry is getting excited. Before I have time to make sense of it all, there's a loud squawk and some thrashing in the tall grass. A huge bird rises like a phoenix in front of us.

"It's a rooster!" Gerry yells. "Shoot it!" I lift the long neck of my gun off its rest on my shoulder and settle its butt into my collarbone before noticing that the entire weapon is upside down.

Bang!

I haven't even turned my gun over and Lori has already popped the bird. It flops into the tall grass and Tessa jogs over to it.

"Wow! Nice one!" The group convenes around Lori. Tessa drops the bird in front of Gerry. He tosses it a few feet in front of the other hunting dog, seventeen-year-old Teesha. She picks it up in her mouth and proudly walks it back to us. Gerry takes the bird from Teesha's mouth and hands it to Lori.

She admires it for a moment, then slides it into a large pocket on the back of her hunting vest. Its talons and the tip of its tail poke out an opening on the side. We spread out again and resume hunting. This time, to be prepared, I carry the gun horizontally.

Ring-necked pheasants don't fly very well, and prefer to run from danger. This makes them particularly easy for dogs to track, because they leave a scented trail on the ground. When the birds do lift off, the wings betray their struggle with loud, clumsy flapping sounds that startle an unsuspecting hiker or a hunter who has allowed herself to daydream for a moment.

Sometimes Gerry sees a bird land in some bushes, far

away, but his dogs don't notice. So he pulls a slender whistle from his pocket and blows into it—its sound is imperceptible to mere human ears, but the dogs stop in their tracks and turn to him. Gerry swings out one arm in the direction of the bird. The dogs immediately head off in the direction he pointed.

I think of my pet dog, Sylvia. I rescued her from a pound just a few months after moving in with Scott. She's a black mutt with lots of retriever instincts; nothing makes her happier than chasing a tennis ball or catching a Frisbee. She knows how to sit, lie down, stay, come and go to her bed on command. She even fetches the newspaper each morning and brings it inside. Yet she still hasn't mastered the correct response to pointing. If I drop a grain of rice on the floor and point to it, she nudges my finger with her nose. I have to lower my hand until my finger is practically touching the rice before she sees and eats it.

So much training goes into hunting dogs, they don't usually hit their peak until five or six years of age. As a dog owner who has poured countless hours into training, I am amazed at what these dogs are capable of. I can't even fathom how I would convince Sylvia to tolerate the bang of a gun. As we hike, I ask Gerry how he trained Tessa and Teesha, who are German shorthaired pointers.

"It's easy," he responds, adding that any dog can overcome gun-shyness; all it takes is patience and—these are my words—tough love.

"There needs to be two of you," he starts. "And then what you do is, you have someone bang two pots or pans together at the exact moment that you place the dog's food on the ground. If the dog flinches or runs away, immediately pick up the bowl. The dog skips that meal. At the next feeding, try it again."

I immediately picture Sylvia, emaciated from fear and starvation. I take a deep breath, trying to prevent my disapproval from leaping into my throat while I ask: What if the dog doesn't get it?

"I've never, in all my years, seen a dog take more than three tries," Gerry says before continuing with the lesson plan.

Once the dog is acclimated to the sound of banging pots, you have someone shoot an air gun or a BB gun when you put down the food. This technique rewards the dog (with a big bowl of food and praise from her owner) for managing her fear as well as for tolerating the bang of the gun.

"I know it sounds mean," Gerry concedes, "but it works really well. Pretty soon the dogs associate gunshots with something really good."

It's Rambo's version of Pavlov's bell, and if Gerry's dogs are any indication, it works. These dogs don't just tolerate guns—they get so excited when they hear shots fired that they drool and whine to be let loose. In fact, the dogs are making this day much more enjoyable than I could have imagined. They are enthusiastic and happy, and when one of them sidles up to me, wagging her tail, I can't help but feel enthusiastic, too. Hunting is their instinct; it's in their blood. These dogs have never known hunting without humans—and guns.

We humans certainly benefit from the dogs. I don't have to try it to know that bird hunting without a dog would be far less efficient. I would have to systematically walk up and down an area that looks like good habitat, covering every square foot and hoping to stumble across a bird. A dog, on the other hand, simply has to stick her nose in the air. She can catch a whiff of a bird who's more than half a mile away. Her stubby tail oscillates slightly, like the meter on a stereo with the volume turned down. The dog trots in large, loose circles until she finds bird scent on the ground. Then she zigs and zags, nose to the ground, with her tail wiggling faster as she closes in on the animal.

The hunter pays special attention to that wiggly little tail. It dictates where to walk and when to pick up the gun and get ready to shoot. Nose to the ground, as the dog homes in on the bird she focuses on one small area, tail still wagging furiously. When the dog's tail freezes, she has found the bird.

A good pointer will stare directly into the bird's eyes, Gerry tells me. If she remains perfectly still, the bird usually does, too. The bird is hypnotized by fear. Once the hunter is close by and ready, she releases the dog, who flushes the bird. The bird hoists its heavy body into the air, flapping loudly, feet dangling, tail cocked. The hunter has a couple of seconds to get off a shot before the bird flies out of range, usually gliding to a landing a few hundred yards away and disappearing under a tree or some heavy brush. So many elements—the bird, the dog, the hunter, the gun—must harmonize for a successful hunt.

We are wading through tall grass and soft mud when Tessa locks in on a pheasant tucked in a ditch beside the road. She holds the bird for what seems like an hour, while Nancy and I make our way over to her.

"Go past those tules," Gerry whispers, his eyes locked on Tessa.

Nancy and I start walking toward the dog. As soon as we're out of Gerry's earshot, I whisper: "What are tules?"

"I don't know," she says.

We both giggle. For the first time, I relax a little. Maybe I do have something in common with these hunter-women, after all.

Gerry holds out his hand, motioning for us to stop. We raise our guns to our shoulders.

"Go ahead and switch your safety off," Gerry whispers.

I place my thumb over the switch and hesitate. I vaguely remember learning, in my hunter education class, to keep the safety on until the bird flies. But Gerry knows so much more than I do. Maybe he's right.

My right thumb works the safety forward to the "off" position, ready to fire. Then I stand, frozen, while Gerry asks, for the umpteenth time, if we are ready.

"Once I release her, you'll just have a second to shoot," he says.

"I'm ready," Nancy and I say. I want to tell Gerry to

hurry up and release the dog already, but saying anything more might break my concentration.

My gun feels so heavy and I'm clenching the barrel so tightly that my left arm starts shaking. My right cheek, pressing the gun down into my shoulder, isn't helping.

Suddenly I hear rustling and frantic flapping. The bird's head pops into view, neck outstretched, eyes open wide. I wrap my index finger as far around the trigger as I can, just as Del taught me. Then I squeeze.

Boom!

The bird drops with a thump, landing not twenty feet away. A few loose feathers drift down leisurely. It was a head shot—a clean, perfect kill.

Burnt gunpowder hangs in the air and I notice, for the first time, its faint sweetness.

"Yeah!" Lori and Debra whoop and holler from the nearby road. "Wahoo!" They jog toward us, the tall grass leaving wet stripes on their pants.

Tessa bounds over to the bird, scoops it up in her mouth and trots a victory lap around us. Nancy and I slap each other high five. Gerry shakes our hands. Everyone is smiling. Gerry tells Tessa to drop the bird. He hands it to me. I hold it by the neck, again thin and warm in my fist. Its feathers are beautiful—green, blue, purple, red, white, brown, black. I can't think of a color that doesn't appear somewhere on this body. The feet, for some reason, are what really amazes me. They are covered in blue-gray skin that is so wrinkled it looks reptilian, but feels soft and taut over perfect, complicated bones. There are talons and toes and ankles, connected by real, working joints. I push my finger against the bottom of one of the claws, and the toe bends gracefully in two different places, a movement I recognize from my own fleshier toes. It is real.

I have imagined this moment hundreds of times. Each time, the act itself was to be simple: Bird sails into view, I pull the trigger. Bang. A single shot. A clean kill. The emotions, I figure, will be the messy part. I brace for a strong cocktail of excitement, pride, guilt, sadness, even disgust. I wonder if I can stomach it.

In the end, it's the other way around. The action, I realize later, was muddled. Nancy and I shot the same bird at precisely the same time, so I will never be sure of that one critical fact: if my shot was fatal. But my own feeling, as the pheasant falls from the sky in a surprisingly slow, graceful flutter, is singular and pure: euphoria.

It isn't until later—weeks after I've eaten the pheasant—that doubt will begin to creep into my head. It starts as a niggling feeling that something isn't quite right. Yet it's not the kind of doubt that I was expecting: remorse over killing the bird, guilt over relishing it. That happens, too, but not until months later, when I am gluing the bird's feathers onto a Styrofoam ball to make a Christmas ornament. No, in the more immediate aftermath, instead I begin to worry that maybe I hadn't killed the bird. Maybe Nancy killed it. Maybe I simply synchronized the impotent pull of my trigger with her fatal one. I'm startled to realize how much I want to be the one who took the bird's life.

★ ★ ★

On the way home from the hunt, I stop at Walmart and buy a Styrofoam cooler and a bag of ice. I place the cooler in my backseat, throw the ice inside and gently lay my bird, double-wrapped in plastic shopping bags, on top. The stacked bird-on-ice is too tall for the lid to close, so I tear open the ice and pour half of it onto a grassy median before putting the bird back on top, faceup, and carefully lowering the lid.

During the long drive home, I replay the final shot, over and over. So many events and lives managed to collide for that one perfect moment. Tessa employed thousands of years of instincts, hundreds of years of breeding and six years of practice to find that bird and hold it in place until I was ready. I was in exactly the right place at the right time. The bird was in exactly the wrong place at the right time. I somehow found it in me to pull the trigger. This is the part that amazes me the most—that I actually did it, I shot the animal. Someone banged two pots together and I didn't run away. I didn't flinch. I dug in.

It feels, in retrospect, like a miracle. Then there's the relief. A post-adrenaline warmth, like stepping off a roller coaster, spreads across my body as I marvel that my entire hunting party escaped unscathed. Every time I think about what I have done, I am amazed all over again. Weeks ago I had never held a shotgun. Now I have killed a bird and will soon eat it for dinner. I roll down the windows, crank up the radio and sing along at the top of my lungs.

★ ★ ★

Euphoria, I learn later, is a common reaction among first-time hunters. Almost every hunter I've ever interviewed has been surprised by it. In *The Omnivore's Dilemma*, Michael Pollan is bowled over by the sensation after shooting a wild boar. He describes a flood of pride, then relief and then, unexpectedly, overwhelming gratitude. "The animal," he writes, "was a gift—from whom or what I couldn't say—but...gratitude is what I felt."

As I hunt more, I will find that no other kill—nearly all of which will be much harder-fought—evokes such pure elation as my first. Still, there is a visceral sense of satisfaction in each of them. And I begin to see how a hunter might crave that special intoxication that oozes from a clean kill.

Some of our greatest writers, from Hemingway to Faulkner and many, many more, have tackled the subject of the hunt. But I never fully appreciated these contributions to literature before I related to this mysterious thrill of killing. In Herman Melville's *Moby-Dick*, for example, Captain Ahab hunts whales in general and the white Moby-Dick in particular with such vicious antagonism that 822 pages cannot explain it. To Ahab, hunting is more addiction than sport or profession or anything else. In one scene, he acknowledges to a mate that his single-mindedness has chewed up his life and spit out his marriage: "Aye, I widowed that poor girl when I married her, Starbuck; and then, the madness, the frenzy, the boiling blood and the smoking brow, with which, for a thousand lowerings old Ahab has furiously, foamingly chased his prey—more a demon than a man!" Before I started hunting, this is how I viewed any thrill-seeking hunter I encountered in books—someone plagued by an addiction that I could never truly understand.

After killing just one pheasant, I begin to think differently about these fictional characters and their reactions to hunting. My own post-hunt euphoria was, I think, rooted in satisfaction and awe. Thanks to the popularity of gardening, many Americans can relate to a version of this feeling. Garrison Keillor, in a story about his fictionalized hometown of Lake Wobegon, describes how a child feels when he grows his own tomato:

> We all go through these terrible things in childhood. Somebody looks at you and says, "You look funny. Your eyes look funny..." And you're scorched, your little heart is scorched by it. And you never completely get over it. But, if you can make this beautiful thing—a tomato, a perfect tomato—that you hold at arm's length and it smells of tomato. And you eat it; you eat it with maybe a little fresh basil and a little piece of cheese. Or you just put a little sugar on it. Or you just put a little French dressing on it. And you realize: this is as good as any tomato anywhere in the world. And all that the best chefs of New York or Paris or London could do for this tomato would only be to cover it up. To realize

> that most of the fancy seasonings and sauces and marinades ever made in the world are orthopedic in nature. That this, you can find the best, you can produce the best of something in Lake Wobegon or Kalamazoo: this is a redemptive experience.

Growing food is an experience that offers pride not only in one's self but, perhaps more important, in one's place. At its essence, hunting is finding food in the wild. Whether hunting for birds or for mushrooms, it evokes an exaggerated form of the satisfaction that Keillor describes. It feels miraculous. Shooting my first pheasant gave me the same exhilaration that I got from reading my favorite chapter books when I was nine or ten years old: The world around me suddenly became bigger than I had ever imagined and, at the same time, it moved closer within my reach.

It's increasingly rare for Americans to relate to this. The continuing decline in hunting mirrors our overall tendency toward spending more time indoors and online, detached from the natural world. Ironically, this has coincided with a sharp rise in meat consumption and, more recently, a growing interest in where our food comes from. Farmers' markets have popped up all over the country—there are now at least 4,385 in the United States. But only about 3 percent of farmers' market vendors sell meat. Most of us still stop at Safeway or Stop & Shop on the way home, to buy steaks to put all of those locally grown vegetables around.

As we distance ourselves from the meat we eat, it's tempting to view any decline in hunting as a reprieve for wildlife. Counterintuitively, hunting can actually protect the health of a wildlife species. When hunting is overseen properly—with scientific study, responsible quotas and enforcement—hunters play an important role in population control. Overpopulation of deer, for example, is a growing problem. The animals have fewer natural predators in the wild. Roads and other development have disrupted their migration routes, which means more deer are crowded into smaller spaces. Deer numbers can skyrocket during years of mild weather and then plummet during a harsh winter, when limited food and habitat can't sustain the inflated population. Say, for example, that a population has grown 30 percent larger than the winter landscape can sustain. This rarely means that 30 percent of the population will die off before spring. Instead, it is likely that the entire population will run out of food before the winter is over. Allowing hunters to cull a certain number of deer helps prevent mass starvation that could collapse the species for decades.

Good hunters are intimately familiar with the land where they hunt, so they can be the first to notice ominous changes in the environment. They value wildlife habitat and will work tirelessly to protect it from urban sprawl. This is the message I hear from one hunter, Lew, who grew up in a part of Colorado where nearly everyone except his own family hunted. His father, he explains, was a pacifist who didn't want to participate in a killing ritual, even though he ate meat on a regular basis. Lew never thought he was missing much until, in adulthood, his writing career led him to a job editing a hunting magazine.

A couple of stories about wild pig hunting in California somehow beckoned him, and when Lew got to the end of one article, he tells me, "I knew I had to do that." He booked a trip to wine country and hired a guide to lead him on his first-ever hunting excursion. As Lew got more interested in hunting, he paid more attention to development news and various issues facing the environment.

"Before hunting," he says, "even though I would hike and camp and fish, I didn't think about those things. I was just moving along with my own life. I never appreciated the precarious nature of the whole ecosystem... With hunting, I started to understand just how limited some resources are in certain areas. I saw the ways that we impact the landscape."

Lew got to know hunters who bemoaned changes to their old hunting grounds. Soon he, too, had a personal interest in seeing valuable habitat spared from development. Hunting had turned Lew into a conservationist.

My reporting job introduces me to Greg, a retired Oregon State Police trooper who spent his career enforcing fishing and hunting laws. Standing more than six feet, two inches tall with thick silver hair and a mustache, he looks the part of a retired cop. Back when he first joined the force, in the 1970s, every police academy graduate wanted to be a game warden. It was a state trooper's dream, the intersection of career and obsessive hobby. Greg tells me that when he grew up hunting in central Oregon, he knew the game wardens personally. To law-abiding hunters, he says, the wardens were superheroes. They were the community's badasses, protecting the game on which every hunter in the state relied.

Wardens have to be experienced hunters themselves, to have knowledge of typical hunters' behavior but also to be able to talk to suspects, hunter-to-hunter (or hunter-to-poacher). Wardens say that poachers frequently admit their crimes, and even brag about them, to other hunters.

Poaching is the illegal killing of wildlife, whether it's hunting without a license, hunting out of season or using illegal means to lure and shoot an animal. The reasons people poach are as varied as the ways in which they break the law. Some are too poor to pay for a license and tag. Some are thrill seekers.

Like any illegal activity, poaching is difficult to quantify. But studies indicate that it's a much bigger problem than even most wildlife biologists would like to admit. In Oregon, not far from Bend, biologists placed radio collars on five hundred mule deer in July 2005. Five years later, 128 of the collared deer had died. Poachers were found responsible for nineteen of those deaths—nearly as many as the twenty-one shot legally by hunters. Cougars killed fifteen. Coyotes killed five. Eight were hit by cars. Five succumbed to disease. Four got tangled in fences or endured some other fatal accident. Fifty-one died of unknown causes, though scientists admitted that poachers could have killed at least some of them. "Sometimes," one biologist told *The Oregonian* newspaper, "we just find the radio collar [lying] out in the sagebrush."

During his career, Greg underwent a change that he says eventually happens to all game wardens. His lifelong love of hunting wildlife was gradually replaced with love for a different type of hunt: "Catching a poacher became much more satisfying than shooting a deer," he says.

This new type of hunting held all the familiar appeals: nervous anticipation, stealth, frustration that eventually cracks and gives way to satisfaction. But the technique was so much more challenging, the stakes so much higher. This was the kind of hunting that animal rights activists joke about: The prey carried guns. By comparison, Greg tells me, "hunting for deer and elk wasn't fun anymore."

In recent years, however, Oregon has struggled to find qualified and interested candidates to become game wardens. Officials chalk it up to low interest in hunting, overall. The children who eventually become state troopers no longer grow up hunting each fall and revering their local wardens. The decline in hunting, in other words, is self-perpetuating.

CHAPTER 7

OFF THE MARK

When I get home from my pheasant hunt, the house is dark. Scott hasn't returned yet from his day of fishing. I leave the cooler on the front porch and check the telephone for messages. There is one from my mother.

"Hi, Lily and Scott, it's Mom calling." She sounds as excited as I feel. Maybe she heard about the bird somehow?

She goes on to announce that I have a niece. Luciana had her baby today—a girl named Sofia.

"I love you both," she signs off before adding, "I'm a grandma!"

I smile as I hang up. This day could not have been more perfect. The cycle of life has not only excused my killing but rewarded it. The bird died and a baby was born. What a trade!

The next morning, I am still on a high. Scott and I

admire the bird, which is now stiff and ice-cold, before carrying it into the backyard to pluck, gut and clean it. Hunters refer to this process using the rosy euphemism "dressing." Neither of us has done this before, but I have a pamphlet with instructions and line drawings titled *How to Field Dress a Bird*. I slip my wedding ring into my jeans pocket—I don't want guts stuck in it—and place the bird on a cutting board perched on top of a trash can.

To start, I pluck small bunches of feathers from the bird's breast and shove them into a plastic bag for future crafts.

This takes a long time, mostly because every part of the bird's body contains a different color or texture of feather. I want all of them, vowing to make the most of this bird whose life I've taken. But I'm also postponing actually picking up the knife and cutting. Eventually, Scott gets antsy.

"Are we going to do this?" he asks gently.

I nod and wipe the down from my fingers. When the knife is in my hand, Scott begins to read.

"'Begin by making an incision all the way through the skin from the breastplate to the vent'—that's the anus."

I shiver at his last word, then take a deep breath. My knife pierces the skin and I glimpse not blood and gore but taut, pink, familiar muscle. I relax a little. It's a sight I've seen hundreds of times, whenever I peel back the plastic wrap on a package of boneless skinless chicken breasts. The remaining steps are simple and reveal surprisingly little blood. It's not much different from pulling the giblets out of a Butterball turkey, really. The organs are small and easy to throw in the trash and then forget about.

Guidebooks call pheasants large birds and they are, at least when you see them bobbing around in the wild. But plucked or skinned, even a mature male pheasant looks terribly scrawny compared with the fryer chickens I'm used to cooking.

We invite Scott's parents, aunt and uncle, all of whom live nearby, to dinner. I spend the whole day preparing our meal. I decide to leave the bird whole because it somehow feels more respectful than cutting it into pieces like fried chicken. I brown the bird on both sides. Then I sauté some sliced onions and apples, stir in white wine and apple cider, put the bird back in the pot and slide the whole thing into the oven, covered, for a couple of hours.

Scott insists on buying steaks to grill, too, because the bird only weighs a couple of pounds, not enough for six adults. I protest at first, not wanting something as mundane as supermarket beef to overshadow my hard-earned pheasant. But I've never had pheasant before, and know that it might not taste very good, so I give in.

Before the guests arrive, I smooth our best tablecloth over our small metal table and carry the chairs from our desks into the dining room. I fold cloth napkins and place them around the table. I light candles.

When I carry the bird to the table, everyone oohs and ahhs. I am too nervous to take the first bite but Scott's aunt Kay spears a bit of breast with her fork, pops it into her mouth and chews.

"It's *good*," she says, her eyes widening so I know she means it.

I taste it. It *is* good. Sweet and tender. Not at all exotic or gamey, it tastes like an especially flavorful chicken. Comfort food. As we eat, I retell the story of the hunt. The guests listen politely as I gush over Tessa's pointing abilities and build to the story's dramatic, execution-style conclusion. At the end of the meal, there are no leftovers.

As Scott and I wash the dishes, I reflect that this is what eating meat should be. The animal was the centerpiece of the entire meal—the entire evening, really. We talked about its life—what little of it I glimpsed, anyway—not just its taste. We marveled at its place on our dinner table, and we felt grateful. We did not take it for granted, like we would a hen raised in cramped quarters and then killed, wrapped in plastic, frozen and tossed into a grocery cart.

Eager to hunt again, I sign up for another workshop, this time for rabbit. One morning in January, I wake up hours before sunrise to drive over the mountains to the Willamette Valley, to a wildlife management area that appears to be one giant mass of blackberry bushes—a horribly invasive species in this part of the state—intersected by a few winding paths of mowed grass. As in the pheasant hunt, we divide into small groups and head out onto state-owned land with the help of experienced volunteers. This time, not all of the hunters are female, and about half are teenagers.

As in the last workshop, we sit through a brief classroom lesson before we bundle up and head into the chilly fog. I'm nervous again, and still uncomfortable with my gun. I'm also excited to get started, mostly because of the dogs, whom I can't wait to see in action.

The volunteer guides each have three or four beagles with triangular bells dangling from their collars. The dogs are actually tied together with a thick rope. There's enough slack that they can run around comfortably, but if one tries to break out on his own, he gets dragged alongside the others until he can get his feet back under him. When I first see them, I can't help but smile. They're adorable, with pudgy bellies and floppy, velvety ears. They look like Snoopy, except that their muzzles are fuzzier and their big brown eyes are wetter and sadder than in the comics. It has never before occurred to me that they're hunting dogs.

When the dogs catch a whiff of rabbit, they start to whine and pull toward the bushes where rabbits are supposedly hiding. The guide lets go of the rope and the dogs dart into the thorny thicket.

One of our guides explains that when a rabbit is scared out of hiding, it tends to run, like most wildlife, in a big circle.

"Nine times out of ten, the rabbit will end up where it started," he says confidently.

This doesn't make sense to me. The place it started is the place it was roused from safety by a yelping dog. Why go back there?

But the guides seem to know what they're talking about.

When the dogs are out of sight, our guides listen to the yelps and peeling bells and order us around: "Run! Get up there!" "Get ready!" "Back up!" "Hurry, you're out of position!" I follow the orders until, suddenly, a rabbit darts out of a shrub to my left.

"Shoot it!" one guide yells.

The rabbit runs straight across a wide path in front of me. I shoulder my gun and line up the sights. I pull the trigger and nothing happens. I fumble for the safety and then pull the trigger again. Bang!

The rabbit somersaults to the far side of the path, then lands on its side.

"Nice!" The guide slaps me on my back and runs over to pick it up. The dogs emerge from the bushes and hurry over to sniff the rabbit.

"The first rabbit of the day!" Another guide shakes me by the shoulders. Other hunters walk over, smiling at me and peering at the rabbit in the guide's hands.

This must be what it feels like to win an Academy Award (minus the gown). Everyone is congratulating me. I feel excited but so stunned that I can't quite believe it's really happening to me. I want to get that rabbit in my hands.

Before I do, one of the volunteer guides takes out a pocketknife and cuts off the animal's head and feet, then tosses them, like softballs, into a nearby blackberry thicket. When he hands it to me, the rabbit is half the mammal it used to be. It's still warm, and it feels surprisingly thin in my hands. I expected it would be chubbier.

Next, he instructs me in skinning the animal. I start

around the neck, tugging a little at a time until the hide peels off like a union suit. The skin is slippery and translucent on the inside, with lusciously soft fur on the other. The guide raises his eyebrows when I wrap it in a plastic grocery bag and tuck it in my backpack.

"Just chuck it into the bushes," he tells me. "You don't have to pack it out."

"No, I want to keep it."

"What? Why?"

As I struggle to come up with a reason, I suddenly picture Scott at home, tying fishing flies on a winter evening. Rabbit fur is commonly used in flies, and sometimes costs several dollars for a three-inch swatch. "My husband ties flies," I tell him. "We'll use it."

The guide nods toward the blubbery wad of skin and fur now inside my backpack and smiles and shrugs, as if to say, *Sure you will.*

Next, he tells me how to gut the animal, which is basically the same procedure as gutting a pheasant. Once I've cut the abdomen open, he reaches in and pulls out the liver, a thin, black, gelatinous sheet that looks far too big to have come from the rabbit in my hands.

A small percentage of wild rabbits carry tularemia, a bacterial infection that can be transmitted to humans. To determine whether your rabbit is infected, the guide says, you should inspect the liver.

"This one looks good," he says. "See how it's smooth and uniform? An infected liver will look like someone sprinkled salt on it."

With that, he tosses the liver into the bushes.

"Go ahead and pull all that stuff out," he says, pointing generally to the offal, "and toss it, too."

He assures me that it will be eaten, enthusiastically, by vultures, coyotes and other scavengers. I start scooping parts out and plopping them onto a weedy shrub next to me. Gravity slithers them down, branch by branch, until they hit the ground. I look down at the animal in my hands. It looks like a skinned cat.

That night, I arrive home with one rabbit and no appetite. I open a bottle of wine and pour most of it into the ziplock bag containing my rabbit. Or what's left of my rabbit, anyway. I pour the rest of the bottle into a pint glass and start sipping. I don't feel particularly guilty about killing this rabbit. But I do feel awfully sorry for myself that I have to eat it.

I go online and identify my rabbit as an Eastern cottontail (*Sylvilagus floridanus*), one of the most common members of the rabbit-hare family. These animals spend most of the day sleeping in a tangle of brush or a burrow in the ground. At night, they come out and eat a combination of leaves and grasses. Rabbits are often considered a nuisance because of the damage they can wreak on crop fields. Oregonians with a valid hunting license may hunt varmints, including rabbits, 365 days a year.

The next day, I debone the meat and cut it into chunks, which I toss in flour and sear. Then I simmer the meat, along with carrots, onions, potatoes, celery and peas, to make a stew. The meat is light-colored and sweet, like pork. And because I accidentally overcook it, it's slightly rubbery, but nonetheless delicious, and Scott and I polish off the whole pot in two meals.

Over the next year, I go on trips with friends and hunt for duck, goose, grouse, chukar, Hungarian partridge and pheasant. I hunt all over Oregon. I take a bird-hunting vacation to Montana, to visit Andy and his girlfriend Jessie, both of whom recently moved there for graduate school.

As I bring home meats that I've never tasted before, I can't help but grow more interested in cooking. I spend hours online, reading through recipes for a dish that will do justice to my latest quarry. I borrow wild game cookbooks from the library. I buy them from bookstores. What used to be a chore—something to hurry up and get through so I could quell my grumbling stomach—suddenly becomes an honor.

One of the things I appreciate most about hunting is that it highlights the uniqueness of each animal, each meat, each meal. Instead of "thighs" or "breasts," a meal is composed of something specific, such as "the American widgeon from that slough in Montana on a cold, cloudy day last fall."

An animal's uniqueness extends to its taste—animals are what they eat, and wild animals eat a more diverse diet than livestock. Farmed animals in the United States feed almost entirely on corn. In fact, if you analyze the typical American diet down to the molecular level, processed corn makes up most of it: from the feed that our livestock turn into meat, to the corn syrup that sweetens our drinks. In *The Omnivore's Dilemma*, Michael Pollan quotes a biologist who refers to North American humans as "corn chips with legs." In my own post-hunting meals, I notice huge taste variations within a single species, particularly duck. I begin paying close attention to what was inside the gullet of each individual. I try to connect a duck's food preferences to the taste of its flesh, which falls somewhere along the spectrum between "fishy" and "beefy."

There are many wonderful cookbooks written about wild game. As with everything DIY, many come with a heaping dose of nostalgia. One of my favorites is *Fish and Game Cookery*, published in 1945. In it, author and outdoorsman Roy Wall laments the decline of wilderness—and Americans' corresponding wildness—in a chapter titled "In Case of an Emergency," which lists recipes for roasted beaver tail and young muskrat. "When America was an expanding frontier, when vast wilderness areas were trackless, it often became necessary for persons to eat anything and everything they could get their hands on," he writes. "Now, with civilization making picnic grounds of these once remote regions, something to eat is much less accidental."

Hunting has changed the way I shop for food. Now, when I saunter through one of several massive groceries in Bend, I take note of the swollen displays of produce flown in from all over the globe. And I realize, yet again, how detached I am from what I usually eat. In the past, whatever I felt like eating, I bought. There were no skills required, no special habitats, no exclusive seasons. When torrential rains ruined the blueberry crop in Oregon, shelves were stocked with blueberries from Maine. When I got a hankering for fresh berries in January, well, I was in luck because a shipment just arrived from Chile.

This could not be more different from the days to which Wall was referring, when our diet depended on where we lived and what plants and animals could survive there. If you didn't own livestock and didn't possess the skill necessary to stalk and kill a wild animal, you probably didn't eat much meat.

I'm not ready to forgo store-bought meat altogether, which would turn me into a vegetarian for all but a few meals a year. But I do start paying closer attention to where my food is raised. I stop buying produce that is grown outside of the United States. I have no trouble finding poultry that was raised in Oregon or Washington. Beef and pork, on the other hand, rarely bear any such labels. I hesitate over the meat cooler, turning over each package of steaks or chops, hoping to find more fine print. For all I know, these cuts of meat could come from the other side of the world.

As I grow more confident carrying a gun in the field, I find I'm still uncomfortable coming home and admitting to certain people that I've been hunting. When we visit with cousins who have young children, and my mother mentions my recent pheasant hunt, I feel cruel. My eyes widen and I shake my head at her, pleading with her not to talk about it. How could I explain to a child what I'd done? All of my rational justifications for hunting seem so meek compared with the basic idea of doing no harm.

I consider what it means to call myself a hunter, and why I'm embarrassed about it. A colleague of mine tells me that he took his son to a Hunter Safety course in Virginia, where the instructor told the students upon graduating: "Congratulations. You are now a member of a despised community."

When friends from college and high school hear that I'm learning to hunt, the first question out of their shocked mouths is usually: "Are you a member of the NRA now?" At first it surprises me that they would even think this. I'm still the same person I was before I started hunting. I still support reasonable gun control. Of course I haven't joined the National Rifle Association.

My friends' question reveals a familiar assumption. Before I moved to Bend, I, too, viewed hunters and NRA members as an interchangeable throng of gun nuts. But opposing gun control wasn't always the primary goal of the NRA, which was founded in 1871 by two veteran Union officers. The men were dismayed by the poor marksmanship of their Civil War troops, so they founded a group to promote safe gun handling and target practice. In fact, it wasn't until the 1970s that the NRA became highly politicized. Rising gun violence—including the high-profile assassinations of John F. Kennedy, Martin Luther King Jr.

and Robert Kennedy—prompted a national debate over the interpretation of the Second Amendment. Amid this discussion, the NRA's voice grew louder and louder. The organization stoked fears that liberal politicians were coming to pry every last gun from our hands.

On its website, the NRA proudly calls itself the "largest pro-hunting organization in the world." Yet, during election season, I notice that the NRA endorses politicians who oppose gun control, labeling them "pro-sportsmen" despite pathetic voting records on environmental issues. Meanwhile, only a fraction of proposed gun bills would apply to the basic rifles and shotguns used in responsible hunting. In fact, even if several of the NRA's worst nightmares simultaneously came true—expanded background checks, mandatory waiting periods, limits to the number of guns purchased by an individual per month, a permanent ban on assault weapons—hunting could continue as it has for more than a century. A hunter has about as much use for a semi-automatic Glock 19 pistol—which can shoot more than one round per second—as a chef does a hand grenade. Why, then, should we allow the NRA to use hunting—the most sane, responsible reason for modern gun ownership—as its argument for weapons that have nothing to do with the fair pursuit of wildlife?

The truth is, many of us don't. The NRA has about 4.3 million members, which is less than 5 percent of American gun owners. At my Hunter Safety class, however, and at every hunting workshop I've attended, I've received a handful of NRA pamphlets and a formal invitation to join the organization. This has reinforced my long-held belief that hunters have adopted gun rights as their leading political cause. (And my own experience suggests that they're succeeding: Buying a gun remains quick and easy to do.) That's not to say that I think everyone who belongs to the NRA is a nut. The NRA runs many admirable programs, including a massive gun safety campaign for children and cheap, accessible basic firearms training for adults. The NRA also does hunters a service by compiling hunting-related news on an easy-to-navigate website.

Every once in a while, a different kind of pamphlet is slipped into the stack handed to me at the end of class. There's one from the Oregon Hunters Association. Or the Boone and Crockett Club. Hunting is, by definition, a mostly solitary pursuit. Still, there are thousands of sportsmen's groups in the United States. And unlike the NRA, most of them are focused not on firearms but on wildlife habitat and hunting access. It is only when I'm flipping through these pamphlets that it dawns on me: The politics of hunting extend way beyond the gun.

Looking to join a hunters' organization other than the NRA, I start researching various groups. But I'm shocked to discover how many of them oppose, for example, sensible bans on toxic lead ammunition. Ducks and other birds ingest small stones to help grind up food in their gizzards. If they swallow lead shot on a beach or at the bottom of a lake, they could eventually die of lead poisoning. Since 1991 it has been illegal to use lead shot for hunting waterfowl. But lead is still the norm in big-game hunting and hunting for other, upland birds.

Nationwide, three thousand tons of lead are shot into the environment by hunters every year, another eighty thousand tons are released at shooting ranges and four thousand tons are lost in ponds and streams as fishing lures and sinkers. The effects of all this lead on an ecosystem can be tragic. The California condor, for example, once lived throughout the southern United States but is now one of the rarest birds in the world—in 2010 there were fewer than two hundred wild birds left, all near the California coast. Condors, unlike other vultures and coyotes, have especially strong acid in their stomachs and can actually digest lead fragments, leaving them particularly vulnerable to lead poisoning. (When a hunter shoots an animal with lead, fragments of the metal usually disperse into the animal's organs, which are often dumped in the field and left for vultures and coyotes to eat.)

With ammunition now made of steel, tungsten and copper, there is no good reason to keep spewing lead into the environment. As one hunter told me, "If you only shoot steel, then that's what you're used to. And you'll hunt very well with it." Lead remains cheaper, but if more people bought non-toxic ammunition, the cost would likely drop because it would no longer be manufactured as a specialty item.

Some of the hunting groups that I research argue for the expansion of off-road vehicle use on public lands, which could damage fragile wildlife habitat and destroy the remote backcountry hunting experience. Other groups complain about responsible hunting quotas, modest fees and game management laws.

The more I read, the more frustrated I become. These hunting groups are arguing against their own long-term interests. My responsibility as a hunter seems obvious: If I am serious about preserving our nation's hunting heritage, I must also be serious about protecting the environment.

I decide to join the Rocky Mountain Elk Foundation, which has a long-standing reputation for conservation- and wildlife-oriented policies. The organization has played a major role in successfully restoring elk populations in North America. But my esteem for the foundation sours in 2010, when the president announces that one of the organization's top priorities is to oppose Endangered Species Act protection for gray wolves, recently reintroduced to the western United States. I am disappointed in the organization's uncharacteristic shortsightedness.

As hunters, we should understand better than anyone the importance of natural predators in a healthy, functioning ecosystem. When wolves were reintroduced to Yellowstone National Park after being eradicated decades earlier, of course elk numbers dropped—that's what happens when a predator is introduced to prey that has lived virtually without threat. But elk numbers didn't drop nearly as much as some biologists had feared. In fact, they noticed changes that helped the entire ecosystem. The elk moved around more, instead of overgrazing certain riparian areas.

This allowed native plants to recover, boosting nearby beaver and fish populations. Coyote numbers dropped a little, due to the competition from wolves. Bird populations flourished. Elk became more wary of humans, too. In short, the presence of wolves brought balance to the ecosystem and made elk behave more like, well, elk.

What first attracted me to the Rocky Mountain Elk Foundation was the priority it placed on habitat protection. That's what makes this anti-wolf vitriol so hard to swallow. Wolves are native to much of elk country, which means that wolf habitat and elk habitat are often one and the same. Moreover, there are millions of people out there who don't hunt and don't care much about elk, but who care very much about wolves. The Rocky Mountain Elk Foundation is squelching an enormous opportunity to ally itself with wolf proponents and protect even more habitat in the long run.

At the same time that he denounces wolves, the group's president pledges that the organization will "become more engaged in the core issues of our time that threaten our hunting heritage and that of our children." But what if environmentalists like me—who wish to see not only ample elk populations but full, healthy ecosystems that include wolves—*are* the future of hunting? I consider joining other hunting organizations and get the same feeling from many of them—they don't represent me.

There is no shortage of conservation-minded sportsmen's groups that specialize in one species: Pheasants Forever, Ducks Unlimited, the National Wild Turkey

Federation. Don't get me wrong, these groups represent a true benefit of hunting: A constituency develops around a particular species to help pay for its habitat protection and vote for its long-term interests. And protecting the habitat for one species nearly always benefits others, too. But from an ecological standpoint, it doesn't always make sense for hunters to focus on one species. That's why I value groups like the National Wildlife Federation, which was founded by a hunter back in 1936 and still reaches out to hunters and anglers today. The National Wildlife Federation aims to protect entire ecosystems. In doing so, it's one of a handful of organizations working to bridge the way-too-distant gap between hunters and conservationists.

In my Hunter Safety class, our instructors spoke frequently about protecting what they—like the Rocky Mountain Elk Foundation president—called "our hunting heritage." E.V. told our class, "You might be the only hunter that a person ever meets. And so you are representing all of us." That's a big responsibility, and one that merits some real thought about how we are perceived by non-hunters. E.V. encouraged us, for example, to cover our dead deer when transporting it, rather than sprawling it over the hood of a car.

"Some people are offended by the sight of a dead animal," he said, "and we need to respect that."

I meet some hunters who boast about being offensive. They enjoy making non-hunters uncomfortable, plastering their bumpers with stickers that say things like VEGETARIAN: OLD INDIAN WORD FOR BAD HUNTER. They brag, as that Hunter Safety instructor in Virginia did, about being despised. But for every American who considers herself a hunter, there are at least nine who don't. In just about every election, non-hunters cast more ballots than we do. Non-hunters decide whether to post NO HUNTING signs on more acres of land than we do. They submit more comments than we do when hunting bans are proposed in public spaces. E.V. was on to something that many other hunters aren't: The future of hunting will largely be determined by non-hunters. It is in our best interest to try to get along.

CHAPTER 8

WILD TASTES

After a day of goose hunting, I bring home a goose, as well as a duck that another hunter shot but didn't want. The next week, I buy a whole chicken at the grocery store. As I cut open its plastic casing, I wonder about its life and how it compared with that of the goose I grilled just a few days earlier.

The greater white-fronted goose (*Anser albifrons*) is a grayish brown bird with orange feet and a white mottled front that inspires its colloquial name, the speckle-belly. Hunters sometimes call them "specks" for short. Whatever you call them, there are more than one million of them in North America. Like most geese, specks are gregarious, social animals that migrate between winter and summer grounds. They travel in large clans called skeins in the air and flocks or gaggles on the ground.

I'm not sure if my goose was a male or a female, as sexing this particular species requires expertise that I don't yet possess. But I'll assume, for the sake of personalization, that it was a she. I'm not sure of her exact age, either, but as a bird with mature coloring and an average weight (somewhere between four and a half and seven pounds) she was at least one year old, and more likely two or three.

My goose hatched from an egg in Alaska, where her flock spends each summer. She was probably one of about six chicks in the nest. She and her siblings fledged quickly but stayed close to both their parents. As summer turned to autumn in her native land, my goose and the rest of her gaggle soared south. They traveled for several months before settling for the winter in California's Central Valley or, perhaps, western Mexico. When spring warmed the air, the flock flew north again, returning to Alaska.

During the second year of her life, my goose found a mate. Specks, like most geese, are monogamous. But unlike Canada geese, who commit until one partner dies, specks are prone to their own version of divorce, and may switch partners after a few years.

On the last morning of her life, my goose awoke on a large farm pond in central Oregon. She and her traveling companions began arriving from the north about three weeks earlier, and found everything they needed—food, water and a lack of predators—within a short distance. The days were getting colder and shorter. Soon it would be time to fly south, but the group was not yet in any hurry. Until this cold mid-October day, that is, when mist wafted off the water as the sun crested over the surrounding buttes. My bird did not know it yet, but it was opening day of goose season.

If she was feeling especially peckish that morning, she might have awoken early and tilted her tail in the air so her beak could reach down for a weedy underwater snack. But she saved her real appetite for a nearby barley field. At about seven thirty, just after sunrise, my bird and the others lifted off the pond and flew to the field for breakfast. Here, she waddled up and down the rows, gorging herself on rich, golden kernels. Swallowing grain after grain, she began to fill the long crop at the bottom of her esophagus. It acted like a silo, dribbling food into her tiny stomach as quickly as it was pumped out into her gizzard.

While my goose ate her breakfast, I pulled a borrowed camouflage jacket over my fishing waders. (I left home the blaze-orange cap that I wear while hunting for rabbits or even pheasant, which rarely fly overhead. Geese—like other birds—have keen eyesight, and the same fluorescent clothing that alerts other hunters to my presence could also tip off a goose in the sky.) Then I crouched beneath some willows at the edge of the farm pond where she had spent the night.

Two hours later, the gaggle had filled their crops, and their overloaded digestive tracts declared an end to the meal. Three particular geese took flight. The skein usually traveled with three vanguards in front, testing a location to make sure it was safe for the others. The birds called to one another as they flew, and from the ground it sounded like manic laughter. My goose hung back with the hordes, waiting until the front-runners floated safely on the pond before setting her wings and gliding in for a landing.

As she approached, I stood up from my blind, shouldered my gun and switched off the safety. I'm not sure why my eyes locked on to her, instead of some other goose in the skein. I lined up the tip of my gun with her neck, to better my chance that one round ball of steel would intersect her slender neck or head. For a second, I held the tip of my gun so it just covered her, making sure my own movement matched the speed of her flight, twenty yards away. Then I pulled the trigger.

She fell into the water immediately. A couple of her feathers drifted behind her in a lazy zigzag, as if in a cartoon. Unlike a cartoon, though, my goose did not pick up her head and roll her eyes at the camera. Stars didn't swirl around her head. She did no double take. She was dead. This time, unlike when I killed the pheasant, remorse wasted no time in finding me. I felt guilty as soon as I picked her up. One moment, this beautiful, well-traveled bird was soaring overhead. The next, she was floating limp in a pond. All because of me.

But her death, I told myself later, was just one part of her life. And her life, though undoubtedly strenuous—if she was just two years old, she had already flown thousands of miles and seen more of the continent than most American citizens do—was not necessarily sad.

My store-bought chicken (*Gallus domesticus*) was another story. As with my goose, I'm not sure whether it was male or female, as both sexes are raised for food. Let's say, for the sake of storytelling, that my bird was a male Cornish rock. My chicken's eventual packaging identified him only as a "broiler," the industry's generic term for a chicken raised for meat (as opposed to a "layer," who provides eggs).

He hatched from an egg and likely spent all but the last few hours of his life in a low-slung shed with about thirty thousand other chickens the same age. Shortly after he hatched, the tip of my chicken's beak was cut off, without anesthesia. For the first week of his life, the lights were left on twenty-four hours a day, to encourage him to eat as much as possible. After that, the lights were turned out for just four hours a night, again to minimize sleep and maximize weight gain. Although he was technically a free-range bird, there is a chance that he never set a claw outdoors or felt a ray of natural sunlight. To earn the "free-range" badge, chicken producers must provide a small outdoor exercise yard. But because these doors aren't opened until the chicks are a couple of weeks old, the birds have no incentive to leave the one home they have ever known. By the time they're full grown, the chickens are so tightly packed—in most commercial operations, each bird occupies a space smaller than an eight-and-a-half-by-eleven sheet of printer paper—that those located near the center of the shed can't easily reach the door.

Unlike most chickens, mine was organic. His feed, therefore, did not contain growth hormones to rush his physical development. But even without performance-enhancing drugs, his breed was developed in the 1940s with the primary goal of growing pornographically large breasts at an astonishing rate. If my chicken was lucky enough to be able to walk until the last day of his life—and let's say optimistically that he was—then it's safe to assume that three of his four closest neighbors were not so lucky. The broiler just to the north of him, for example, its wing touching my chicken's wing, one day felt its leg snap under the weight of all that breast meat. This injured bird was not removed from the shed, but left to lie and wait until the entire flock had reached slaughter age, about seven weeks. Lame chickens are slaughtered and sold along with the sound ones. (Dead chickens, however, are removed from the shed every day or so, and are not sold as meat.)

The last day of my chicken's life began as all the others had, with the shed's bright lights switching on. He opened his eyes and began to eat. At some point, the doors at one end of the shed opened and a group of people wearing white hazmat suits entered. They grabbed chickens upside down by their legs. Clutching several birds in each hand, they thrust them into transport crates. Some of the birds were dead by the time these workers reached them. They were tossed in a pile and discarded. The live, crated birds—including mine—were loaded onto a truck and transported to a slaughterhouse. There, my bird was unloaded by a different worker, who slid his ankles into shackles. As he hung upside down, a conveyor system dragged him through an electrically charged water bath that paralyzed most of his body. His eyes and beak might still have been able to move as he reached the next station, where a mechanized blade sliced through his throat. My bird's two-month life was over.

To me, the story of my chicken's life is an undeniably sad one, although I can't know what my chicken thought and felt. All of animal ethics hinges on assumptions about what it means to be a species other than human, something we can imagine and guess and study but can never really know.

Here is what I do know: I bear responsibility for the death of my goose. But I bear responsibility for the entire life and death of my chicken. And one of those scenarios is more bearable to me than the other.

My goose probably endured near-misses by other hunters and possibly cars. She endured hard, hungry winters. She may have lost a mate to hunting or to disease. Some of her goslings were probably snatched up by a fox before they could fly. But none of that was my fault. My chicken would never have lived at all if not for demand by meat eaters like me. No matter how little I saw of it, everything about my chicken's life and death was my doing. A natural death was out of the question because his entire life was, in a way, unnatural.

At times, I catch myself romanticizing the idea of a "natural" death. I think of it in terms of a peaceful human death: *He died in his sleep.* But nature can be as brutal as any slaughterhouse worker. If my goose hadn't been shot, she could have suffered a broken wing or leg and eventually starved to death, slowly and painfully. She could have been torn apart by a coyote or chopped to pieces by an airplane engine.

As far as I was concerned, both my goose and my chicken had tasty, nutritious afterlives. I skinned and gutted my goose within an hour of her death. We took one bone-in breast to the house of some friends and grilled it, then ate it with fresh salmon that one friend had recently caught. The meat was dark and rich, more like steak than chicken. A few days later, I deboned the rest of the goose and cut the meat into long, thin strips. I marinated it, in the fridge, in a bowl of soy sauce, garlic, brown sugar and dried red pepper flakes. We got out a small metal smoker that I recently bought used. I stoked a charcoal fire in a pan at the bottom, then laid the strips of meat on a grill on top. Hours later, the jerky was stiff and dry. We nibbled on some, and saved the rest to pack on long ski trips.

I unwrapped the chicken's plastic suit and tossed the giblets in the garbage can. I patted the bird dry, then filled its cavity with a paltry amount of stuffing. (I've often wondered why these carefully engineered birds aren't bred to have larger cavities, since I'd take a second helping of stuffing over more breast meat any day.) I roasted the bird and served it as the centerpiece of a Thanksgiving-like feast.

But unlike with the goose, there was no talk of the chicken's life, no heartfelt toast. Just dinner as usual.

Or, rather, what has only recently become usual. My interest piqued by imagining the lives of my chicken and goose, I seek out books about the industrialization of food, which I am startled to discover is such a recent phenomenon. Agriculture itself is only about eleven thousand years old. In *An Edible History of Humanity*, Tom Standage tells us that if all of man's 150,000-year history were likened to one hour, "it is only in the last four and a half minutes that humans began to adopt farming, and agriculture only became the dominant means of providing human subsistence in the last minute and a half."

Today more than nine out of ten land animals killed for food in the United States are broilers. Fast-food chain KFC alone buys nearly one billion per year. On top of that, more than 250 million chicks are destroyed each year, most of them layers who—oops!—turned out to be males. Michael Pollan writes: "The industrialization—and brutalization—of animals in America is a relatively new, evitable, and local phenomenon: No other country raises and slaughters its food animals quite as intensively or as brutally as we do."

I read through stacks of books on modern food production, and the more I learn about the treatment of livestock, the more enthusiastic I become about hunting as a compassionate, alternative meat source. These animals are free-range, and if their lives are not exactly easy—life in the wild is full of hardship—well, at least they are free. Before it lands on my dinner plate, a wild duck gets a chance to be a real duck, to dabble and splash and migrate and do everything that other ducks have done for thousands of years.

The meat that I buy from the grocery store cannot be

as special to me as the meat I hunt and kill myself. But those few, extraordinary meals that result from a hunt can—and do—change the way that I feel about even boring old livestock that I still buy from the grocery store. As I nick the plastic with my knife and unwrap the seal around a package of boneless, skinless chicken breasts, for example, I recognize the taut, quivery flesh of my pheasant. I recognize the graceful curve that was once covered by skin and adorned with feathers. My experiences in start-to-finish butchery have granted me the ability to mentally rebuild the bird, bone by bone, slice by slice, pluck by pluck. I can picture its former self, like a ghost. Until even the floppiest, most unrecognizably boneless, skinless cutlet is no longer just a piece of meat. It's a piece of an animal. A piece of its life.

I used to forgo packages of thighs or breasts in favor of a whole animal—a chicken or turkey—only once or twice a year. Just rinsing it, patting it dry and placing it in the roasting pan felt daunting. Its animalness was too obvious, too on display. I felt guilty and clumsy while handling it raw. I hurried to get it in the oven, so it could be transformed into something easier to sit with—a browned, aromatic meal. But with each animal that I gut and dress, I gain confidence. I begin to buy whole chickens from the supermarket, instead of packages of breasts or thighs. It's cheaper this way. And there is less waste—something that is becoming important to me.

A 2009 study found that more than 40 percent of all

food produced in the United States is thrown away instead of eaten. Waste occurs during food manufacturing and distribution, but most of it is attributable to consumers who buy the food and then throw it away. After shooting a bird and seeing it as not only a piece of meat but a whole life, I find myself loath to throw away even a scrap of meat.

Something else happens—unrelated to hunting—that makes me even more conscientious about wasting food. Scott and I become friends with an elderly man named Raymond who lives next door to us. Raymond lives in the same seafoam-green bungalow where he was born. It has two bedrooms and one bathroom. Raymond was born with some sort of developmental disability, one whose name we never learn. When he speaks, he slurs and is hard to understand. With nothing but meager Social Security and disability payments to keep him afloat, Raymond must pinch pennies to make sure he has enough to eat each day. Soon Scott and I find ourselves driving Raymond to various grocery stores so he can take advantage of coupons. Each outing turns into a long ordeal. He doesn't read very well, so it takes him several minutes to discern whether a sale price applies to a particular can of stewed tomatoes. Sometimes I can't bear to look at the canned chili and stale donuts in Raymond's grocery cart. Because he is always seeking the best deal, he rarely ends up with the makings of a healthy diet.

I think of Raymond when we finish eating a well-balanced meal. If we have leftovers, I pack them into Tupperware containers and carry them next door. Every once in a while, we invite Raymond to eat dinner with us.

One night, Raymond sits at our table and tells us a story we've already heard dozens of times: After his father got into a car accident and was unable to work at the lumber mill in Bend, the government awarded him a scholarship to go back to school to learn a new trade. The whole family moved to Klamath Falls, about 130 miles south of Bend, for two years.

"Did you like Klamath Falls?" I poke around for a new layer in this oft-cited tale.

"Eh," he says, shrugging. "It was okay. Not as good as Bend."

"Why not?"

"The water didn't taste as good."

Occasionally, Raymond does this: He reduces something to such a basic level that he ends up saying something profound. After all, when considering how to rank two small eastern Oregon towns, what could be more essential than comparing the quality of our most basic need?

Living in Bend for more than four years, I've thought a lot about what makes a place worth calling home. I've grown frustrated at work, and I'm toying with the idea of applying for jobs in other states. Reporters move a lot, which is just a fact of the profession. And I've been here so much longer than I ever planned. Maybe it's time to move up to a bigger, more respected newspaper. I browse the dwindling job ads, but it's a tough time to be in the market for a reporting gig. In my head, I turn over the name of each place—Anchorage, Salt Lake City, Sarasota—and wonder if it would be worth it to move there.

I've always seen myself as a city person, but now that I'm faced with the choice of staying in Bend or moving to an only slightly larger city, I find myself reconsidering what it is that appeals to me about cities. The diversity interests me most, the hodgepodge of languages and backgrounds, the variety of ethnic foods, the array of nightlife. More people means more chances to make friends, right? But I can't shake this niggling feeling that maybe these things aren't what I want, after all. Maybe what I'm finding right here—in Bend and the surrounding landscape—is just as valuable.

I hash out my dilemma with Scott. He tells me that he will go wherever I want, but I can tell that deep down, he wants to stay in Bend. For the umpteenth time, I run through the pros and cons of moving to a larger city, a larger paper. When I ask for his opinion, he sighs.

"Lil, I don't think about these things the same way you do."

"What do you mean? You don't ever think about leaving your job for something bigger and better?"

"I don't know." He pauses. "I guess I don't want as much as you do. I mean, yes, I want a job that gives me some amount of satisfaction. But at the end of the day, it's always going to be just a job to me. More than anything, I want to come home to you. I want us to go camping on the weekends, maybe fishing, maybe skiing. Maybe go to a movie, maybe stay home and read. That's all I want."

I'm embarrassed that I've spent so little time thinking about how much good is in my life right now, right as it is. The things that Scott just framed as modest desires, aren't they really the biggest, most important things in life? Someone to love, a decent income, hobbies that make us happy. Why should I—why should anyone—want more than that?

In the fall, our friend Betsy takes us mushroom picking. Mushroom hunting—like animal hunting—had a bad reputation in my family when I was growing up. We actually knew of one person who hunted mushrooms on the way home from the subway station, which was fine in theory—there were no guns involved, after all, and no animals were harmed. But still we concluded that this man was crazy. All mushrooms looked pretty much the same to us. What if he plucked the wrong kind by accident and died of mushroom poisoning? How could one ever know *without a doubt* that a particular wild fungus is safe?

As with animal hunting, my life in Oregon has dulled these concerns. For thousands of years, people have foraged for wild food and survived. There is a body of hard-earned knowledge out there, practically begging to be passed on to other generations. The people I know who forage for mushrooms have reassured me: If you know what you're doing, it's very safe. (And if you're ever in doubt, they add, you don't eat it.) Only a few known species of mushrooms are, if ingested, capable of killing an otherwise healthy adult. Most poisonous mushrooms cause run-of-the-mill food-poisoning symptoms: diarrhea and vomiting but nothing close to death. Besides, after hunting with loaded guns, mushroom hunting sounds relatively stress-free. So one Saturday morning, we pull on comfortable, layered clothing and sturdy boots and gather a few canvas tote bags to carry our bounty, in case we get lucky.

In the car, Betsy turns to us and delivers a stern lecture.

"The biggest rule in mushroom hunting is that you *never tell anyone where you went*," she says. "I'm serious. I'm showing this place to you because I trust you."

This code of secrecy is common among practitioners of all different sports in Oregon. When I first moved here, it annoyed me. Anglers complained when a prominent outdoors magazine profiled their favorite "secret" fishing holes. Skiers lamented any published ballyhoo of their favorite backcountry trails. Hunters bemoaned that a guide led a paying customer to the clearing where they, unguided, shot a buck three out of the last five years. I understood the fear that a favorite, seldom-visited spot could suddenly become overrun with outsiders. What bothered me was the sense of ownership underlying that fear. I want to say to each of these whiners: You think you discovered this snow-covered slope? This deep bend in the river? This patch of browse? Sorry, bub, it's the twenty-first century. People have scaled this hillside, fished this stream, hunted this piece of ground for thousands of years. You are not the first. This world is everyone's to explore.

But when I try mushroom hunting, I sort of get it.

There are exceptions, but the general rule is that mushrooms tend to grow in the same spots, year after year. And they don't fruit year-round. More than one person can ski the same slope, fish the same waters, hunt the same forest. But if you return to a familiar, fertile ground and another forager has just picked it clean, you're out of luck. Betsy has spent years hunting mushrooms with a friend's Japanese American family. Her friend's grandmother knew, for example, that one particular pine tree was surrounded by one mushroom species at a precise time of year. One year, she went to the tree and other pickers had not only removed all of the mushrooms but raked the ground. Some pickers do this to churn up every last mushroom, but raking can destroy the underground roots that would produce next year's fruit. Betsy saw this woman mourn a true loss that day.

As Betsy recounts this story, I realize: What is the harm—other than grouchiness—in laying claim to a place? Perhaps that's our best hope for protecting it: a constituency of people with real connections to the land.

Mushrooms are not technically plants, and they're obviously not animals. They are the sole occupant of a third kingdom of life: fungi. The capped mushrooms that we're used to—the kind you could slice and sprinkle on a pizza—are actually the reproductive structures of an organism that is largely invisible. It's off-putting to think of these delicacies as genitalia so we use the more polite, poetic term: *fruiting bodies*. Unlike plants, which produce their own food, mushrooms must ingest nutrients from another source. The mouth and digestive system—the gut of the fungus—is a web of thread-like fibers called mycelium. In most of the fungi we eat, the mycelium is buried underground. In some, it weaves through a tree trunk.

Betsy takes us to her secret spot, and we hike around looking for chanterelles and matsutake. She quickly finds one of each and explains how to identify them. The chanterelles (*Cantharellus cibarius* and *C. subalbidus*) look like trumpet-shaped coral. Some are white with amber edges; some are yellow or orange all over. All are beautiful. There are a few inedible (only one is poisonous) mushrooms that could possibly be mistaken for chanterelles, but Betsy, an experienced forager, offers to confirm our finds for us.

The matsutake (*Armillaria ponderosa*) is rarer than the chanterelle. It grows primarily in the Pacific Northwest, though a closely related species blooms in Japan. It is white and unremarkable-looking, similar to the cultivated mushrooms sold at any grocery store. But then Betsy turns it upside down and instructs us to inhale, our noses to its gills. The spicy fragrance is unlike anything I've ever smelled. In what will become my favorite mushroom guidebook, *Mushrooms Demystified*, David Arora describes it as "a provocative compromise between 'red hots' and dirty socks." To me, the scent is an earthy-citrusy combination, like fresh grapefruit squeezed over rich, loamy soil.

We spread out and begin walking across the forest, eyes trained on the floor. Betsy has an almost superhuman ability to spy mushrooms from several yards away. But to my eyes, a few dried pine needles and the dim forest light manage to camouflage even the brightest chanterelles. The matsutake are even more difficult to spot because you can rarely glimpse the mushroom itself. Instead, you look for a raised clump of duff and then use a stick to nudge it over, hoping to unearth a white mushroom cap. Betsy advises us to crouch down, to better notice any promising divots.

I find myself enjoying this search even though I'm not picking anything. I notice parts of nature that I would normally overlook—lacy bits of lichen, fluorescent green and robin's-egg blue, strewn along the ground. And though I'm not seeing either type of edible mushroom that we're targeting, I do see mushrooms galore. Small purple toadstools poke out of black soil. Glistening smooth caps perched atop slender stems pepper a mossy log, as if slimy mushrooms were cast in the artistic reenactment of an ant colony. Shelf mushrooms jut out from the trunks of live trees.

I bend over to examine yet another mound of duff that is probably just that. A white ridge seems to protrude from the clump of soil. I take my stick and flip over the clump. And there, lying exposed, is a giant white mushroom cap. I gasp.

"Betsy," I yell, "I think I found one!"

I drop my stick and use my hands to shovel soil away from the stem. Next I grasp the thick stalk and rock the whole mushroom back and forth until it tears away from the ground. It's magnificent—the top is more than six inches in diameter, and its whole being reeks of that strange, fruity scent. But it's bigger than the matsutake Scott and Betsy have found so far, most of them just a couple of inches wide. And the cap is less spherical, more convex. Could this really be the same species?

Betsy appears beside me.

"That's huge! Is it a matsutake?"

"I'm not really sure."

I'm almost shaking as she takes the giant mushroom, turns it over and smells it. "Mmmm... It's definitely a matsutake. Good find!"

For the rest of the day, I feel giddy.

Mushroom hunting as recreation has developed a sort of whimsical persona. The species bear common names that are anything but—angel wing, man on horseback, ma'am on motorcycle, shaggy parasol, poor man's slippery jack, dead man's foot. When mushrooms pop up in naturally occurring circles, the formations are called "fairy rings." Certain mushrooms, called candy caps, have a sugary taste when dried. Foragers pick them, dry them, chop them up and bake them in cookies or sprinkle them on ice cream.

I haven't admitted this to Betsy but the truth is, I don't like eating regular mushrooms. The texture and the taste are just a little too unusual, although I will choke them down to be polite or sometimes out of laziness. With these wild mushrooms, however, I can't wait to dig in. Finding them was such a joy that I expect their taste will be, too.

We cook only the chanterelles the first evening. Some people have allergic reactions to wild mushrooms, so Betsy warns us to take precautions. We eat just a small amount of one type of mushroom at a time. And we don't drink much alcohol with this meal, as the interaction can compound allergies. Betsy tells us to wait a day to see if we have any reaction—intestinal distress or even a rash—before eating more.

We opt for the simplest preparation: sautéed in butter. Instead of chopping the chanterelles, Betsy shows us how to peel them apart like string cheese. They are mild-flavored with a firm, fibrous texture similar to chicken breast. They aren't bad, exactly, but I don't enjoy them as much as I'd hoped.

The next day, I discover that the matsutake have a pungent, otherworldly flavor, as their smell would suggest. Unlike the chanterelles, they can't be gummed and swallowed quickly. You have to really chew them, like calamari, and that just releases more of the strange taste. I get the first slice of sautéed mushroom down but gag on the second. I push my plate away and watch as Scott relishes the rest of what we cooked. I'm jealous of him, and disappointed in myself for not having a more sophisticated palate.

I once wrote an article for the newspaper about the Pacific lamprey, a parasitic fish that hatches in streams, then swims to the ocean and eventually returns to fresh water to spawn. These eel-like animals are one of the oldest species still living on earth—an estimated hundred million years older than the earliest dinosaurs—and they've been an important food source to local Indian tribes for thousands of years. But lately, tribal elders have had to coax children to even try a bite. Their palates are accustomed to cheeseburgers and fries, not dried or char-grilled lamprey.

Our palates change over time, depending on the foods that are most available in our society. In his book *Putting Meat on the American Table*, historian Roger Horowitz traces the change on American dinner tables from cured to fresh pork. Our taste for salted barrel pork—a staple that any reader of *Little House on the Prairie* will recognize—dissolved in the late nineteenth century as fresh beef appeared in more urban markets. As the cultural preference tilted toward beef, pork producers were forced to reexamine their offerings and switch to products such as fresh chops and loins, which had a flavor and texture more similar to beef. Two hundred years ago, when more of us ate wild game, Americans undoubtedly had a different palate than we do today. Although humans could eat hundreds of animal species, only a few are readily available to us. Just 14 of the world's 148 large terrestrial mammals have been domesticated. As someone who grew up eating chicken, pork, beef and occasionally turkey, I found even duck meat difficult to eat at first, more flavorful than what I was used to. Not knowing how to describe it, I fell back on that generic, derogatory word: "gamey." But with time, I learned to love duck meat. Befitting of waterfowl, it tastes to me like slightly fishy-tasting turkey.

I once read an article about picky eaters, which I vaguely remember mentioned that a child's palate changes so much while growing that she must taste a food at least seventeen times before knowing for sure whether she likes it. Though I'm fully grown, I wonder if I've given mushrooms a fair try. If I eat them twenty, thirty, forty times—will I start to appreciate them? On New Year's Eve, I hatch a resolution: I will eat mushrooms at least once a week this year.

During some of the early meals, I almost resort to holding my nose while I chew. But over time, it gets easier. I find myself tossing mushrooms into my stir-fry or sprinkling them on pizza by choice, rather than a sense of obligation. And by the fall, I am eager to get out and pick mushrooms. I want to feel that thrill of finding a miraculous mushroom again. I also want to bring it home, cook it up and savor it the way it deserves.

The rush of finding a wild mushroom reminds me of what I felt after shooting my pheasant. It also reminds me of a camping trip that Scott and I took earlier that fall. We hiked miles into a wilderness area to fish in a remote pond that we'd spotted on a map. When we got there, the pond was surrounded by wild huckleberry bushes. I never even strung up my fly rod, just squatted down and began picking berries. Fortunately, I had a few large ziplock bags in my backpack. The minuscule fruit was dark purple and tangy sweet, as if a blueberry and a raspberry had a tiny, delicious baby. We froze most of the berries. Each month, we scooped out a small allotment to sprinkle on pancakes as they sizzled, or to cook into a sweet, sticky sauce to pour over ice cream.

Foraging for wild food gives me an almost religious feeling of serendipity: When I stumble upon a hillside of gleaming, ripe huckleberries, or unearth a fragrant mushroom, the universe is confirming that I am in exactly the right place at exactly the right time.

CHAPTER 9

GOOD DOG, BAD WOLF

The same fall that Betsy takes us mushroom hunting, I continue to get out and hunt for birds. Of all the things I enjoy about hunting, my favorite is watching the dogs. I jump at any invitation to join an expedition that includes a trained dog. These animals are incredible, first because of their physical abilities: to sniff out a bird and then startle it and hold it still. Next, they astound me with how well they work with humans. They read their owners' body language and anticipate their slightest movements. I notice their happiness in the field, their satisfaction while retrieving a bird their owner has shot. This makes me rethink my relationship with my own pet dog.

I adopted Sylvia three years ago in April. Like so many

modern-day love affairs, ours began online. After browsing Petfinder.com for months, I stumbled across a profile for Missy, a one-year-old female who had been picked up as a stray and was thought to be a flat-coated retriever mix. I'd read that the breed was well mannered, playful and large enough to go running or skiing, but not as unruly as the hundred-pound Great Pyrenees whom Scott already owned when we met. One Saturday, we drove Scott's dog, Bob, out to meet this mystery gal, and the two ignored each other completely. Even when we took the pooch home with us, and she and Bob rode together in the back of Scott's car, they avoided all eye contact. The next morning, they finally acknowledged each other by playing wildly in our backyard.

Back then, I'd had no real interest in hunting. But in hindsight, if I had thought to expose her to birds right away, Sylvia probably could have become a fine hunting dog. Instead, we raised her with little access to non-human or canine animals—with one notable exception: fish.

Scott and I started taking Sylvia on fishing trips immediately after we rescued her. She was good company and, unlike Bob, she stayed close enough and obeyed commands well enough that she rarely caused trouble. I can't imagine exactly what was going through Sylvia's diminutive head the first time she watched Scott hook a fish, but I know that it was a life-changing event. We were standing on the bank of the Williamson River, next to a deep pool of crystal-clear water. The fish chomped Scott's fly and leapt into the air. And Sylvia lost her head. She surged into the current, swimming in circles where the fish had breached a few moments earlier. Periodically, she plunged her entire head underwater to glimpse the fish. Sylvia's ancestors may have been bred to find and retrieve birds, but from that moment on, Sylvia has seemed perfectly happy dedicating her life to fish.

She knows more about fly-fishing than most humans do. When fish are rising to feed off the surface of the water, she often notices before I do. If I'm fishing with a dry fly—an imitation of a floating bug such as a mayfly—she stands aquiver and stares as it drifts downstream. If I'm fishing with a nymph—a weighted, sunken fly—she watches the neon foam strike indicator that I attach to my line. And she goes bonkers when it pops underneath the surface. She can barely control herself when she hears the whir of a reel. And when I get my fly caught on a piece of submerged wood and reel in a stick instead of a fish, Sylvia cavorts around the bank with the "catch" (after I've unhooked it and handed it to her) in her mouth, triumphant. On the rare occasion that I do hook a fish, Sylvia watches rapt as I try to reel it in, occasionally becoming so impatient that she tries to swim out and greet it. When I let the fish go, she bounds over to the spot in the water where I last held it and plunges her head in, hoping to catch a whiff. After a few years of fishing, Sylvia knows where to stand to be close to the action when a fish is hooked and landed. When I'm nymphing, she knows at what point in the drift a fish is most likely to strike, and her excitement builds as the fly approaches.

Sylvia loves fly-fishing more than most humans I know do, including me. We once took a long-weekend camping trip in which both Scott and I fished non-stop for three days and caught *nothing.* By the end of that kind of weekend, I was barely paying attention to my line anymore. The cast was automatic, mere background to my vivid daydreams. But Sylvia ran back and forth along the banks between us, checking our lines and watching, her muscles tense with optimism. She knew that any cast could be the one that resulted in a squirming fish.

Once, Scott and Sylvia were fly-fishing alone along a small, remote stream. Scott flogged the water all day and caught nothing. In late afternoon, they turned to hike back to the car. After a mile or so, Scott noticed that Sylvia had stopped a little way ahead of him. He could only see her tail and back end, as her head was thrust into some tall reeds on the river's edge. As he tiptoed closer, he saw her whole body. She was standing on point, perfectly still and at peak attention. Scott peered past the reeds and into the water, which dropped off steeply from the bank. There, about six feet below, in a deep, calm pool, was a giant rainbow trout. Just as Scott spotted the fish, Sylvia torpedoed into the water, headfirst, with her mouth opened wide. She popped up, stunned and disoriented, with an empty jaw. When Scott came home, he looked as if he'd discovered time travel. His eyes were wild and he couldn't wait to tell me, in perfect detail, about what Sylvia had done. He was amazed—first that she had somehow spotted the fish and, second, that she dove after it.

"What would you have done if she came up with that fish in her mouth?" I asked.

"Are you kidding?" He shook his head, and I realized that Sylvia's disappointment was no match for Scott's. "I would have bonked it on the head, grilled it up and fed it to her. She earned it."

When Scott pats Sylvia during a post-fishing slumber, he turns to me and says: "I'll probably have a lot of dogs in my life. But none of them will be as interested in fishing, none will love to do what I love to do, as much as Sylvia." He's right; how often does a family pet share our own hobbies?

I love my mutt to an embarrassing degree. Scott and I have coined dozens of nicknames for her (Wiggles, Wigs, Sylvester and Peeps McGoo are just the beginning). And though she can't speak for herself, all indications are that she loves me back. But her affection for me stems, I think, from my feeding her, walking her and playing ball or Frisbee with her. I am her caregiver. We are companions. Yet I wonder how much deeper our connection would be if we were also hunting partners, if I could draw out her latent instincts and plumb the long-bred abilities she doesn't know she has.

Wolves and dogs are the same species, and their genomes are almost impossible to distinguish. Even a Chihuahua is just one taxonomical subcategory away from a wolf. This bald, pint-size version of the wolf has simply been through an unnatural breeding process, carefully controlled by humans. This alone is what qualifies it as *Canis lupus familiaris* instead of *C. lupus.* And even though we created dogs, we don't have good records of how we did it. It's only in the last decade or so that scientists have turned serious about their studies of the history of the dog.

Other than humans, wolves were once the mammal with the most varied geographic distribution on earth. They were also one of our biggest predatory competitors, which is why we managed to wipe them out of so many different regions. But as this widespread annihilation was taking place, we were also taming some members of the same species into our best friends. Evidence suggests that humans domesticated wolves simultaneously in several places around the world. There are various theories about exactly how wolves were tamed. In one particularly odd theory, wolf pups were stolen from their dens at one or two days old, and lactating women nursed them from their own breasts. Regardless of the specifics, each theory rests on some level of consent from the wolves themselves. For example, some scientists believe that Siberian huskies stem from a population of semi-domesticated wolves who flocked to nomadic tribes of people when hunting became difficult during harsh winters. The people tied the dogs up and taught them to pull sleds. In return, they fed the dogs. Then, when the snow melted, the dogs went off and lived in the wild, again hunting on their own, until winter.

Although the "how" of domestication is still a mystery, researchers tend to agree on the "why." Nearly every theory involves a symbiotic relationship that includes food and hunting. Humans and dogs have been hunting together for as long as forty thousand years, according to some estimates. Many of the dog breeds that we recognize today were developed much more recently to specialize in one category of game or one stage of the hunting process.

Thick-coated dogs such as Labrador retrievers, for example, are naturally equipped to brave icy waters while fetching downed waterfowl. Dogs who pursue upland birds—birds living on dry land, including chukar, grouse and pheasant—are usually either pointers or flushers. Pointers, such as German shorthairs, run way ahead of the hunter to track prey and then freeze, holding a bird in place until the hunter draws near. Flushing dogs, such as cocker spaniels, stay close to the hunters and scare birds into the air as they find them. Both categories of dogs are easily trained to retrieve birds who've been shot. Today the vast majority of American dog owners do not hunt. But their four-legged companions still have these hunting abilities coursing through their veins.

Armed with a couple of books, I try to teach Sylvia to retrieve birds. I drag feathered bird wings around our yard, then let her out and encourage her to sniff them out of hiding. But she's not used to following her nose, and doesn't notice anything different about the yard.

I tell her to sit and stay while I throw a duck decoy into the water. But after years of dashing madly after balls and Frisbees the moment they leave my hand, she struggles to wait. For a hunter, this is a liability. You don't necessarily want a dog leaping into the water as soon as you down a duck: Other birds at whom you want to take aim might approach. My next pet, I resolve, will be a bird dog. For now, though, I am happy with my ball-chasing furry friend.

In 2009, wolf hunts open in two Western states for the first time in decades. Personally, I can't imagine shooting a wolf—it is too closely related to a dog, the animal with whom I have forged my deepest human-animal friendship. But I don't entirely oppose wolf hunts, either, because they could build tolerance—among conservative ranchers, especially—for a self-sustaining wolf population. As I've mentioned, when hunters are allowed to pursue an animal, a constituency develops around that species. It reminds me of controversial, high-priced hunts in Africa and Asia for endangered animals. As a conservationist, it pains me to think of someone pulling the trigger on a rare tiger. But if the astronomical price tag for hunting one tiger raises enough money to protect hundreds of them, isn't it doing more good than harm?

Wolf reintroduction raises ethical questions that I would not have considered before I started hunting. As an environmentalist, I believe in the principle of reintroduction. We should right the wrong that we committed by removing the species from giant swaths of North America more than sixty years ago. But this reintroduction comes with caveats to which the wolves cannot possibly agree. For example, state and federal management plans call for an individual wolf to be "fatally managed," or killed, if it preys on multiple livestock animals. Wildlife managers assume that wolves are capable of quickly learning the difference between wild and domestic prey. Scientists who study wolf and dog behavior, on the other hand, aren't sure that's a realistic demand. Human activity, particularly ranching, now takes up the vast majority of wolves' available habitat.

Wolves have no shortage of enemies. Ranchers seem to universally despise the animals for preying on their livestock. And hunters are surprisingly eager to join the wolf hatred, too. To combat all of this, some environmental groups have launched campaigns to discourage fear of this animal that humans have loaded with symbolism for millennia. Non-profits in Montana and New Mexico, for example, actually bring wolves into school classrooms. The message from such groups is that wolves are gentle creatures, not fierce ones, and so we should be kind to them. But wolves are not gentle, they're wild animals. There is a real danger in demanding that animals be good, kind, friendly—or any other anthropomorphic trait—to deserve respect or protection.

Sadly, wildlife management is a political sport. Interest groups—including sportsmen and environmentalists, who often have opposing aims—wield power over the politicians who allocate funding for state agencies. This means the wildlife managers who set hunting seasons and quotas are beholden to politicians. Worse, they could permit irresponsible levels of hunting in an effort to collect the maximum number of fees, bolstering their short-term budget.

Many hunters, I've found, think about individual animals as members of a larger population. As long as the population is healthy and stable, a certain number of individuals can be ethically harvested for food. In some cases, these killings are actually beneficial to the health of a population. Well-managed hunts can cull a population that has outgrown an ecosystem and is causing damage to the land. Hunting discourages animals from becoming tame or moving into developed areas, reducing the number of human-wildlife conflicts. In a recent study, well-regulated hunts were found to help migratory animals adapt more quickly to habitat changes.

Occasionally, we even think of ourselves in such a way that the good of the population overrules the health of one individual. Take clinical medical trials, for example. Doctors willingly, and in some cases knowingly, administer placebos—substances that they know will not slow disease or save lives—to their patients with fatal diseases. But they do this with the hope that the one patient's experience—even one that fails to slow death—will eventually save more lives.

Before I began to hunt, I thought of animals as individuals, with families and emotions and a whole slew of anthropomorphic traits. This strikes me as the environmentalist, vegetarian, animal lover's approach: Any death of any individual being is painful and bad. The trouble is, I now think of animals both as members of a population *and* as individuals. It makes for a lot of hand-wringing.

But maybe it's a necessary paradox; it's what makes me a responsible hunter.

When summer arrives, we add to our usual camping and fishing trips a new activity: scouting for birds. We look for small ponds and streams where ducks congregate. We peruse rocky river canyons for signs of an upland game bird called a chukar partridge. And then, in the fall, we camp in southeast Oregon and hike through steep canyons in search of these elusive birds. Before this trip, I have never seen a chukar in person, and everything I know about the species I learned in books.

The chukar (*Alectoris chukar*) is native to Eurasia, and especially common in the western Himalayas. A chukar is about the size of a pigeon and mostly black and white, with a dark stripe around the eye and bright red legs, eyelids and beaks.

These avian homesteaders prefer habitat from which other game birds would flee: steep, rocky cliffs in harsh, arid climates. Because they live in such steep terrain, chukar hunting is grueling. This is especially true without a trained dog to sniff out the birds. Scott and I cover miles of steep ground before I see one lone chukar. It flies out of sight before I can recognize it and shoulder my gun. Because they live in areas with few trees, chukars spend most of their time on the ground. When startled or roused, they run or fly quickly uphill and over the canyon edge.

Despite my lack of traditional success—we come home with an empty cooler—I enjoy each day's outing. I listen for the birds to call to each other—chukka chukka chukka—as if they're cackling their name. I look for insects and seeds, including the small kernels of native Oregon bunchgrasses, that the birds might munch on.

I notice how much more satisfying it is to climb a steep canyon in pursuit of an animal than to hike the same feature for the vague sake of recreation or exercise. Scott and I don't hike much, but we have started cross-country skiing together, which is like hiking on skis. Scott has skied his whole life. I went downhill skiing a couple of times in high school and college, but haven't cross-country skied before.

On one of our first outings, we drive to the Edison Sno-Park, a network of trails with electricity-related names. We peruse the map and decide to ski to the AC/DC shelter because it has the best name. Most of the route to the hut is uphill, which is difficult on skis but not impossible. Directly under each foot, the base of my skis is etched into a fish-scale pattern. By stomping it into the snow, I can grip the trail. Or I can point the tips of my skis outward and walk straight up, leaving a herringbone pattern in the snow. Or, on steeper terrain, I can turn my skis perpendicular to the hill and sidestep up it. Most of the time, no matter what I choose, I still manage to slip backward, catching myself only with my poles.

Just as I start to get tired, we ski past a sign declaring that the shelter is one and a half miles away. About ten minutes later, we pass a second sign saying the same thing. By the time we reach the third sign, I am imagining myself in a horror movie. It would end with me collapsing, dehydrated and exhausted but still one and a half miles from the AC/DC shelter. I imagine myself as a modern-day Sisyphus and also the boulder, doomed to futilely push myself uphill.

Shortly after the third sign, I slide off the trail and land on my back in a pile of snow, soft enough that it doesn't hurt. It's so soft, however, that I can't stand up—there's no surface to push off. I thrash and wiggle, trying to find a magical position that somehow gives me leverage. I grunt. I sweat. I swear. I grunt some more. Eventually, I pack down enough snow to stand up. Tears and snot drip off my chin. Sweat glues to my skin all the clothing layers under my waterproof jacket.

Since my first day of skiing, I have struggled to notice when I start to feel warm or chilled. Scott seems to have an internal thermometer with an alarm attached. He stops skiing occasionally and peels off a layer, or yanks a sweater out of his pack and pulls it over his head. "If you're already hot or cold," he has warned me, "it's too late. It takes too much work for your body to warm itself back up, or dry off sweat." But my climate-controlled life has stripped away any sensitivity I might have had to these dips and rises in body temperature. I notice only when I am hot enough to sweat, or cold enough to shiver.

I stomp up the hill. Scott is out of sight, which is for the best. I whimper as I slip and slide along the path. Why do people do this to themselves? The trail weaves around a tree and there is Scott, waiting for me.

"How're you doing?" He sounds so cheerful I could throttle him.

"I hate this." I look away from him, instead focusing on the empty trail in front of me.

"Do you want to stop? Take a break?"

"No. I want to hurry up and get to the damn shelter." If I say any more, I could start crying again.

He glides—how does he do that?—behind me and says gently, "Hey, are you okay?"

"No. I just want to get there." I shuffle my skis a few more times and then ask, "How much farther is it?"

"Probably not too far."

Probably? Tears flood my eyes. Then Scott says: "You can do it, Lily."

"No," I mumble, "I can't."

He's still right behind me, and I know that he is about to deliver a pep talk. I can practically feel the obnoxious encouragement on the back of my neck. He's going to say that yes, I can do it. I'm stronger than I think. We're almost there. When we get there I'll be so glad I did this.

Instead, he says: "We can turn back if you want. But either way, it's more skiing. You're in Fortitude Valley, Lil."

For a moment, I forget how miserable I am. I look around, as if he has told me, literally, where we are. But this is the AC/DC trail. I already know that. I've seen the map.

"Where is that?" I finally ask. "Fortitude Valley? I don't know where that is."

"Yeah you do. Everyone does. Fortitude Valley is..." He waits for the right words, perhaps aware of how short my fuse is. "You don't choose to come here, but when things get tough, it's where you are. And you just have to get through it."

Gray sky. Mounds of tricky, evil snow. Scrawny lodgepole pines judging me as I limp past, sniffling. Behind us, more obnoxiously cheerful skiers who say things like "What a lovely day!" when I step off the trail so they may glide past. Scott is right: This *is* Fortitude Valley. Or hell. He's also right that there's nothing to do but keep going. I resolve to save what little energy I have left for skiing, and not waste it on anger or whining.

And then, just like that, the heavens part and the angels sing. A stout log cabin appears before me, with a few pairs of skis poking out of the snow in front.

"Shelter ahoy!" I pump my fists in the air, poles dangling from my wrists and slapping my sides.

Scott snaps my picture and kisses me. He's so proud of me that I am ashamed at my grumpy, childish behavior. It shouldn't be such a big deal to ski to the top of a ridge. All around me children are doing it, for goodness' sake. We drink water and eat granola bars next to the woodstove, where other skiers have already stoked a hot fire. The downhill is fast and treacherous. But I gladly accept the fear of flying down a slippery hill over the miserable slog of shoving myself up it.

As the winter wears on, we ski almost every day that we don't have to work. For me, the physical challenge is just one part of the hardship of skiing. The other part is figuring out how to occupy my mind during a seemingly endless trudge.

Early in the day, when I'm feeling energetic, I daydream. Once I get too tired to be anything but practical, I make mental lists of articles I want to write. Next, I compose pneumonic devices to remember my lists. Eventually, exhausted, I resort to the only thing I can think to do: count my steps. One, two, three, four, five, six, seven, eight. Around 250, I start getting sloppy. And then something else happens: nothing. If I ski long enough, my mind becomes so exhausted that it almost shuts off. The novelty of not thinking about anything is sort of fun. John Cage, the composer and poet, once wrote: "In Zen they say: If something is boring after two minutes, try it for four. If still boring, try it for eight, sixteen, thirty-two, and so on. Eventually one discovers that it's not boring at all but very interesting." At the tipping point of exhaustion, I veer as close to meditation as I ever have, and it feels like I've reached Nirvana.

In the spring, we return to the site of my infamous breakdown. I have knots in my stomach as I click my boots into the bindings. I tell myself that whatever happens, however hard it gets, I will not cry. Fortitude Valley, here I come.

But this time, the trail to the AC/DC shelter is apparently diverted around Fortitude Valley. We arrive—both of us—in good spirits. I am tired and of course I've fallen more times than I wanted to count, but I remain positive. I feel as if, in one winter, cross-country skiing has transformed me from a whimpering child into a capable adult. As we sit down on a rustic wooden bench next to the woodstove, a family with two young children—probably eight and ten years old—gets up to leave.

Children who willingly cross-country ski amaze me. There is no grand finale, no reward, no intoxicating sensation of flying like you get downhill skiing. It is hard work one way and hard work on the way back. I watch as the youngest child, a boy, snaps his boots into his ski bindings. He looks so matter-of-fact, so calm. Never mind that he is about to slide down a hill that reduced me—an adult—to tears.

"I think I would be a better person," I whisper to Scott, "if I'd grown up cross-country skiing."

He swallows a mouthful of water and laughs. "How so?"

"I don't know, exactly... I guess I'd be more patient, more able to live in the moment, enjoy the journey. Something like that, anyway."

Although I've always liked the abstract idea of hiking, I never really understood the point of it before. But now, bird hunting, I find myself enjoying these long treks through difficult terrain. Even on days when I don't see a single chukar, I take pleasure in the hike. My heightened senses occupy my mind, so there's no need for games or counting. With no trail to follow, I stay focused and scrutinize each step. I look for possible nesting sites or the flash of a red beak. I listen for calls and cackles. I know I am nearing the top of a steep canyon when I smell the citrusy sagebrush that grows above, on flat ground. I feel completely present; taking in the land around me and interpreting it, in real time, with no distractions. I don't daydream or compose lists or check my BlackBerry.

Perhaps what I needed to become a hiker was a specific goal, a reason to traipse for miles. Hunting and gathering are, after all, the primary reasons that humans developed stamina to travel long distances on foot. Daniel Lieberman, an evolutionary biologist at Harvard, believes that the uniquely human capacity for long-distance running (think 26.2-mile marathons or even longer ultra-marathons) is a vestige of our ancient method of "persistence" hunting, or chasing a wild animal to exhaustion and eventually death. Hiking was once an integral part of the human experience of the hunt, just as fetching was key to the dog's experience of it.

Yet our lives have moved so far away from hunting that we no longer recognize the origins of these daily routines. Even our language is filled with words and phrases derived from hunting. "Buck," the word for a male deer, for example, is slang for one dollar because in the nineteenth century, Americans could buy one deer carcass for a dollar. A "sitting duck" is the easiest shot a hunter will ever get (so easy it's considered unsportsmanlike).

When you remove hunting from human life, this is what you get: aimless hiking and "hanging out" with our dogs, who are bred to catch Frisbees or to not shed fur. How bizarre, then, that something as passive as hiking has become the domain of environmentalists. After all, we assume that any solstice-worshipping hippie loves to hike. Hunting, on the other hand, is for rednecks who couldn't care less about the health of the planet.

The truth, of course, is that many hunters care greatly about the environment. In fact, most of the hunters I know go hunting in search of an outdoor experience first and wild, healthy meat second. Drinking beer in the woods or nabbing a giant set of antlers—these possibilities don't enter their consciousness.

Until I moved to Oregon, I made no distinction between hunters who pursued animals for different purposes. Now I know: There are hunters and there are trophy hunters. The people who aim to bring home tasty meat don't usually worry about the size of an animal's rack. In his cookbook, Roy Wall writes, "The sportsman who bags a noble head, a monarch of a wilderness glade, has a just right to be proud, but in doing so he imposes double duty upon the camp cook, for, in most cases, the finer the head, the tougher the meat."

Scott and I don't get cable television at home, but whenever we stay in a hotel, I flip through the cable stations to find a hunting show. In some parts of the country, there are around-the-clock channels devoted to hunting. But most of these shows display little if any footage of the actual pursuit, instead focusing solely on the kill. A typical program tells the story of one hunt in about five minutes.

It begins with a quick introduction of the hunter and the guide, then shares a little information about the guns each is carrying. The bulk of the footage is a rapid sequence of spotting the animal, setting up the shot and taking it. The hunter and guide spend a few seconds admiring the animal and remarking on its gargantuan size. Then the show switches to a new hunt. There is no map to situate viewers, no insight into the species' behavior, no indication that the pursuit took hours or days or weeks. Just wham, bam, boom, cut to commercial.

Serious trophy hunters spend hundreds of thousands of dollars traveling to remote countries and hiring guides to help them stalk the biggest animals. The Boone and Crockett Club, co-founded by Theodore Roosevelt, is still a major force in habitat conservation. It also plays another, more nefarious role in modern big-game hunting: record keeping. The club has a formula for determining the score of any antlered animal killed, based on the width of the horns and the number of tines, or points, jutting out of them. In keeping and promoting these records, the club perpetuates an obsession with gargantuan size that is likely detrimental to wildlife. A study in 2009 found that the horns of Canadian bighorn sheep have shrunk because of hunting. By targeting individuals with the largest horns, trophy hunters are systematically altering the gene pool of the entire species. The lead researcher of this study told *National Geographic*, "Human-harvested organisms are the fastest-changing organisms yet observed in the wild."

In central Oregon, every outdoors supply store has a

"brag board" near the entrance. This is a bulletin board where customers can tack photos of themselves with their prey. I used to scurry past these displays, averting my eyes. It seemed so cruel and twisted for people to pose, grinning, with a dead animal. The creature's teeth are often stained with the blood that dribbled out of its mouth during its last breath. Its tongue hangs out, looking unnaturally long and seeming to mock the dignity that the animal possessed during life.

In a way, I get it now. I understand how much work goes into tracking and killing wild game. Where I used to see a grisly image of mockery, I now see a memento of a hard-fought victory. Still, some of these pictures tug at my insides.

That fall, after buying my duck stamp and other tags needed to hunt birds, I stand at the brag board of a local outdoors supply store, entranced by one photo of a man straddling a prone bear. There's something about photos of dead bears, in particular, that startles and saddens me. In life, bears are so elusive. And they seem, to me at least, more dog-like than most wild game. A bear's forehead slopes to its nose at the same angle as my Sylvia's. This particular bear is a grizzly. Its paws are the size of dinner plates, with claws like switchblades. No doubt it was a huge, threatening animal during life. But it's been reduced now to an impotent mound, disarmed and stretched out, with a burly, grinning man standing over it. It's not even cold and already it looks more rug than bear.

CHAPTER 10

FRIENDS FOR DINNER

In July 2008, after weeks filled with planning, Scott and I take a fishing vacation with four friends in the Alaskan wilderness. On the Fourth, Scott and I land in Anchorage at midnight. Fireworks ignite around the city, though the night sky never gets dark enough to see them very well.

The next morning, armed with a long list, we purchase food for the next eight days. Then we take a cab from the grocery store to a tiny floatplane operation, where we meet our friends. There's Andy and Jessie, from Missoula. There's our friend Ryan, who lives in Bend and is a regular fishing buddy of ours. And then there's Evan, a friend of Andy's from Minnesota whom we haven't met before.

Andy was a fishing guide in this area during college,

so he knows the river well and has made arrangements to hire the planes and rent the rafts. We load our clothing and fishing gear, along with two white coolers and two inflatable rafts, onto two tiny airplanes. We fly for an hour before landing near the shore of a small lake. We wade to shore, assemble and inflate the rafts and stuff our gear into dry bags.

I'm nervous as the pilots wave and their planes lift off to return to Anchorage, leaving behind a whitecapped wake. We're on our own in a veritable wilderness, and if anything goes wrong we will have no chance of getting help. This will be the first time I have been in clear danger from predators; grizzly bears are plentiful here, and will fish the same river we do, for the same king salmon. Andy has told us that for the most part, bears try to avoid humans. But any bear, particularly a protective mama with cubs, could become aggressive if startled.

Before we left Anchorage, Scott and I purchased a fifty-dollar can of aerosol bear spray, a spicy concoction that could temporarily disarm an attacking bear if sprayed directly in its face. Andy has packed a 12-gauge shotgun, too, which means that for the first time I (or at least our group) will be armed for self-protection. Andy gives us a gentle reminder that anyone who heads into the woods—even just to go to the bathroom—needs to carry the gun or spray. My pulse quickens when he says that. I'm not sure which I'm more afraid of: the threat of a snarling bear or the idea of using a gun for self-defense. What really makes me nervous, though, is that I'm not sure whether I'll enjoy myself with nothing to do but fish for eight days.

The sun is warm, though, and the mood jovial as we push off from shore. We laugh and whoop, rowing across the lake and into a tiny stream: Talachulitna Creek. The weather quickly turns awful—a cold rain moves in on the second day and follows us for the rest of the trip. Andy says he's never seen the mosquitoes so thick. We wear head nets all day, every day, and I shudder to think how I would fare without one—bugs buzzing in my ears, up my nose and in my eyes. The river is blown out by all the rain, so the fishing is slow.

And yet, every day is adventure. Each curve of the river brings new scenery and a new stretch of fish habitat to try to solve. Where will we stop for lunch? How will we find enough dry wood to start a fire? I bond with Jessie, and sometimes it only takes one word to make us erupt in laughter. I start to cast like a professional, flinging out more line and heavier flies than I ever have before. We manage to catch rainbow trout, arctic grayling and, on occasion, massive king salmon. We eat fresh salmon for dinner almost every night, and even make grayling sushi. (So much for all the food we've packed in coolers.) The biggest surprise of the trip is how much I enjoy it.

Here's something else that surprises me: We aren't alone, after all. Several times a day we hear the buzz of airplanes overhead. And the most popular fishing holes usually have a raft of anglers nearby. Some are even flown in by helicopter from a local fishing lodge.

These other anglers tell us stories about a misguided pair of New Yorkers paddling one or two days ahead of us. They have mishaps with grizzly bears. They flip their raft in a rapid. They run out of food. I start to feel as if I'm following a ghost version of me—the person I could have been if I hadn't moved to Bend and met Scott. The pilot who eventually flies us out of the bush will tell us that he found the New Yorkers while he was picking up another group. They paddled downstream to the take-out days before they were supposed to leave. They were wet, cold, hungry and terrified. They couldn't wait to go home.

I sympathize with these New Yorkers, and understand how they got in over their heads. I have so much more respect for the power of nature than I did four years ago. Almost every year, someone in Bend drowns in the Deschutes River, usually wearing a bathing suit and floating in an inner tube. Until I spent time boating with Scott, I never understood exactly how forceful rushing water could be. Also every year or so, a deer turns on a dog who chases it and gores a beloved family pet to death. A mountain lion moves near the city limits and feasts on house cats and small dogs. Back when most of my wildlife interactions happened via television, I didn't appreciate the power of a wild animal leaning into its instincts.

Daylight shines twenty-four hours a day here, and the whole landscape roars with life as if every being, from the mosquito to the grizzly bear, is trying to make up for the lost time of winter. Giant king salmon swim past us upstream, led by their noses hundreds of miles from the ocean to the exact stretch of fresh water where they first hatched from a pink egg. As soon as these behemoth fish leave the sea, they stop eating. Mating is their only goal.

As they migrate upstream, their silver skin turns as red as an autumn maple tree in Vermont, with the color peaking just as they spawn.

The salmon are the lifeblood of this river; countless other species depend on their summer pilgrimage. Trout wait for them to enter the river and lay eggs the way impatient schoolchildren wait for their mothers to put dinner on the table. Some of these trout trail just a few inches behind the female salmon, at times slamming into their swollen bellies in hopes of releasing a few delicious eggs.

After a salmon spawns, it rots and dies. This is a gradual process, the opposite of getting thwacked on the head by a fisherman or clawed open by a bear. Instead, a salmon's flesh fades from scarlet red to translucent. Even as chunks of its body fall off and drift downstream, the fish remains aggressive and vigilant. Half dead, the zombie-fish hovers over its own eggs, ready to muster a vicious snap of the jaws at any greedy leech or trout that tries to gobble its young.

For the last few months, Scott and I have been talking about starting a family. As I watch these salmon, I marvel at their drive to procreate. It elevates these otherwise normal fish into something verging on supernatural. It makes me wonder: How would parenthood change me? Would I become something unrecognizable, some snarling mother-beast, a translucent ghost of my former self?

Some of my friends have had children, but not my closest ones—nobody with whom I could sit down and talk about these fears honestly. Nathan is an obvious choice, but he's so far away and we talk on the phone so infrequently. Besides, I still haven't met his new daughter, or seen Nathan in person in his new role as a parent.

When I think about having children, all of my fears center on what would happen to me—to my marriage, to my career, to my life. This underlines what I see as the uncomfortable paradox of parenting: Deciding to have a child is, at its core, a selfish decision. But raising that child is a never-ending act of selflessness. Scott and I talk circles around this, until I meekly declare: I'm not ready. Scott is patient and understanding. Yet even after the subject is dropped, I can't help but fixate on the questions he is too polite to ask: What am I waiting for? What could possibly make me ready?

The month after we arrive home, Republican presidential candidate John McCain names Alaska governor Sarah Palin as his running mate. Palin quickly rouses just about every controversy imaginable, including some centered on her family's hunting traditions. She is lauded by some for her ability "to field dress a moose," but criticized by others for her remorseless attitude about killing animals. Hunting is back in the spotlight.

Until I moved to Oregon, the only time I heard about the sport was during presidential campaigns. Every four years, it seems, candidates don crisp vests, visit shooting ranges and bow before the almighty NRA to show their support for the Second Amendment. But what of wildlife management? Or habitat conservation? I'm not voting for Sarah Palin but I am frustrated by the lack of civility and deep consideration that accompanies our national debate about hunting. It seems we Americans can't move past two polarizing issues—Guns! Killing animals!—to dissect the meatier and, dare I say, more relevant issues begging to be discussed. The reasons why a person hunts, the ethical guidelines she chooses to obey, even the technology she uses or refuses to use—say a lot about who she is. With rare exceptions, media coverage addresses hunting in a yes-or-no manner. Does the candidate hunt? Yes or no. Next question. We have missed a tremendous opportunity to learn more about who these candidates are, why they believe what they do and whether their actions live up to their beliefs.

In the fall, the generosity of other hunters continues to amaze me. Andy and Jessie invite me to go bird hunting. Andy and his dad, Hank, offer advice. The hunting and fishing writer for my newspaper, Gary Lewis, gives me books on hunting. Others offer to loan me guns or other equipment, too. Psychologist and philosopher Erich Fromm argues that hunting has long been connected to our social tendency to work together and share rewards. "Luck in hunting was not equally divided among all hunters; hence the practical outcome was that those who had luck today would share their food with those who would be lucky tomorrow," he writes. "Assuming hunting behavior led to genetic changes, the conclusion would be that modern man has an innate impulse for cooperation and sharing, rather than for killing and cruelty." Hunting and sharing continue to go together today through groups like Sportsmen Against Hunger, a nationwide food bank that accepts donated game and distributes free meals to the needy.

Still, I don't want to become a barnacle on these people who have helped me so much. Besides, I figure there's no better way to test my hunting skills than to go it alone. So in October, Scott and I take a weeklong camping and bird-hunting trip. (Well, I hunt; he opts for the more modern, gun-free hike. Over the last two years, Scott has grown increasingly interested in my hunting adventures, but he still isn't ready to pick up a gun.)

We set up camp and then start scouting the area for places to hunt. I notice a cluster of three small ponds on the map, so we drive there in hopes of finding some ducks. There are generally two ways to hunt for ducks. One is to spread decoys on a pond or lake, hide in a nearby blind and occasionally toot on a duck call to dupe birds flying overhead into landing on the water. The second is to try to sneak up and flush ducks that are already sitting on a small pond or stream. I'm going to try the second method.

I get out of the car, hike toward the first slough and then squint into the sun. I can't quite tell if anything is on the water. Hopeful, I creep closer—heel, toe, heel, toe. A row of pine trees shields me from the view of any animal that might be on the water. When I reach the last tree, there are still ten yards between the pond and me. I don't own any camouflage, but this morning I made sure to pull on neutral green and beige clothing, to better blend in with my surroundings. Birds have keen eyesight, and wild birds in particular tend to avoid blocks of solid color, the telltale sign of something man-made.

I crouch down and take one long step from behind the tree trunk. I freeze, watching the surface of the pond. No new movement, just the same windblown wrinkles. Another step, then another pause. Another step. I continue this until I get just a few feet from the water and notice a wave ripple out from a spot to my left. Something splashes, and the tall grass at the edge of the pond flickers. A duck thrashes into the air. Yes! I stand up straight and shoulder my gun. I switch off the safety, line up my sights and pull the trigger. Bang!

But the duck flies on, its wings not missing a beat. I pump the next shell into the chamber and line up my sights again. This time, the duck is farther away. Bang!

Again, the bird keeps flying. I reach into my pocket for another shell, but it's empty. I left the rest of my ammunition in the car. As I lower my gun, the duck circles the pond and flies directly over me, then rises higher into the air until it disappears.

Disappointed, I head back to the car to fill my pockets with shotgun shells. Then I hike to the second pond from the road but find no birds. I scoot under a large willow, so a bird flying overhead doesn't see me. I am ready. I wait. And wait. Then wait some more. I daydream a little. But mostly, I fret about what I would do if a duck flew overhead, spotted the water and opted to descend.

When I am actively pursuing an animal, I don't have time to think about these qualms. I focus solely on the chase and assume permission to take the life of my quarry. The kill is the goal. There is no time for doubts. Instead, it's during times like this, sitting quietly with no animal in sight, that these ethical questions bubble up.

The question hunters most often ask themselves and one another is: Is this shot *sporting*? Simply put, a sporting shot is one that gives the prey a reasonable chance at survival. There are no carved-in-stone definitions of what's reasonable—a hunter must analyze each situation and sort out the ethics for herself. But there are some widely accepted rules, such as: It's fair to shoot a duck in flight but not one sitting on the water. Though it's legal in most states, baiting deer, or setting out piles of corn and salt licks to draw in herds, is believed to give the hunter an unfair advantage. In addition, many hunters develop personal codes of ethics such as *I will shoot bucks but not does.* (Depending on the individual, this vow could stem from a wish to minimize one's impact on the deer population—or could reflect a machismo attitude like the one taught to young boys: Never hit a girl.)

Sportsmen's groups have their own written credos for "fair-chase" hunting. The Boone and Crockett Club's definition is considered the national standard for big-game hunting: "The ethical, sportsmanlike, and lawful pursuit and taking of any free-ranging wild, native

North American big game animal in a manner that does not give the hunter an improper advantage over such animals." In 2005, the group adopted a statement condemning so-called canned hunts, in which large animals are bred in captivity and then "released" in a fenced area for high-paying clients to shoot. One short-lived but highly publicized ranch tried allowing clients to operate a gun over the Internet and, with the click of a mouse, take the life of a captive hog. To me, the excitement of hunting lies in gleaning enough knowledge—about the prey and about the landscape—to find the animal in its own habitat. Baiting animals or shooting them within an enclosure not only raises ethical concerns, but negates the main challenge of hunting.

My own definition of fair chase is to interfere with an ecosystem as little as possible while participating as a predator. I didn't settle quickly on this general rule. Rather, it's the result of many months of consideration and research. I've had long talks with other hunters and with Scott. I've read books on ethics, as well as short stories and essays about hunting. Like so much related to hunting, it is subject to change depending on unique circumstances or on some new piece of insight. It remains impressively, frustratingly difficult to make generalizations about an activity that has so many variables.

When I first moved to Oregon, for example, I was wowed by the bow hunters I met. It seemed the ultimate act of sportsmanship to creep within thirty yards of an animal (the range of most bows) and then kill it with a well-placed arrow. But the more I talked to bow hunters, the more horror stories I heard about animals being maimed instead of killed. I know it is possible to make a clean kill with a bow and arrow because I've met people who have done it. But others have had to chase down the wounded animal for twelve hours, hitting it with three arrows before it finally dropped dead. My newspaper has published photos of deer that appeared in backyards and parks with arrows sticking out of their necks. It takes a lot of practice to be able to shoot a rifle or a shotgun consistently. It takes much more experience to make a fatal shot with an arrow.

Today, however, as I sit under a shrub and ponder the ethics of hunting, there is only one question dogging me. It is perhaps the least likely one, considering that I've been hunting for two years already. It is certainly the biggest, most overriding question: Is it wrong to kill animals?

•

As strange as it sounds, hunting doesn't have to involve the death of an animal. There are, it turns out, some nonlethal forms. I met my first "catch-and-release" hunter—a woman—earlier that fall, when presidential campaigns were in full throttle. I was profiling a rural voting precinct, so I went door to door to gauge the local political climate. It's the kind of story I love to write because it's an excuse to knock on random doors and—more often than not—get invited into people's homes. There is no job that gets you inside more living rooms than being a newspaper reporter. If you're as nosy as I am, this alone is enough of a reason to go into journalism. You get to see what kind of sweatpants and slippers a couple change into when they get home from work. Meet their pets and their children. See their wallpaper. Smell what they're cooking for dinner.

Alas, this particular woman was standing outside when I pulled up, so we did all our talking in her front yard as the sun dipped below the mountains behind her farmhouse. She was in her early fifties, with short brown hair and an athletic build. She explained that she's a staunch Republican and fears Barack Obama would take away her guns if he were elected. Then I asked if she hunts.

"I do." She smiled. "I'm a bird hunter, but mostly I do catch-and-release."

"What?"

She laughed. "I love that reaction—I get it all the time." She went on to explain that she trains pointing dogs, and when she takes her dogs out in the field, her only goal is for them to find a bird and hold it on point. She commands them to release the bird, the bird flies away and she rewards her dogs with treats.

Turns out she's not the only person to enjoy all aspects of the hunt up to the kill. In 2010, the Whitetail Pro Series began hosting deer-hunting tournaments in which contestants stalk deer, zero in on them using digital scopes equipped with memory cards and then fire blank shells. Hunters earn scores based on the size and number of deer that they would have killed humanely had their guns been loaded with traditional ammunition.

And in England, since fox hunting with dogs was outlawed in 2004, dedicated hunters on horseback have eliminated the fox altogether and taken to pursuing a designated human. When they catch up to their quarry they don't harm him or her but take what satisfaction they've gained from the chase and call it a day.

These so-called humane methods of hunting offer new potential for the sport by eliminating all of its most contentious aspects: human danger, impact on wildlife populations and, of course, the ugly reality of death. But this represents a tiny fraction of the hunting that occurs on the planet. And because it avoids all of the slippery ethical questions that plague fatal hunters, including me, isn't it, in a way, missing a main point? Spanish writer, philosopher and hunter José Ortega y Gasset rebuked an early-twentieth-century British version known as photographic hunting, in which hunters captured prey and photographed it before releasing it unharmed. "One can refuse to hunt," he writes, "but if one hunts one has to accept certain ultimate requirements... Without these ingredients the spirit of the hunt disappears. The animal's behavior is wholly inspired by the conviction that his life is at stake, and if it turns out that this is a complete fiction, that it is only a matter of taking his picture, the hunt becomes a farce and its specific tension evaporates."

My discomfort with catch-and-release hunting begins to bleed into my feelings about catch-and-release fishing.

Fishing is more complicated, though. Unlike hunters, anglers can't always target individuals, and often must reel in a fish and identify its species or measure its length before determining whether it's even legal to kill. In other words, every sport fisherman must practice some catch-and-release. And as Scott points out, when anglers release their catch, it allows more people to participate in the sport. Still, I find myself feeling guilty for trying to trick fish into biting my fly when I know I'm only going to release them. It feels like teasing or bullying. Perhaps cooking the fish and eating it—putting it to use—would be more respectful. I worry that, as Ortega y Gasset argued, there is something farcical about engaging in a life-or-death struggle minus the death.

Of course, whether or not we hunt, we all kill animals on a regular basis. Our entire society is built upon the sanctity of human life at the expense of animal life. Geese are killed to prevent collisions with the airplanes that transport us and our goods. We perform medical research on animals, to enhance and prolong our own lives. Each day in the United States, six thousand acres of open space, including working farms and forestland, are developed. Much of our sugarcane is harvested from fields that were shorn from life-teeming tropical rain forests. The roads we drive on, the manicured lawns we play on, the stores we shop in—all of it used to be wildlife habitat. Power plants, oil and gas wells and even wind farms deal enormous blows to animal populations.

In Bend, the local park district decides to euthanize

109 resident geese. Instead of migrating between seasonal habitats, these birds have made themselves at home in Bend, year-round. And they've reproduced, despite the park district's multi-year efforts at birth control, hazing by dogs and even relocation to a bird sanctuary more than a hundred miles away. Fields of feces create a nuisance and a health hazard. And the aggressive geese displace other native birds. So district officials do the right thing by having the birds killed humanely, then donating the meat to local food banks, which struggle to feed the needy during a long, deep recession.

But the decision sparks furious dissent. Mourners hold an earnest memorial for the geese in one of their favorite, goose-shit-ridden parks. Angry protesters show up at the soup kitchen after the birds have been served and threaten the manager. I'm baffled by these responses. Bend is not a town where it's common to see public protests of any kind. Where do these activists think meat comes from?

Few of us bother to question our habit of eating animals, even as we question the killing of animals for other, less frequent purposes. Anthrozoologist Hal Herzog writes that Americans "kill 200 food animals for every animal used in a scientific experiment, 2,000 for each unwanted dog euthanized in an animal shelter and 40,000 for every baby harp seal bludgeoned to death on a Canadian ice floe." Animal rights protesters target fur-coat-wearing fashionistas but let slide the thousands of grocery stores that crowd live lobsters into miserable, algae-covered tanks. In the United States, vegetarianism is, by all accounts, more popular than ever before. Still, the lifestyle is rare. While the exact number of vegetarians is difficult to discern, most surveys and polls produce estimates of seven to eleven million. That means we have about as many vegetarians as residents of the state of North Carolina. This is a tiny minority, a blip in the demographic chart. There are roughly two million more hunters than vegetarians in the country. Not to mention, one survey found that 60 percent of the people who identified themselves as vegetarians admitted they had consumed meat in the last twenty-four hours.

Animals kill other animals, of course, and we don't judge them for it, we simply call them carnivores. Most vegetarians I know continue to feed meat to their pet dogs. Some wild animals kill with no intention of eating their victims. Wolves and elk, for example, are both known to slay members of their own species while jockeying for status during mating season.

Some animal rights activists argue that because humans no longer *need* meat to survive, we should all become vegetarians. It is an understandable, even noble desire to avoid unnecessary death or suffering. Yet even a vegan diet kills animals. "The grain that the vegan eats," Michael Pollan writes, "is harvested with a combine that shreds field mice, while the farmer's tractor wheel crushes woodchucks in their burrows and his pesticides drop songbirds from the sky; after harvest whatever animals that would eat our crops we exterminate... If America was suddenly to adopt a strictly vegetarian diet, it isn't at all clear that the total number of animals killed each year would necessarily decline, since to feed everyone animal pasture and rangeland would have to give way to more intensively cultivated row crops." Cattle, elk, sheep and antelope, for example, can chew grass and turn it directly into protein. So in mountainous, rocky regions that are better suited for grazing than farming, raising animals for meat is often the most efficient way to reap food from the land. In other words, if our concern is, as Pinchot put it, the greatest good to the greatest number for the longest time, then eating animals may sometimes be the most ethical decision.

Arguments for or against meat eating almost always involve some definition of otherness, the drawing of a line that separates "us" from "them." The precise geography of this boundary is determined by the individual eater. To some pescivores, for example, cows and pigs are too intelligent, too fuzzy or too reminiscent of humans to eat. Fish, on the other hand, are fair game.

Novelist Jonathan Safran Foer draws a comparison between the arguments about meat eating and abortion. In both cases, he writes, "it is impossible to definitively know some of the most important details (When is a fetus a person, as opposed to a potential person? What is animal experience really like?) and that cuts right to one's deepest discomforts, often provoking defensiveness or aggression. It's a slippery, frustrating and resonant subject. Each question prompts another, and it's easy to find yourself defending a position far more extreme than you actually believe or could live by. Or worse, finding no position worth defending or living by." One person's definition of an ethical meal can—and does—bow and sway throughout one's lifetime. Perhaps that's why, in the United States, vegetarians are outnumbered, three to one, by ex-vegetarians.

Philosophers pose stark hypothetical questions to help each of us understand where, exactly, to draw the ethical boundaries of our own eating habits. On one end of the spectrum, philosophers Peter Singer and Tom Regan argue that "speciesism," or the exploitation and oppression of non-human animals, is equivalent to sexism, racism and the exploitation and oppression of humans. If we justify our treatment of animals—namely, eating them—by pointing to their inability to speak and their incapacity to think rationally, then why not, they ask, eat brain-damaged humans who are also unable to speak or think rationally?

But philosopher Cora Diamond counters that the *rights* of a person or animal are not the real issue. Her problem with the Singer-Regan argument for not eating animals is that it implies a vegetarian would have no qualms about eating the cow that has been struck by lightning. She writes, "There is nothing in the discussion which suggests that a cow is not something to eat; it is only that one must not help the process (of turning a living cow into food) along." Diamond concludes that Singer and Regan are ignoring a critical difference between human beings and other animals. It is something else that makes so many of us willing to eat cows but not people (even people who die in car accidents), something that extends beyond eating. For example, a formal funeral and a published obituary are appropriate for a human baby who dies at two weeks old, but not for a dog—not even one that has been loved by a family for many years. Or, from another angle, just because an animal is capable of suffering doesn't mean that we should do everything we possibly can to avoid its suffering. "That *this* is a being which I ought not to make suffer, or whose suffering I should try to prevent," Diamond writes, "constitutes a *special* relationship to it."

As a newspaper reporter, I am acutely aware that we measure tragedy, at least in part, by its relative closeness. The summer that I worked at the *Hartford Courant*, I wrote about a group of local firefighters who flew to Oregon to help fight a raging wildfire. The fire had ravaged thousands of acres, which seemed like a shame. It wasn't until I moved to Oregon that I understood the serious threat of such a fire. A rampant wildfire isn't just a shame, it is terrifying. The first time I interviewed a family who had to be evacuated from their home because of a wildfire, I saw tears well up in a young mother's eyes and I understood what the fire really was: a tragedy. This shift was all because of closeness. A special relationship. That same reasoning explains why I can eat steaks from a cow and feel nothing, then fall to pieces when it's time to euthanize a beloved pet dog. I never met the cow. But I loved that dog like a member of my family.

It also explains the ire over the deaths of 109 geese. Bend residents feel too close to these animals—why, a goose strutted past me in the park just the other day—to approve of their being cooked and eaten. But what does it say about us that we're only comfortable with our meat being raised out of sight and out of mind? Thinking about it too hard points out our hypocrisy. "No other people in history," Pollan writes, "has lived at quite so great a remove from the animals they eat."

This makes hunting an especially complex endeavor: The chase fosters a special relationship, a direct connection between predator and prey. Yet that closeness does not infringe upon the predator's willingness to eat the prey. Instead, killing and eating the prey becomes an expression of that relationship's specialness.

Hunting regularly reminds me of an enormous category of animal-human relationship that I otherwise ignore in my day-to-day life. My relationships with animals tend to fall under two broad headlines: Friends (my pet dogs) and Enemies (vermin mice).

But there exists a vast world between the two. The black bear who picked huckleberries from the same shrub just hours before I did. The chicken who lays eggs for my breakfast. The squirrel who lives in the tree in my backyard. The cow who produced milk for my cheddar cheese. The family of rabbits displaced by the construction of a new house on my block. The shrimp who wiggled into a net and landed in my stir-fry. The fish who also wiggled into that net and was thrown back, dead. These animals' lives are intertwined with mine, whether or not I acknowledge them.

Most animal species on earth probably fit somewhere

in this middle ground, and yet I have no words to express how I feel about them, how I treat them, what role they play in my everyday life. I have no category for them. Every once in a while, a pretty hummingbird who sips from my feeder slips into the Friend category. The squirrel who breaks the feeder scoots onto the Enemy side. The rest of the time, they don't exist.

The problem with this binary formula is that, for more than twenty years, it has tricked me into thinking that I face an equally stark choice in my own treatment of animals. As a meat eater, I could:

1. Eat friends for dinner, or
2. Convince myself that dinner was an enemy deserving of death.

What reasonable person would choose the first? And after meeting an animal—almost any animal!—and looking into its eyes, how could anyone believe the second? For most of my life, I did the only logical thing: I chose instead not to think about my meat as animals at all. But to hunt is to confront this overlooked category of animals on earth. It is to stand up and admit: This is how the world works.

In one scene in his book, Foer witnesses the slaughter of a pig. The pig looks at him during its last seconds of life, and Foer is moved by this. "The pig wasn't a receptacle of my forgetting," he writes. "The animal was a receptacle of my concern. I felt—I feel—relief in that. My relief doesn't matter to the pig. But it matters to me." Indeed, our respect for animals is, in some ways, all we can offer them. Animals die whether we acknowledge them or not, whether we eat them or not, whether we participate in their deaths willfully or indirectly.

To me, hunting my own meat feels like saying grace before a meal and really, for the first time in my life, meaning it. I grew up in a household that said grace before supper: *God is great, God is good. And we thank Him for our food.* As a kid, I thought it was a little silly. It was the only time we ever mentioned God. And the imperfect rhyme bothered me. It is only since I started killing my dinner—watched it switch, in an instant, from living to dead—that I have felt truly grateful for a meal.

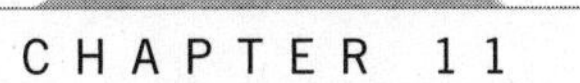

CHAPTER 11

YEAR OF DEATH

In October I grow nervous because back in early summer, I bought an elk tag. This means I have the right to kill one male elk in one particular area about a hundred miles south of our home. I don't feel ready to put myself in a situation where it's even possible to shoot something as big as an elk. The guilt might be too much to bear, not to mention—oh God—the guts.

"Elk guts," one hunter tells me, "are the real test."

I look for an excuse to call off the four-day hunt, and soon find one in our dog Bob.

Every evening, Scott and I come home from work and take our dogs on a walk. One day we are only a block from the house when Bob starts hopping on three legs. The next day, I take a long lunch break and drive Bob to the veterinarian. He is diagnosed with osteosarcoma, a bone tumor, in his right hind leg. The vet says this disease usually causes a swift but painful death. She prescribes a mild opiate and reassures me that "you'll know" when Bob is in too much pain to go on. I drive Bob back home and call Scott at work to tearfully deliver the news.

Bob's decline is steady. As October wears on, our walks get shorter, and then, to save Bob's strength, we start dividing them up. I take Sylvia on an energetic jog around the neighborhood. Scott takes Bob on a slow, elderly stroll, letting him savor the smells that have provided the high point of each day since Scott adopted him nine years earlier. Elk-hunting season arrives and we stay home to care for Bob.

One Monday in mid-November, I arrive at work and open my inbox. The first email informs me that one of my co-workers, Jim Witty, died of a sudden, massive heart attack earlier that morning. Jim was the outdoors writer, and just fifty years old. He was beloved by the paper's readers, some of whom will later tell me that although they never met Jim they felt he was their hiking buddy. Scott and I had just bumped into Jim three days earlier at a pizza parlor, with his wife and two friends.

A couple of days before Thanksgiving, we learn that our sister-in-law's father, also named Jim—who helps run the non-profit where Scott works—has been hospitalized. The whole family was on the Oregon coast when he started feeling ill. His wife drove him back to Bend. A few miles from home, he fell unconscious, so she drove straight to the hospital.

"It doesn't look like he's going to make it," Scott tells me.

On Thanksgiving Day, Jim's children and wife gather around his body as a flight deck of life-support machines are switched off, one by one. He was an attorney in Bend whose primary client was the Confederated Tribes of Warm Springs. He was also a U.S. Marines veteran, so his funeral is elaborate, with American Indian blessings and a three-volley salute.

On our drive home from the funeral, Scott and I talk about the spiritual ceremonies we just witnessed.

"It really makes you stop and think...," Scott starts.

"I know."

"...about what we'll leave behind when we die."

"I know."

"I need to figure out what I believe in."

"Wait...what are you talking about?"

"I don't necessarily mean religion, but..." He trails off and pauses before adding, "What would my funeral look like? I wasn't in the military. I don't go to church. I'm not Native American. So what will you do at my funeral, just throw me in the hole?"

I can't help it. I let out a laugh.

"What's so funny?"

"You're right—you are." I shake my head and try to turn serious. "It's just, I thought we were talking about something else."

"What?"

"I was just thinking, when you said 'what we leave behind,' that you were talking about kids. That, you know, we need to hurry up and have some...or there won't be anyone at our funerals."

Scott laughs.

"Good point," he says.

On December 11, I come home from work to find Bob lying on his stomach, unable to stand. We've already arranged for the vet to come to our house the following afternoon, a Friday, to euthanize him. Now it's clear that's not soon enough. I call Scott in tears.

"It's Bob," I say between sobs. "You need to come home."

He carries Bob into the backyard, which is where he appeared to be heading when he fell. We stroke his soft fur and then I go inside and call the vet.

While we wait for her to arrive, we pet Bob and feed him pieces of sausage that we had in our refrigerator. We tell him what a good dog he is. That we will always love him. That we were so lucky to have him.

Later that night, I tell Scott that it feels like this recent string of deaths is consuming our lives. I have no idea that it's just beginning.

For Christmas, Scott and I fly to Washington, DC. We gather with my mother's extended family at my parents' weekend home near the Chesapeake Bay. The celebration is hectic and crowded, full of small talk with some family members and long, involved conversations with others.

My cousin Donna and her husband, Seth, are spending the night here with their daughter, Audrey, who was born a year ago in August. I haven't seen Donna since before

Audrey's birth, and I'm curious to see her as a mother. For several years now, my life with Scott has paralleled Donna's with Seth. We married within a week of each other. We both recently bought houses. Audrey represents this huge step that they have taken and we have not, a step that still makes me nervous.

When the other family members leave and my parents go to bed, Scott and I stay up talking in the living room with Donna, Seth and my sister, Gretchen.

I can't wait to get into bed with Scott, and have some time away from the rest of my family. I lift the covers and crawl in, stretching out alongside him and letting out a big sigh.

And then I hear a sharp shriek.

Scott and I look at each other.

"Did someone just scream?" I whisper.

Scott nods, his brow furrowed.

We sit up and hear it again. This time, it's louder and unmistakable—pure agony. It's Donna.

We leap out of bed and fling open the door.

Donna is running into the dining room with Audrey limp in her arms.

"Call 911!" she screams, then bends down and places her ear to Audrey's chest. My parents and sister appear, stunned, in the doorway on the other side of the room. My father has the cordless phone in his hand. He has already dialed. "We need an ambulance."

By now, Donna—who is an EMT by profession—is performing CPR, rapidly pressing her palm into Audrey's tiny chest and then stopping to breathe into her pink mouth. Seth sits on the floor with Audrey's feet in his lap, rubbing her legs.

My father gives our address to the dispatcher, then Audrey's age. Donna lifts her mouth off Audrey's and says, through tears, "I can't do this."

"Yes you can," my mom says. "You're doing it, Donna. Keep going, you're doing a good job."

Donna doesn't miss a compression or a breath, even though it's clear her heart is breaking. We can hear it in her wailing, see it in her tears. While all of this is going on, I'm standing helpless on the side. My body is in a state of hypervigilance; I shiver a little, acutely aware of how the scene looks, how my own saliva tastes, how my parents' dining room smells. Scott returns to our bedroom and yanks on his clothes and shoes. I follow and pull on a pair of boots over my pajamas. We walk outside, to the end of the driveway. The neighborhood is dark and the streets are wet. In the distance, we hear sirens.

I run back in to tell everyone that we can hear the ambulance—it's coming. Then I run back out to wait with Scott. We don't say anything, just stand nervously. The sirens are loud and clear but there is still no vehicle in sight. Minutes pass before lights finally crest the hill in front of us.

We wave our arms and a fire engine pulls up and parks on the street, next to the driveway.

"Hurry! She's in here!"

Two paramedics stride in and seize Audrey in their blue-gloved hands. I watch from the front hallway as they run her outside, her body jiggling in their arms. An ambulance pulls into the driveway. We all get dressed and drive in two cars to the hospital. Two people—police officers, paramedics, I can't remember—stay behind because they are required to investigate the house.

I park the car and walk into the hospital with Scott. In the emergency room, Donna and Seth stand outside a room where doctors and nurses are circled around Audrey. Donna pleads with the doctor to try using defibrillator paddles, to restart Audrey's heart. I pace through the white halls of the hospital, feeling like a ghost, until my dad sees me. He ushers Gretchen, Scott and me away, into a private waiting room.

I sit and stare at the lotus pattern of the upholstered chair in front of me, trying to memorize every curve, every color. My mind is racing and I have no idea how to slow it down, how to calm my thoughts. I grip Scott's hand. At first it feels like something to do, although I quickly grow afraid that if I loosen it, Audrey will die.

My parents come in to give occasional updates, but there is no good news. Audrey is not responding to any medical treatments.

A short, middle-aged woman comes in and tells us she's a volunteer trauma counselor. She stands and stares at us, probably wishing, as we all do, that she knew what to say. Then she leaves.

Eventually someone—my mom? my dad?—comes in and, crying, breaks the news. Audrey has been declared dead. Donna and Seth are holding her, saying good-bye.

Later they both enter the room, red swollen eyes, blank stares of grief. They look like zombies. And in a way, they are. Audrey's death has robbed them of their identities as parents, their dreams for their daughter, all those birthdays yet to come, the dress-up games, the dances, the wedding, the grandchildren. So much more than one life has been lost.

Scott, Gretchen and I drive back to the house to gather everyone's belongings. My parents drive Donna and Seth to our family home in Takoma Park, the house I grew up in. For days, we hole up there, trying to comfort one another, trying to figure out what just happened. How could this blue-eyed, curly-haired toddler—beaming with the promise of a full life—just *die* for no reason? We will never get an answer. A months-long autopsy and extensive genetic tests will yield no explanations. Because Audrey was over a year old, her death will not be ruled sudden infant death syndrome. Instead, it will fall under an equally vague but lesser-known category: sudden unexplained death in childhood.

I have never heard my parents' house so quiet before. The silence is punctuated by faint pulsations of crying, from behind a closed door. Or the gurgling of an empty stomach. Nobody sleeps. We sit or pace, our eyes glassy, our faces pale.

In January, I work nights and weekends to finish a series

of articles about a young woman I met in the fall. Summer Stiers is thirty-one years old but has gray hair and walks with a cane. She was healthy most of her life until, at age eighteen, she began having regular seizures. In her twenties, her health deteriorated further. She suffered intestinal bleeding, muscle atrophy and brittle bones. Her kidneys failed and she relied on nightly dialysis treatments to stay alive. Doctors believe she is dying of an unknown genetic illness, and they are studying her in hopes of someday helping any others who turn up with her rare, unnamed disorder. I enjoy spending time with Summer—who somehow remains upbeat and grateful for what she still has—but she is also a reminder that death lurks around the corner.

I find myself obsessing over the animals I've killed while hunting. Was that shot I took at the goose last fall truly sporting? How young was the pheasant I killed during that women's workshop? Had it lived long enough to enjoy any of its life? I worry that I have caused surviving animals to feel grief and pain similar to what I have endured during the past few months, and the guilt is almost unbearable.

For the first week of February, Scott and I go on a backcountry ski trip with friends. It should be relaxing, a chance to get away from work and the trauma of Audrey's death. But I see danger everywhere I look. When we ski, I worry about avalanches. When we get back to the remote cabin where we're staying, I fear Scott could die in his sleep.

On February 10, Scott and I sit down on the couch

after a long day of work. The phone rings, and Scott walks into the dining room and answers it.

"Hi, Mel," he says.

It's my dad. I glance at the clock—a little after nine, past midnight for my parents on the East Coast. Shit. This isn't a good phone call. Next, I hear Scott say, "She is, just a second. Hey Lil?" I immediately think of my dad's father, who is eighty-eight and getting weaker by the day.

I take a deep breath and pick up the phone.

"Hello?"

"Hi, Lily, it's Dad." I have never heard his voice sound so sad. Whatever he's calling about is even worse than I'd thought.

"What's wrong?" My heart races and I sweat, just like the night Audrey died.

"I have terrible news. Nathan killed himself tonight."

The past tense is what hits me first. *Killed.* He's gone, it's already too late. Later, my father will tell me that I shrieked. That the sound made him feel like he'd punched me in the gut. But in the moment, I don't even realize that I'm making noise. If he had punched me, I wouldn't have noticed. If I'd been run over by a truck, I wouldn't have noticed.

In the exact moment that those horrible words slip out of my father's mouth, I feel every ounce of pain and sorrow that will take turns pummeling me over the next hours, days, months, years. There will be times when a single memory or a sudden pang of guilt hurts so deeply that I fall to the floor in sobs. But there will never be another moment when all of it hits me at once: The gaping wound left by my only brother, my companion since birth. Co-holder of the memories that my parents sometimes fudge or forget. The one who introduced me to Top 40 music and later encouraged me to reject it, explaining that it was so much cooler to find my own taste than to accept what's popular. The person who convinced me to eat pig tail—it tasted like an especially fatty hot dog—during one of my visits to Brazil.

The details of Nathan's death are relayed to our family secondhand and through a translator. During an argument with his girlfriend, he grabbed a pair of scissors and cut through a net that stretched above the railing of his balcony. Then he plunged seventeen stories to his death.

I spend the rest of the night on my living room rug. I panic when I try to remember my brother—anything, any scene with the two of us, any lesson he taught me—and I can't. Nothing. Scott rubs my back and tells me that the memories will come back when I'm not in shock anymore. Every once in a while, I call my sister and we cry together, she in Los Angeles, me in Bend. Even her sniffling comforts me. When we hang up, I curl up on the floor and convulse in sobs. My stomach twists and churns. I picture him jumping. Even in my own head, I can't stop him; he leaps again and again. I imagine what he must have thought as he fell. I hope—as I always will—that his last feelings were not remorse or panic but liberation and calm. I hope that somehow he found peace. How could he have lost control of himself so completely? Nathan always had a fiery temper, but I never imagined it could kill him. How did I not think to call him? The last time I saw Nathan was when Scott and I visited him before we were married. Between then and now, a few simple words could have made all the difference: *Are you okay? I love you. I want you to live.* The distance that grew between us these last few years becomes a talisman of self-torture.

The next day, Scott and I walk Sylvia along the Deschutes River near our house. It's sunny and unseasonably warm. Hordes of geese and ducks bob along the river. We stop midway across a footbridge. Scott peers into the water to look for fish. I watch the birds, which sit on the surface but are as blurred through my tears as fish under water. I am reminded of the birds I have killed, the rabbit I have killed, the fish I have killed, and another wave of sadness batters me.

Nathan's daughter, Sofia, was born on the same day I shot my first pheasant, and at the time I was thrilled to have both occasions coincide. Now the pheasant's death and my brother's will be forever linked, and I couldn't be more ashamed of the ecstasy I felt over killing that bird.

I've long thought of death as something that happens during old age. But the truth is, of course, that children like Audrey die. Young adults like Nathan die. Middle-aged people like my co-worker Jim die. People die at all ages, of all sorts of causes, many of which are "natural."

The same truth holds for other animals. By hunting, I have cut short the lives of my prey. Though I don't know their ages, all of my birds and rabbits appeared to be in the prime of life when I killed them. Again I ask myself: Was the killing worth it? Were these deaths really justified for a few special meals? I feel like a monster.

That night we take a red-eye to Washington, DC. When we land, it takes all the strength I can summon to put one foot in front of the other and walk off the plane. I know as soon as I see my parents that Nathan's death will become even more real. I'm not sure I can bear it. Yet somehow, I do.

Over the next few days, my parents scramble to plan a funeral. Those questions that Scott raised during our ride home from Jim's funeral swirl through my mind. Our immediate family has no religion to pave the way. We celebrated almost every holiday when I was growing up, the fortunate result of my Catholic-raised mother and Jewish-raised father. My parents had left behind their respective religions by the time they married, so our celebrations were unburdened by Church or Synagogue or even much of an explanation. On Easter we hunted eggs; on Sukkot we built an outdoor fort; during Chanukah we ate latkes; on Christmas we praised the miracle of Santa. This is not to say that we weren't spiritual in our own ways. Nathan explored Judaism during high school and college, then devoured books about Sufism as an adult. But he never, as far as we know, reached any confident conclusions. His beliefs, what ceremonies he would have wanted—these details are all left for us to guess. Friends and family descend on our house, bringing food and kind words, sharing our tears. They weave a sort of human cocoon around us, one that I am reluctant to leave.

Two weeks later, Scott and I return to Bend. We get home just in time to attend the funeral of Scott's favorite uncle, who died ten days after Nathan, ending a long battle with multiple myeloma.

Another cousin goes into labor four months early and her baby boy dies just after his birth.

Summer Stiers dies.

When I call my father to wish him a happy Father's Day, he tells me that one of my uncles has died of Alzheimer's disease.

Raymond, our neighbor, falls from a chair and dies of a head injury.

I struggle with anxiety. At times I am overwhelmed by fear, terrified that I could lose anything and anyone at any moment. This horror creeps into my life in strange ways: I buy travel insurance whenever I fly, and develop an intense fear of heights. Perhaps because I read so much fiction and watch so many movies, I have a ridiculous tendency to see my own life as the plot of a novel. So much of life is determined by luck, I know this. And yet I am quick to spot literary devices such as foreshadowing and symbolism in the random everyday. Maybe these deaths are trying to tell me something about what to do, or what will happen. I worry that somehow I have caused all of this death and despair, that these people would still be alive and well if they hadn't worked with me or become friends with me or been related to me. Should I just bid my friends and family adieu and lock myself in my house until the dying stops?

"It's not you," Scott tells me. "It's life. People die."

Working at the newspaper each day is a struggle. My cubicle is near the police scanner, and when news of a fatal car accident buzzes through the air, it stings me in a way that it never used to. My imagination is lean and agile from the interval training of these past few months. It strides easily to the conclusion that the driver—the one found dead on arrival—is Scott. I stare at my cell phone and steel myself for an ominous ring.

Reporters deal with tragedy on a regular basis, and we develop coping mechanisms to separate our own lives from those we write about. That's not to say that certain stories don't move us. Just that sometimes we have to deal with three or four tragedies in one day of work. We find ways to laugh through them, to detach from them and get through the business of writing. But grief has sloughed off my calluses.

For fleeting moments, I feel angry at my brother, which experts say is a common response to suicide. How could he do this to all of us? To his daughter, especially. And to my parents. Almost immediately, pity displaces my anger. Even battered by grief, our lives go on. Nathan, on the other hand, has lost everything. We remain close with Luciana, and when she emails us photos of Sofia, opening each file feels like I'm being stabbed in the chest. Sofia looks so grown-up, less like a baby and more like a real person. Yet Nathan will never get to see her like this. Her first joke, her first day of school, her first bike ride—he is missing everything. As are Donna and Seth. Now that Audrey is gone, they face a future unlike anything they have imagined since her birth.

In the coming year, once joyous holidays will become overwrought with sadness. This is particularly true of Christmas, which will roust all the trauma of Audrey's death. Anxiety and depression loom over all of us. I mourn not only these lost family members, but also the family that we were before the deaths. I will never again see Audrey or Nathan. Nor will I ever see that earlier, happier version of my parents or my cousins. Whenever I stop and really think about that, well, it still sucks the breath from my lungs. And for some reason, I hope it always will.

Meanwhile, friends of mine begin to bear children. Two of my closest girlfriends in Bend tell me, within weeks of my brother's death, they are pregnant. I spend the summer knitting baby gifts and planning showers. All of these deaths and births start to feel like the circle of life is pulling tighter and tighter around me.

What is the universe trying to tell me? Here I go again.

I question the decision that Scott and I have made to start a family of our own. The way I imagine myself as mother is nothing like it was a few months ago, before the dying began. I am so much more fearful now, more aware of how ephemeral everything important really is. Where I once pictured baby clothes, giggly baths and finger painting, I now see miscarriages, birth defects and spinal cord injuries. If Scott and I do have children, how could I bear the constant anxiety that something bad could happen to them?

Life is full of risks that could bring death: Driving down a highway with thousands of other drivers, any one of whom could be drunk, exhausted, distracted or enraged. Flying on a plane when a storm hits or an engine fails. Crossing the street while a bus runs a red light. Riding an elevator that was overlooked during its last inspection. Eating tainted food. And that's not to mention internal dangers. Aneurysm. Cancer. Sudden cardiac arrest.

Intellectually, I have always known this. But now I know it viscerally. I can close my eyes and see Donna performing CPR on Audrey's perfectly plump little body. I can hear the agonizing scream she let out when she found her baby. I can hear the tone in my father's voice when he said, "I have terrible news." I can feel my pulse quicken as I brace for his next words. These events, these sensations, they have changed me to my core.

Fatal hunting accidents are extremely rare when compared with, say, car accidents. But compared with other, more popular pastimes such as playing video games, hunting is downright perilous. I don't usually consider myself a physical risk taker. I have no interest in skydiving or bungee jumping. Looking back on it, though, I appreciate the danger involved in hunting. It has been meditative to me, spending so many hours so aware of life's high stakes.

Human life used to be full of risks like these. Not too many generations ago, in places like Oregon, the majority of residents killed animals for food on a regular basis. Wild predators, weather events and even minor infections posed serious threats. Today we have complex safety laws to prevent deaths that just a couple of generations ago were commonplace. Science has provided all kinds of treatments and prevention, even extreme interventions that can keep a body pumped full of blood and air when there's no brain left to dictate how.

We may live longer than any generation in human history, but eventually we all die. Death is an essential part of life. Yet we acknowledge it so seldom. Most of the time, we don't even call death by its real name. We prefer euphemisms—passed on, crossed over, gone to a better place or, simply, gone.

The last few years of hunting have forced me to take a rare, honest look at mortality. Hunting has everything to do with death. You can't kill an animal—watch its eyes flicker for the very last time—and not think long and hard about the finality of it. Nothing could have prepared me for these losses, but I feel grateful that I have, even in some small way, spent the last couple of years facing death. Of course, each kill has been the relatively easy, bearable loss of a wild animal. But deep down I have known that the next death in my life could be of someone I love.

I obsess over the guns I keep locked in an upstairs closet. Gunshot is the most common method of suicide in the United States. In fact, gun-inflicted suicides outnumber gun-related homicides and accidental deaths combined. My brother did not shoot himself. If he had, I would probably get my guns melted down and molded into some peaceful symbol. My heart aches for every victim of gunfire, and I worry that as a gun owner, I am somehow complicit in this violence.

Despite all of this, I am shocked to find myself wanting a child more than ever before. My fears about parenting have shifted suddenly: I no longer worry about what a child would do to my career or my social life; instead, I obsess over what could happen to the child. In a strange way, this comforts me. This more selfless fear is more crippling and yet it feels more appropriate, more parental. As my cousins and parents fall into a void of grief, the depth of which terrifies me, I realize all that I could miss out on if I succumb to my fears. Maybe all of this loss will make me a better parent in some ways, too. I won't take things for granted. I will appreciate the good days and happy moments because I will know how fleeting they are. Here's the thing: If anxiety is one side of a coin, then the flip side—the brighter side—is appreciating each moment. Because if the next breath I draw could be my last, then it had better be a deep one.

All of these deaths have focused me. I waste less time

browsing the Internet. I watch fewer bad movies. I shrug off minor annoyances and feel silly to think that just one year ago, I would have wasted time being upset. I work hard at my job, but I no longer fret about the size of my newspaper or the title on my business card. When we go camping and fly-fishing on weekends, I find myself better able to let go of work and other stress and simply enjoy the moment. I examine the glistening fly as it drifts down the stream in front of me. I revel in the sunlight and the cool water swirling around my legs. I look at Scott, standing downstream from me, and I smile. I am here, right now, paying attention.

These losses have left me lean and urgent, the way elk in Yellowstone must feel since their predator, the wolf, returned. This is why it's important to be reminded of death: so that you never forget that life is temporary and every day matters.

Some days, grief knocks the wind out of me. But just as often, I am bowled over by love. At night, I lie down in our bed and nestle my face into Scott's warm neck. Something—life?—grabs me by the shoulders, looks me in the eyes and tells me: *Cherish this. You only have this. You only have now.* I inhale my favorite scent in the world—his. And I feel truly, completely lucky.

CHAPTER 12

KILLING BAMBI, REVIVING ARTEMIS

A few years ago, I spent a day on the banks of a roaring Deschutes River waterfall with a Warm Springs Indian named Roland. He worked as a "creeler" at the time, counting the salmon and steelhead caught by tribal members at this ancient fishing site. He was short and round, with a wide grin framing a chipped front tooth. I was interviewing him about his job and we got off track, talking about his own passion for fishing. He mentioned that in his culture, it's a tradition to cease fishing for months or even years when a close friend or family member dies. He didn't explain the custom fully at the time, but I think about it now because suddenly it makes sense to me. There's the obvious reason: that in mourning we lose our desire even for things that once made us happy. But I suspect another reason, too. Perhaps it's as simple as needing a break from death, even the death of fish. One reason that hunting is so uncomfortable to non-hunters in the first place is because of its connection to death. After my brother's death, unable to bear another modicum of guilt or sadness, I stop hunting.

This hiatus does not, however, mean that I manage to avoid killing anything. In the fall of 2009, I take a leave of absence from the newspaper and move across the country to Ann Arbor for a nine-month journalism fellowship. Scott stays in Bend but flies to Michigan every month or so to visit.

One afternoon in September, I am pushing a rickety electric mower across the lawn of my rented house when I feel a thump against my shin. I look down. Something small and gray convulses at my feet. I gasp, let go of the mower and jump back, but this little animal flops and I accidentally step on it. It pops straight up and lands on its side, unable to run forward like a normal... whatever it is. I must have cut off one of its legs. It lurches, as if compensating for its lost limb—limbs? I shriek and run toward the house.

From the porch, I glance back, bracing for another pained hop, or at least a twitch. Instead, I see nothing. The grass is still. I take a deep breath and creep back to the lawn mower. The animal lies dead, a few feet away.

I decide to finish mowing. The death was unfortunate, I tell myself, but half the lawn remains unmowed and I have nobody to do it for me. Lawn maintenance is written into my lease. So I resume, slowly pushing the grumbling mower and glancing back every few feet at the dead critter. Should I move it? Put it in a bag and throw it in the garbage? Leave it for some hungry scavenger? As I turn the mower and head back toward the scene of the death, I notice a patch of fur on the ground, a few feet from the dead animal. Suddenly, another creature pops out of the fluff and clumsily hops away.

I flinch but mow on, my eyes glued to the tiny (live) animal. It's still unsteady as it crosses the lawn and ducks under a spruce. Its ears aren't long yet, but they stand up. Though there's no ball of fluff on its rump, I know exactly what it is: a baby cottontail. It's also the spitting image of the thing I just killed. When I cross the lawn again, I give the patch of bunny-bearing fur a wide berth, but two more hop out of the fuzz-covered hole. They scamper toward the street. I blink back tears but keep mowing, eager to finish the job so I can spend the rest of the day indoors.

As I put away the lawn mower, every pinecone I step on makes me jump. I eye every squirrel and sparrow with paranoia. In my own yard, I am terrified of all these tiny creatures squatting around me, and angry, too, as if they've framed me for a crime I never intended to commit. I'm equally afraid that I will accidentally kill something else. I waffle between feeling like an invader who has no business being here and wondering: Can't a person be left alone in her own yard?

In the safety of my locked house, I calm myself down. It doesn't take me long to conclude that I should go back outside and pick up the dead rabbit. Running in here to avoid it doesn't change the fact that I killed an animal. The least I can do is bury it out of respect.

I unlock my door and tiptoe outside. But the bunny is gone. Less than half an hour has passed since I went inside, and it's nowhere to be seen. I walk all over the lawn to make sure, focusing my search around the fluff-covered burrow. No rabbit. I know what probably happened: A neighbor's cat got it, or a bird swooped down and carried it away in its talons. Nature is everywhere, even in this suburban neighborhood. Still, an unrealistic worry creeps into my brain and won't let go: Maybe the grief-stricken mother dragged the baby back into her burrow, to mourn over the body. I start to think of the bunny as Audrey, and Donna as the mother. I am a monster, a murderer, a destroyer of life and the happiness of all the bunnies who loved this one. Tears well up in my eyes and I run back inside.

For the rest of the day, I am haunted not only by guilt but also by my own distress over this death. Why did the death of this rabbit bother me so much? The first time I killed a rabbit, during that hunting workshop with the beagles, I was thrilled about it. What's the difference? Yes, that was an adult and this was a baby, but there must be a more significant distinction, too.

I have felt nervous and ambivalent approaching every hunt I've ever gone on. In retrospect, these feelings were part of my mental preparation. Each hunt has precipitated a new reckoning of all the thoughts involved in my initial decision three years ago to try hunting.

In fact, every one of my hunting experiences could be broken down into a series of questions that I ask myself. It starts with: Do I want to hunt? Then, do I want to hunt for this species? At this time? In this place? Using this weapon? With these hunters? Other, subtler questions make up each of these: Am I ready to bear the guilt that may follow this killing? Will this experience be one that I can retell and feel proud of? Eventually, this flow chart culminates in one particular shot at one individual animal. And so by the time I pull the trigger, I have decided that I do want to kill *this* animal in *this* place at *this* moment. The kill is undeniably purposeful.

With this baby rabbit, however, I asked myself no such questions. I had no time to prepare for the emotions, to weigh the magnitude of that life against my own intentions. After several years of narrowing so much purpose onto each life I take, I am even more horrified by this thoughtless, careless death than I would have been before I started hunting.

Those of us who hunt, who kill animals on purpose, open ourselves up to a lot of criticism, including from other hunters. Since I started hunting, I have joined in the condemnation of so-called road hunters, who lean out their windows hoping to shoot their prey without getting out of their trucks. Hunters who hike miles away from roads and backpack into wilderness areas criticize these hunters as being lazy. But their laziness pales in comparison with the hundreds of millions who buy their meat already killed, butchered and shrink-wrapped, right? Well, not necessarily.

It's a big deal to kill an animal, and an even bigger deal to do so on purpose. As hunters, we carry a grave responsibility. The questions we ask ourselves—Is it sporting? Did I give fair chase?—matter. Not everyone believes hunting is justified, and that's okay. But each hunter must justify it to herself. And these questions matter even more when you consider that not everyone can hunt.

There is not nearly enough habitat or wildlife left in the United States to sustain three hundred million hunters. Unlike gun ownership, the Constitution—no matter how you interpret it—does not guarantee our right to hunt. Several states, however, have amended their own constitutions to protect citizens' right to hunt. Though I have become a staunch defender of hunting, I don't think of it as an inalienable right, nor do I think it should be considered one. Hunting is an immense privilege. And one of its greatest values is that it requires near-constant reassessment of the situation and surroundings. Just because you can take a shot at an animal, for example, doesn't mean you should. Likewise, responsible agencies should continuously, realistically assess state wildlife populations, habitat health and hunting quotas. It's not a stretch to imagine that wildlife habitat in a particular area—or even an entire state—could become so degraded that hunting there is no longer feasible or responsible.

Even though not all Americans can hunt, there are ways for non-hunters to support responsible hunting. The owners of large acreage could allow ethical hunters to hunt on their property as the law permits. And imagine if environmentalists partnered with hunters to turn conservation-minded sportsmen's groups into juggernauts more powerful than the National Rifle Association. Perhaps most important, greater acceptance of hunting could go a long way toward promoting sound environmental policies. Hunters and non-hunters alike share ownership of local wildlife, which are property of the state. Non-hunters can help by supporting sound wildlife management policies. Agencies that regulate hunting should be held to high scientific standards and should be adequately financed. These goals require all citizens—not just hunters—to pay close attention and vote responsibly.

Even during my hiatus, I can't stop thinking about these issues facing the future of American hunting. After all that I've gained from my hunting experiences, I feel a strange responsibility to pick up my gun and hunt again. It's as if, by taking a break from hunting, I am turning my back on the very tradition I once vowed to help resuscitate. Feeling too guilty to keep hunting, I am startled by this new guilt caused by *not* hunting.

Interestingly, many experts believe that women are the key to reviving hunting and fishing in the United States. The idea is that if the mother of a household hunts, her children are more likely to embrace the sport. Most states now offer Becoming an Outdoors Woman workshops, like the pheasant hunt I joined, to introduce women to hunting and fishing. The effort seems to be working. One in ten

American hunters is female—our gender's highest participation rate in history—and we are the only demographic of hunters currently on the rise.

Today, with reliable firearms and other improved technology, there is no physical reason why women can't hunt as capably as men. I know a woman who shot a six-point buck when she was almost eight months pregnant. In fact, female newcomers might be exactly what the sport needs for one obscure reason: Many female hunters learn the sport as adults. Making a conscientious decision to hunt—rather than doing it because your parents want you to—requires serious ethical deliberation that is likely to creep into other aspects of hunting, too.

My friend Jessie, for example, never considered hunting until, in her twenties, she fell in love with a hunter. At first, this still wasn't enough to convince Jessie to pick up a gun. But she enjoyed cooking and eating the wild game that Andy brought home. Soon she was hiking behind Andy and his parents as they hunted.

"Eventually, I thought that if I was going to eat it, I should be okay with killing it," she told me. "And I wanted to see what goes into the death of animal... There is no way to really understand it until you do it yourself."

Jessie had contemplated the philosophy and psychology of hunting before she ever loaded a gun. All of this consideration made her more likely to become a committed hunter, continuing the tradition well into the future. It also made her more likely to recognize the far-reaching implications of hunting, and to do her part by joining sportsmen's organizations and closely following the politics of wildlife management. In short, she is exactly the kind of new hunter that is needed to keep the tradition alive.

In a sense, hunting is a final frontier of feminism. As women make up a growing percentage of American hunters, we quietly lay claim to a part of humanity that has been dominated by men. Women born to hunters or in love with hunters weren't always allowed to participate. Tina, a hunter in her early sixties who lives in La Pine, grew up with two brothers. Her father took the boys hunting and sometimes let Tina tag along but didn't let her hold the gun or take a shot. One day, one of the boys killed a deer. The father instructed the children in how to dress the deer, but both boys were too scared to take the knife to the hide. Eventually, Tina grew frustrated of waiting.

"I said, 'Give me that, I'll do it,' and I grabbed the knife," she told me. "And then I cleaned the whole deer."

It's strange to think that for millennia, men have done almost all the hunting while women focused on other roles such as gathering and child rearing. Of course there are exceptions—Artemis the huntress, for example, is a feminist symbol from ancient Greece. On a whim, I pick up a book of Greek mythology at a library in Ann Arbor. Almost immediately, I am captivated by the stories about strong, mysterious Artemis.

For this goddess, the Greeks spared no meaning. She represented Nature, Wilderness, the Hunt, the Moon, Virginity and Fertility. To some, she embodied Death and Vengeance. She was as complicated as her broad assignments would suggest, a goddess of contradictions. Armed with a silver bow and a quiver of arrows, she inflicted illness and death on those she also protected. She had great respect for animals and also hunted them, leading a posse of nymphs through the forest in pursuit of game. She wore a short tunic like men did at the time, instead of the long gowns worn by women, to allow freer movement while hunting. She loved wild things but she held captive a group of stags who pulled her golden chariot. She was usually depicted with one of these deer or with one of her many hunting dogs.

Combing through the stories of Artemis, it's clear that different hunts meant very different things to her, and to the Greek people who worshipped her. She could be vengeful. Some versions of mythology claim that Artemis killed Orion—the only hunter whose skills matched her own—as punishment for boasting that he would track down and kill every wild beast alive.

Artemis was virginal and fiercely protective of her purity. A hunter named Actaeon once happened upon her and her nymphs bathing nude in a secluded pond. Stunned by their beauty, he hid and watched them. But Artemis saw him and showed no mercy: She turned him into a buck and then spurred his own dogs to chase it. They tore the stag apart before finally recognizing him.

The goddess could also be kind. As she was a master archer, death by her arrow was considered a gift, a blessing. Everyone must die somehow, after all, and this death was swift and painless. She also healed and protected. As the goddess of childbirth, she represented the entire life cycle. But rather than create life of her own by bearing children, she asked her father, Zeus, for the gift of eternal virginity.

Reading this makes me stop and think, again, about my ongoing hesitance over whether to have children. It also makes me more aggressive in my Artemis research. Yet I find nothing to explain the goddess's reasoning for her request. Would gaining a child have meant losing something even more precious to her? Was she, too, afraid of all the heartache lurking in the depths of parenthood? Did she know something that I don't?

When I stop my own navel-gazing and consider Artemis as a public icon, I am stunned that one deity could embody all of these diametrical forces. She was light and dark, life and death. She was not a Disney princess, all good or all bad. She was—all the mythology notwithstanding—realistic. No character in today's culture would be asked to represent such a complex collection of properties. We much prefer to compartmentalize our symbols. This is particularly true in our portrayal of animals.

Take, again, the movie *Bambi*, for example, which reinforces the myth that hunting and eating prey is equivalent to killing and eating one's friends—Bambi, Faline or Thumper—for dinner. Hunting has taught me to rethink this. I know that the truth is murkier and more complicated. Yet since this string of deaths began last fall, I have found myself trying to believe again in Disney's version of pain-free nature. After so much loss in such a short period of time, I've wanted to separate myself as much as possible from any death or suffering. I haven't wanted Artemis, I've wanted Bambi.

Gradually, in the months after my brother's death, I begin to feel less raw. And I haven't forgotten my epiphanies about the values of hunting. Hunting has changed the way I think about the food I eat and my pet dog, not to mention the animals that live out of sight but all around me. It has given me a deeper connection to the fast-growing community where I live. It has changed the way I follow politics. Still, I have only brushed the surface. I have not yet wrung all the meaning I can out of this new adventure.

Autumn is under way, and my temporary home in Michigan offers a new landscape and a new species to pursue. It's time to reload my gun and hunt again. This time, I will try my hand at deer.

CHAPTER 13

DEER DIARY

In Michigan, some people talk about deer the way New Yorkers talk about rats. They are varmints and they are everywhere. Munching azaleas in the backyard. Sprawled out, bleeding, on the side of the highway. Bounding through the woods past my favorite jogging trail. Despite their ubiquity, shooting a deer will be the ultimate test of my decision to hunt; even a small buck weighs more than I do. I wonder, yet again, if I will be able to look into its eyes and pull the trigger. I wonder if I will be able to gut it without vomiting and then wanting nothing to do with its meat.

As hunting season approaches, I start researching rifles. I brought my shotguns with me to Michigan, just in case I decided to do some bird hunting. But for big-game hunting, a rifle offers longer range and more accuracy, which adds up to a more humane kill. I do some research online, talk to as many hunters as I can and spend hours in front of the gun counter at a large outdoors store. Eventually, I settle on buying a Weatherby 7mm-08. It's a generous caliber for the small white-tailed deer that populate Michigan. But, in an effort to limit my accumulation of guns, I want something big enough to shoot larger mule deer and possibly even elk back in Oregon.

I ask locals for instructions and then drive nearly an hour to a shooting range where a volunteer teaches me how to load and shoot the gun for the first time. I go back and practice a few more times before opening day.

Even in liberal Ann Arbor, it's impossible to ignore deer-hunting season. That's not to say that everyone here hunts. Far from it. But everyone I meet here *knows* someone who hunts. And nobody is shocked by the practice. Unlike Oregon, where the population doubled between 1965 and 2010, Michigan residents have been leaving in droves, due to a weak economy and the decline of the auto industry. That means most Michiganders who remain have lived in the state for many years, so local traditions such as hunting don't shock them. Small towns including Dexter, just ten miles outside of Ann Arbor, still post "buck poles" each fall. These large wooden beams are erected in a prominent spot in town, where successful hunters hang their gutted stags for all to see and admire.

In Oregon, as in many other parts of the country, the buck pole would attract more protesters than admirers. Not here. When Scott and I visit one during opening weekend of deer season, we are the only people laughing nervously as we step between two rows of hanging bucks, swinging gently in the breeze. Nearby, a stockpot of venison stew has been set atop a grill, and we smell the rich meat but are informed, apologetically, that it won't be ready to eat for a couple of hours.

"That's okay," I tell Scott as we walk back to our car. "I'm not really in the mood for venison."

But a couple of weeks into deer season, I find myself holding a rifle and sitting perfectly still on the ground beneath a maple tree, watching for deer. My mentor at the university, Charles Eisendrath, also owns a cherry farm in northern Michigan, and has generously offered to let me hunt on his property. Scott is here, too, to help if I shoot something.

White-tailed deer (*Odocoileus virginianus*) are sometimes called Virginia deer or simply whitetails. They are one of three closely related deer species native to the United States. Deer live just about everywhere, from cities to suburbs to rugged wilderness areas. They are particularly attracted to farms because they love many of the same foods we do: wheat, oats, corn and soybeans, not to mention fruits, vegetables and even the grasses we grow to feed livestock. If large numbers of deer are allowed to browse these crops freely, they cause astronomical amounts of damage. With most U.S. farms operating on slim profit margins, farmers can't afford to let deer (or, for that matter, geese or rabbits) gobble their crops unchecked. And neither can we, the mouths who rely on these farms for our food. To discourage unwanted grazing, some farmers employ hazing techniques including noise machines, mechanical scarecrows and explosives. But the most common deterrent is hunting. Every state in the Union offers some form of special hunting tags to reduce crop damage. "Everyone in North America who lives each day on agricultural foods," writes cultural anthropologist Richard Nelson, "belongs to an ecological network that necessarily involves deer hunting... In this sense, the blood of deer runs through our veins as surely as we take bread and wine at our table."

Instead of hunting amid rows of cultivated cherry trees, however, I'm sitting in a sixty-acre stretch of forest that separates the orchard from a large lake. I don't have a problem with hunting deer on farmland—that's what Charles is doing, just a few hundred yards away—but one of my goals is to learn to track deer.

Almost every American has seen a deer. But going out and finding one is a different story altogether. It feels daunting at first, like finding a needle in a haystack. Except this needle can move where it likes, so I should be able to learn its preferences and predict its location.

Deer favor the margins between two types of habitat—areas where thick trees edge up to fertile clearings, farmland or prairie. Like cattle and many other mammals, deer are ruminants, which means they have four stomach chambers. A doe eats by chewing up plants and swallowing them into her first stomach. This chamber gradually softens the food. Once that first stomach is full, and before the food moves to the second chamber, the animal beds down somewhere and regurgitates its semi-digested cud, re-chews it and swallows it again. When Nelson sneaks close enough to a doe to watch her ruminate, he writes that it "looked as if mice were running up and down inside her esophagus." In short, deer spend their days alternately eating and then lying down, hidden in dense shrubs or trees, to digest. Deer tend to spend the middle of the day bedded down, and they are most active at dusk and dawn.

The trick to deer hunting is finding out where the animals eat and where they bed down to sleep or ruminate. Deer become attached to certain places, and they tend to travel between them along well-worn routes. A common strategy is to plant oneself along such a path and wait for an animal to arrive.

"You're looking for a deer superhighway," Charles says as he points out a couple of spots that have proved promising in previous years.

I have a buck tag, which means I may shoot a deer with visible antlers. This is the deer tag that anyone can buy. But Charles, as a farm owner, also has a handful of doe tags. Doe hunting is more closely regulated to protect deer populations, so doe tags are much harder to come by. (Only a handful of bucks are needed to inseminate dozens and dozens of does, ensuring plenty of fawns next spring.) Charles gives me one of these tags, which I tuck in my backpack next to the buck tag.

"The does are fatter, and have better-tasting meat," he adds.

At first, I make up my mind that I will only shoot a buck. I feel a little guilty about the other advantages afforded me by hunting on this private land, next to a farm—namely, no competition from other hunters and all of Charles's knowledge of the place. A doe tag—which, combined with my buck tag, enables me to shoot any deer I see—feels too lenient, almost decadent. But then a couple of days go by and we don't see a single deer.

Each morning, before sunrise, we creep out to one of the supposedly high-traffic areas that Charles has pointed out. We hide and wait for the sun to come up, hoping that dawn will bring deer rush hour. When I get too cold, or tired of sitting, we walk around as quietly as possible, looking for prints or scat or some other sign that deer are in the vicinity.

"If a doe walked right in front of us, right now," Scott whispers to me on the third day of hunting, "would you shoot it?"

I don't hesitate: "Oh yeah. Definitely."

Deer have acute senses of sound and sight, but they're even more attuned to smells. To get anywhere near a wild deer, you have to pay attention to which way the wind is blowing, as deer can detect ribbons of human scent a mile away. Hunters gain an advantage when deer season coincides with the rut, or mating season. This is when hormones cause otherwise wary bucks to let down their guard.

Unfortunately, our mid-November outing is a week

or two before the rut picks up. It's also unseasonably warm and dry, which makes it pleasant to sit still in the woods, but the worst possible weather for tracking big game. In cold weather, animals must eat more to stay warm. When it's mild, they can bed down and postpone their meals until darkness falls and the threat of predation drops. Also, dry leaves and twigs make it almost impossible to sneak up on a species armed with superhuman hearing.

As Scott and I creep around in the woods, all of these challenges overwhelm me. When I accidentally snapped that twig, was it as loud as I think it was? Did the wind just change direction? If I turn north and there is a deer in that gully, will it be able to smell me? Stay down, a buck in that thicket might be able to see me if I stand upright on this ridge.

I begin to understand why hunting is often compared to war. Aside from the obvious commonality of a pursuit to kill, there is the need for a physical strategy. Sometimes when I hear myself speak, I sound like I'm commanding a battalion. Sitting on a small ridge overlooking a promising-looking thicket of shrubs, for example, I decide that I need to inform Scott of my new strategy. But I don't want any deer that might be hidden below to see me, so I army-crawl to Scott, trying to keep my knees from rustling the dried leaves too loudly.

"New plan," I whisper. "I'm going to go up this hill, past the orchard and then come back down over there." I point across the thicket to another treed hill. "You stay here. Then, when I'm in position—I won't motion to you, that might give me away—walk slowly through the valley toward me."

"Why?"

"If something comes out of those shrubs, I'll at least be able to see it. And then we'll know that's where they are."

"Got it."

For days, we walk from one of Charles's hot spots to the next. Then we sit as still as possible in these strategic locations, and wait. This is how I always pictured deer hunting, and it's why I expected the experience would confirm my preference for bird hunting. Sitting perfectly still in the woods always seemed, well, boring. But I quickly discover that it's not. In fact, it's amazing what I get to see. Songbirds skitter right up to me, unaware that I'm here. Squirrels dart across downed logs, performing their pre-winter chores, I suppose. This is life in the forest, and for the first time I have a front-row seat. As with skiing or hiking, it is a satisfying mental exercise to keep my mind present, to stay engaged in the subtle entertainment that unfolds before me.

On our last evening, after three days without seeing a single deer, I hear something crunch the dried leaves. Crunch. Crunch. I suck in my breath. Through sweaty palms, I tighten my grip on my rifle and peer through the scope. The forest is quiet again. Perhaps the deer saw me move a little? Then crunch, crunch, crunch. It resumes walking.

This time, it's a deer, I'm certain. I sit as still as I possibly can, but my heart thumps louder than a bass drum. Will the deer hear my thundering heart and be scared away? Every muscle in my body is flexed. The stepping sounds are coming from below the hill where I sit, on the other side of some thin, bare maple trees. I refocus my scope to get a better view. Deer are masters of disguise; their gray-brown fur blends into just about any backdrop. I scan the area slowly, but see nothing. Just a fat gray squirrel hopping through the leaves. Crunch. It hops again. Crunch. It takes me a moment to realize that the crunching is synchronized with the squirrel's movements.

Shit. It's not a deer at all, just a stupid gray squirrel.

Darkness falls and soon I can only see a few feet in front of me. Still, I am hesitant to leave my perch. Scott, who was sitting against a tree trunk about ten yards to my left, stands up. He's a human cacophony: crushing leaves, shattering branches and crackling layers of clothing. I glare at him and he shrugs. Then he tiptoes toward me, still making a racket.

"Lil, it's dark."

"I know." I sigh and stand up. I almost fall over, my legs are so stiff; my butt is completely numb. As we walk back to the farmhouse, I lament the outcome of my first-ever deer-hunting trip.

"Didn't that sound like a deer walking across those dead leaves?"

"Yeah, it really did."

"Never in my life have I been so amped up about a stupid squirrel."

He slips his arm around my shoulders.

"Well," he says, "now you know how Sylvia feels."

Two days later, back in Oregon for Thanksgiving, we are on the way home from my in-laws' house when Scott hits a deer with my ten-year-old Toyota. We gasp as a trio of deer bounds across the highway, one after another. Scott slams on the brakes and slows the car to about twenty miles per hour by the time we intersect their path, but he can't quite avoid the third deer. Our right-front bumper hits its hind leg. The deer keeps running, no doubt from a surge of adrenaline, and we keep driving. The car is fine but we both fall silent, worrying about the doe. It reminds me of my rabbit-assassination-by-lawn-mower. Yes, I just spent four days attempting to kill a deer. But this isn't how I wanted to do it.

The next fall, I am back in Oregon and more determined than ever to bag a deer. In Oregon, there is one brief statewide deer season followed by a series of controlled hunts, in which a limited number of tags are awarded by lottery for a particular part of the state. This system allows biologists to more closely manage deer populations because they can survey deer numbers and even size, then adjust as needed the number of tags awarded the following year. But it makes things much more complicated for hunters.

As the deadline for lottery entries approaches, I cave

and spend fifteen dollars on a statistical guide to the lottery system. It explains how hunters have fared in each unit during past years: in drawing a tag and in "filling" the tag (killing an animal). I manage to emerge from this long lottery process with buck deer and bull elk tags for the same unit, about a hundred miles south of Bend. It doesn't have a spectacular reputation. When I unfold a map of the unit, I notice immediately that a cobweb of small dirt roads stretches over the entire region. Deer and elk live here, but there's no remote wilderness area where I can eke out an advantage over other hunters by my sheer willingness to hike.

I read everything I can find about deer hunting. I interview every hunter I can think of. When Scott and I go fishing during the summer, I practice looking for deer tracks. When I find some—cloven hoofprints—I realize that I don't know if the cleft comes to a sharp point at the front of the hoof or the back. This seems important, so I find a pasture of grazing cows and, assuming they are closely related enough to be indicative, inspect their hooves. The notch is at the front.

As the season approaches, I also prepare by worrying. I'm not too nervous about failing to locate a deer—that feels mostly out of my control. But I am scared of what to do if I somehow manage to find and shoot one. Gutting a deer sounds like a monumental task. Also called field dressing, the process involves removing the organs to prevent bacterial contamination of the meat, or muscle. It helps cool the meat more quickly, to keep it from spoiling.

And gutting it enables a hunter to cut the animal in quarters and transport it out of the woods. While doing all of this, however, you need to make sure the meat isn't contaminated by urine or feces. Deer urine contains strong-smelling hormones that can alter the taste of the meat, and the feces and other intestinal contents contain loads of bacteria.

Field dressing is a process that begins with this vile instruction: *Make a deep incision all the way around the anus and tie it closed with a piece of string.* I shudder whenever I read this and skip ahead to the easier-sounding steps, like slicing open the abdominal cavity. Another list-topping fear: What if I shoot a deer in the evening, and have to track it, gut it and pack it out in the dark? It's legal to be on USDA Forest Service property after sundown, but I can't fathom having to hike across uneven, trail-less land in the dark, not to mention trying to spot droplets of blood to find an animal I've wounded but not yet killed. Or carrying fifty pounds of venison on my back.

For several weeks before deer season opens, we head south and camp in my unit each weekend. We drive all over, pulling off small logging roads to hike up hills, my eyes trained on the ground for deer tracks or poop (for which hunters use another pleasant euphemism: sign). Fresh deer tracks are still sharp around the edges, not yet disturbed by wind or rain or dew. The print is darker than the surrounding ground because the damp under-layer of dirt hasn't dried yet. Very fresh deer scat looks like a glistening pile of dark-roasted coffee beans. As it ages, it looks more like a pile of chewed-up bits of grass and leaves.

One day, as we drive down a paved forest road in the pouring rain, we see a pair of deer jog up an open hill. Scott slows the car and I peek at them through my binoculars. One is a doe, the other a buck, his antlers still covered in their spring velvet.

Male deer shed and regrow a new pair of antlers every year. Hunters tend to think that the more tines or points on the antlers, the older the deer is. But diet and genetics also play a big role in antler size and formation. In January or February, a buck will rub his antlers against a tree until they fall off, one at a time. I imagine that they start to feel like loose teeth, and their release brings relief even though it sometimes draws a little blood. In late spring, the antlers sprout up again, this time coated in velvet. As fall approaches, bucks rub against trees to remove the velvet in long, bloody strips. During the late autumn rut, when they compete to mate with does, bucks spar with one another, antler-to-antler, until one surrenders. Occasionally one buck will gore the other. Rarely, two bucks lock antlers during a fight. If they can't separate, they will both starve or die of exhaustion or be killed by a predator.

The day before deer season opens, Scott and I work all day, then pack up the car and drive two hours south. We pull into a small campground just before midnight and don headlamps to set up our tent. I set the folding travel alarm clock for four thirty, then change my mind and reset it for five. I've had a long week at work. No sooner do I close my eyes than the alarm is beeping next to me. Several snooze rounds later, I switch on my headlamp and start pulling on layers, still inside my sleeping bag. We were planning on boiling water and brewing coffee but it's too cold to stand around waiting for the stove to heat up, so instead we jump in the car and drive to a spot that we visited while scouting.

It's almost six when we pull off a dirt road. I take my gun out of its case and load it with one bullet, then slip a few more in my pockets. In a hurry to get in place before the sun rises, we make our way noisily to a wide, rotting stump. It's still plenty dark, so our headlamps bob as we pick our way there. I chose this landmark because it will shield our silhouettes while offering a 180-degree view. As we hike, I wonder if deer can see our lights. I shrug off my backpack and sit down against the stump. Scott sits against another downed tree about ten yards behind me. I try to stay as still and quiet as possible.

Daylight builds and soon I can actually see the landscape around me. At one point, a chipmunk races along a downed log and nearly bounces off my back. I take it as a compliment: I'm sitting perfectly still, as far as this animal can tell. No sign of any deer, though.

By midmorning, I am frozen stiff and I motion to Scott that it's time to get up and leave. We hike back to the car and drive back to our tent for a nap and some long-awaited coffee. That afternoon, it gets warm—over seventy degrees—and Scott fishes a nearby creek while I re-read a book called *Deer Hunting*, by master hunter Gary Lewis. The chapter I've opened is all about the importance of stealth. He recommends taking one step and then waiting, listening. Humans are the only animals who stride with such regular cadence, he writes, so walking to one's normal beat is an alarm to the sensitive ears of wild deer. I feel guilty when I think back to our groggy, clumsy hike this morning.

We head back up to the chosen stump a little later, this time parking farther down the butte and hiking—more slowly and quietly this time—up the ridge. We sit next to another tree stump in a slightly wider clearing. Scott reads and I scan the clearing, peering into dark spots for the flick of a tail, the crook of an antler. This time, I'm sure we'll see a deer. I wonder, too, if I really want to shoot it. How guilty will I feel? Do I really want to deal with the guts and skin and—shudder—tying off the anus? After another two hours of backbreakingly still and silent sitting, however, I have my answer: a resounding *yes*. It's still opening day of deer season and I already feel that I've put in too much effort to go home empty-handed.

When it's dark, we switch our headlamps on and tiptoe back to the car. We return to camp, heat up a quick dinner of leftover meat loaf and go to bed at eight thirty. Two men driving a sedan with New Jersey plates pull into the campsite next to ours and proceed to chop wood for an hour and a half, carrying flashlights past our tent as they drag logs back to their chopping block.

The next morning, the alarm buzzes at four thirty. I'm

already awake but wishing I were asleep. I hit the snooze button a couple of times and stay nestled in my sleeping bag. Plink, plink. It starts to drizzle on the tent. We wake up and drive back to what we've started to call our "parking spot." Again wearing headlamps, we tiptoe up the butte and out to the same giant stump where we watched the sun set yesterday. Nada. A few hours later, we tiptoe back out to the logging road and start heading down to our car. As we creep—heel, toe, heel, toe—we notice several fresh-looking hoofprints underfoot.

"Were they sneaking past us on the *road*?" I whisper.

"Looks like it," Scott says.

We head back to camp, where we fry up some potatoes, onions, peppers, mushrooms, cocktail weenies and eggs together in a skillet. As we cook, we remark how strange it is that we haven't heard any shots. I admit to Scott that yesterday evening, sitting quietly in the clearing, I actually wondered if I had the dates wrong. But I didn't.

"Maybe we're just in a really bad place," Scott suggests.

"Or maybe"—I fumble for a more optimistic angle—"it's so warm that other hunters aren't having any luck, either."

We drive to the closest town, Chiloquin, to fill up our tank, and I ask the gas attendant if many people have brought deer through town.

"I've only seen one," he says. "Everyone else has bad reports. Two weeks ago, when it was colder, they were down here. But once it got warm again, they went back up."

I nod. When I recount this conversation to Scott, back

in the car, he's full of questions: Where did the successful hunter bag his deer? Where is this "up" place that the deer go when it's warm? We're on top of a butte and we're not even seeing any hunters, much less any deer. I have no answers.

Later that afternoon, we head back to our spot, this time hiking up to a little opening that overlooks the road. Again, we notice fresh tracks on our approach, which I take to be a good omen. The sun is behind me and I see a long shadow of myself carrying a gun. It's startling how the gun changes my appearance. I look dangerous, which I guess is how the animals see me.

Scott reads a book under a thick stand of fir. I sit out, more exposed, for a better view. I can't stop fidgeting. My stomach grumbles. My back is sore. Where are the deer? I get up and hike slowly around the top of the butte until sundown, again seeing no sign of deer.

The next morning, we awake to howling wind, gushing rain, thunder and lightning. According to my readings, the deer will be bedded down somewhere, not moving around, in a storm this severe. I switch off the alarm and we fall back asleep.

That afternoon, the sky brightens and we hike through thickets of trees to a small clearing. Scott sits against a tree trunk, behind and below me.

I've been sitting for what feels like hours, daydreaming and trying not to fidget. My gun is in my lap. Staring into this clearing is becoming meditative. I start to notice everything about it. I recognize which chipmunks are running along certain routes, and what they sound like when they stop and pick up a nut or seed. I peer across the clearing and imagine that dozens of deer are waiting in the shadows, watching me. That crook, could it be part of an antler? I check through my binoculars. Nope, just a branch.

And then, like magic, a doe appears in the clearing in front of me. I mean, *appears.* I don't hear her arrive or watch her step into the light. It's as if a special-effects engineer just beamed her into view. Right as I see her, she notices me. We stare at each other for what feels like several minutes, her giant funnel ears scooped toward me. (The species that inhabits this part of Oregon, mule deer, or *Odocoileus hemionus*, are named for their giant ears.) I try to stay as still as possible, hoping that my impossibly loud heart won't scare her away. I narrow my eyes, trying to will a pair of antlers onto her head as magically as she popped into my clearing.

But she doesn't have antlers. And she has had enough. She bounces away, flicking her hooves backward with each graceful leap. I take a deep breath. Even if she had been a he, antlers and all, there's no way I could have raised my gun to my shoulder fast enough to shoot before she bounded away. Lesson learned: Don't hold your gun in your lap; hold it where you can use it.

Later, Scott tells me that the doe was looking straight at *him*. It reminds me of that Disney World ride where the ghosts all seem to be making eye contact with you, no matter who else is in the haunted house.

At the end of the day, we head back to camp, pack up

and head home. Scott needs to check in at his office. I need to restock our cooler.

Two days later, we return to what we've dubbed Deer Camp. We awaken at five (deer hunting is turning out to be more of a marathon than a sprint, so why wear ourselves out with masochistic wake-up calls?) and drive back up what Scott has started calling "Buck Butte." He's an optimist.

We settle under some trees facing the small dirt road that we hiked. It's a little warmer this morning, above freezing, and foggy. Once the sun is up, I creep around to the various clearings that I've identified on this butte, looking for deer. Then Scott and I hike up behind where we sat. We notice a thin game trail that might be fun to investigate later. By now, I'm ready for breakfast. We go back to camp, eat and then decide to explore the low-lying area around the butte. All day today, we have heard gunshots all around us. Unlike experienced hunters, I can't hear the difference between a shotgun and a rifle. Duck-hunting season has opened, and we're not far from a popular bird-hunting area, so it's possible that some of the shots are being fired at birds. Still, I can't help but worry that others are bagging deer left and right.

We hike a long loop, locating another promising game trail. This time, we follow it. I enjoy the new scenery and feel excited that this could be where the deer are. Where my deer is. We see fresh deer and elk scat, which is embarrassingly exciting. I'm so thrilled by this poop that I feel like a proud new parent. We also notice strange marks in the duff, as if a buck (or a bull elk) has pawed at the ground or dragged a hoof. Pine needles are bunched up and the bare ground is revealed underneath. Later, I learn that sometimes when a buck is startled, it paws at the ground.

That evening, during our headlamp-lit hike back to the car, I put on a cheerful face.

"I know this sounds crazy because I haven't even seen a deer, but I feel like I'm getting better at this," I tell Scott. "I can focus longer. I'm quieter. I'm more prepared to switch off the safety and take a shot if I do see a deer... I think."

On the drive back to our camp, I pray the only way I know how: by wishing on a star. *Please, please let me shoot a buck tomorrow*, I think as I look at the dappled sky. Then I add: *Safely. Humanely.*

The next morning, it's dumping rain when the alarm sounds. I switch it off and roll over. We sleep in and finally emerge from the tent at nine thirty in the morning. In hunting time, this is already afternoon. We drink our coffee and eat our breakfast in the car, trying to stay dry. We drive southeast and discover that in the farthest corner of my hunting unit, it's not raining. We are not too far from where we saw the velvety buck while we were scouting, either.

This is classic mule deer country. It's open, especially since a wildfire burned the area years ago, with rocky outcroppings and steep canyons. We park and hike to the top of a steep rock formation. The hiking is tough. We thrash through snowbrush and at one point I even tumble backward off a rocky ledge. I'm not hurt, but I also can't imagine sneaking up on anything here. Too bad, because there are signs of deer everywhere. We walk past bedded-down shrubs that actually reek of wild animal. We step over piles of poop so fresh that they still gleam with moisture. There are sharp tracks and well-worn game trails beaucoup.

Next we hike through a low-lying draw that we spotted from a perch on a tall outcropping. Again, deer sign is everywhere here. We creep along, stopping frequently while I peer through my binoculars to scan the low-lying vegetation. Again, we can't help but sound loud and clumsy. Again, the deer stay hidden. Again, gunshots are firing all around us. After a few more hours, we head west to our old familiar area and find that the rain is gone and only a thick mist lingers.

We head back to what we've dubbed the low country, surrounding our butte, and hike out to the game trail. We follow it to a clearing that looks like the intersection of eight or ten different trails. An interchange on the deer superhighway, perhaps?

I sit as still as possible between two snug snowbrushes. As water drips from trees, the forest sounds like a bowl of Rice Krispies: snapping, crackling, popping. It keeps my attention piqued, no time for daydreaming today. Was that a dollop of water dripping from a ponderosa branch onto a manzanita leaf? Or was it a doe stepping carelessly on a pinecone? Was that a chipmunk scurrying across dry bark or was it the creak of a giant buck's knee? I am so alert that I notice out of the corner of my eye when a pale yellow currant leaf drops to the forest floor. I watch a black ant scale the branch of a snowbrush next to me. This scene is still and almost silent, yet I feel as if I am surrounded by garish displays of life and movement.

Again, darkness falls without so much as a teasing flash of deer. I click on my headlamp and start the long, cold trudge back to the car. My whole body feels heavy and sore. I am pissed off. Frustrated. I don't know what the hell I'm doing. How could I have ever been duped into thinking I'd get lucky enough to shoot a buck?

Sunday morning we wake up early and hike back up Buck Butte, to the end of an overgrown logging road and down a brushy slope to a small clearing. We sneak behind a log and wait for the sun to come up. The stars are incredible. But we're both sleepy. Scott starts to snore and I don't even bother poking him. What's the point? I'm too uncomfortable to stay still, too fidgety to keep quiet. I wonder if I'm getting louder—worse at this—or if I'm just more attuned to how much noise every little movement makes.

Once the sun is up, we tiptoe across the clearing to a steep patch of trees where I heard—or thought I heard—something big lumber through the dark. As we pick our way downhill through the fallen branches, we see piles of giant poop pellets. Some of the piles are very fresh. My stomach flutters. There is at least one big buck traveling through here on a regular basis.

Later in the day, we come up with a new plan. We start at the base of the butte and creep up the side of a clearing, quickly and, for once, quietly. Then we cut across the clearing and tuck ourselves behind an old snag. We find some gelatinous, neon-orange mushrooms sprouting from a stump, and Scott finds a piece of wood that looks like the face of an owl. As the sun sets, still seeing and hearing nothing, I creep along the side of the slope until I reach another small drainage. I hide there with a good view down the clearing. The wind direction is perfect—anything walking up or down the hillside wouldn't catch a whiff of me. But again, I see no deer. I leave frustrated and discouraged. The animals are only moving at night, it seems. Almost a week has passed since I saw that doe. This is getting pathetic.

The next day I hide among the snowbrush at the edge of a small, flat clearing at the bottom of Buck Butte. I keep hearing buck-like sounds that quicken my pulse but turn out to be squirrels or, in one case, a black woodpecker with a white head and red patch. Later, I consult a guidebook and discover that this was a rare white-headed woodpecker (*Picoides albolarvatus*). Instead of waiting for the sun to go down, I scribble notes in my journal, right here, right now. I'm sure it's a dead giveaway to the deer, but fuck it. After today, there are two days left in deer season. It's time to look for silver linings.

So I didn't get my deer. All is not lost. For one thing, I have enjoyed spending so much time with Scott. It's surprisingly romantic, being alone with him in the woods, seeing him embrace this new activity, this new goal.

Two, everywhere I go now, I see game trails and deer beds and all kinds of animal scat—coyote, deer, elk, rabbit. One night I even dreamed of deer poop. These woods feel alive in a new way to me. As with fly-fishing, I feel like I am learning a foreign language.

At dark, Scott and I head back to the car and then camp. We drink whiskey and reheat leftover meat loaf in a skillet. We build a campfire and sit close to it, poking the wood and talking about our families and the family we hope to have of our own someday. Nothing promotes deep, philosophical conversations like an open fire does. Even though this hunt has been frustratingly death-free, it's impossible not to think about the circle of life. Dead deer or not, it's what hunting is.

That night, I am drunk and my stomach is so upset that I barely sleep. I curl into the fetal position and hatch a strategy for tomorrow. When the alarm buzzes at five, I am already awake and grateful to get out of the tent.

We spend the morning exploring the top of Buck Butte, again seeing no deer. In the afternoon, we walk along a small stream and hunt not for deer but for mushrooms. Success! We collect a box of white chanterelles, savoring the instant gratification that deer hunting has so cruelly withheld. As it gets dark, we drive to a highway diner for supper, with plans to go to bed early and give it our all tomorrow, the last day of deer season.

But this day comes and goes like most of the others did: without a deer in sight.

It reminds me of a moral that anglers often repeat: There's a reason it's called *fishing* instead of *catching*. I know this. Big-game hunting is very new to me, and I must be patient. But after two fruitless deer seasons—first in Michigan, now in Oregon—my patience is waning.

CHAPTER 14

BIG GAME

I return to work for a week before elk season opens. Everyone in the newsroom asks about my deer hunt and I relay the disappointing news. Or, rather, non-news. I try to get myself excited about elk season but I can't muster much enthusiasm. If I couldn't track down a buck deer, what hope do I have of finding a bull elk? Biologists estimate that the average American elk hunter kills an elk only once every eight years.

Elk (*Cervus canadensis*) are closely related to deer, and the two species' habitats overlap, but they behave quite differently. To hunters, these behavioral differences matter. Deer prefer to nibble on shrubs and young trees, whereas elk, like cattle, favor grass. And while deer usually travel alone or in small groups, elk live in herds—sometimes called gangs—that range from just a few animals to hundreds. Biologists predict that this behavior stems from a time when elk lived on flat, open plains.

Deer hunters can succeed by locating prime deer habitat and then waiting, hidden, for an animal or two to show up. But because elk live in larger groups—and because there are fewer elk than deer—hunters usually have to cover a lot more ground to find them.

"I expect to see a deer every day that I spend deer hunting," Andy tells me, "but it's a good day of elk hunting if I just find fresh tracks or scat."

This further deflates my hopes as I recall my less-than-fruitful deer season.

To make matters worse, elk season in Oregon is shorter than deer season. I actually have two tags for my wildlife unit, but each tag is only good for a span of about four days. During each season, I am only permitted to kill a male elk, or bull. As with deer, hunting for female elk is tightly regulated for the protection of the overall population. As a result, there are as many as seven females (called cows) for every male, further diminishing my chances.

At the start of the first elk season, we drive down to my unit and wake up at four thirty in the morning. We have almost three hours before sunrise. I pull on layer after layer to withstand the cold. Scott tries to encourage me as we drive toward Buck Butte.

"Maybe we've had it all wrong, and it's really called Bull Butte."

Only a couple of weeks have passed since deer season, but winter has made significant headway. We are soon greeted by a thin layer of snow. This is a welcome development. Snow will dampen our footsteps as we creep through the woods. And it will highlight any recent hoofprints that we stumble across, too.

A large brown animal darts across the logging road ahead of us, flashing in our sights only when it's in the headlights, then vanishing into the dark.

"Oh my God, an elk!" Scott is genuinely amazed. He turns to me. "I didn't think we'd see one."

"Me neither."

He rolls the car to a halt. I peer into the dense woods through my window on the passenger's side, where the elk disappeared. All I see is blackness.

I jump out and tie some fluorescent orange flagging on one of the branches nearby. If we don't see anything promising this morning, we could come back here and try to follow the tracks in the snow.

I get back in the car.

"Well," I say, "this is already more successful than deer season."

We roll over the snow-covered dirt road, the tires barely making a sound. All those days of deer hunting come back to me, and a familiar combination of frustration and disappointment edges into my stomach. I take a deep breath and remind myself that this is a new day, in pursuit of a new animal. Anything can happen. This time, at least the weather is cooperating.

The road barely climbs the butte when we decide to park and travel the rest of the way on foot, to avoid getting stuck in the slush or mud. The sky is dusty black, with no stars, as we pull on all of our layers. I hoist my backpack—stuffed with a raincoat, snacks and a water bottle, not to mention rope, canvas bags and knives, just in case—onto my shoulders. I pick up my rifle and load it.

We hike quietly. The snow offers another perk—it illuminates the forest. We pick our way up a steep slope, hugging the edge of a thicket. Afraid that the sun will rise pretty soon, I motion to Scott that we should settle down here just in case an animal travels along the swath of thinned trees that stretches before us. We both sit against tree trunks, facing north. We listen. We wait. I peer into the darkness, trying to make sense of the shadows and blotches. I notice some noise uphill from us, which of course I imagine is a herd of elk traversing the butte. Daylight takes its sweet time. When the sun does come up, we stay put for a while longer. The air is still and foggy. I'm starting to get cold and that promising window of dawn, when elk, like deer, are supposedly most mobile, has almost passed. It's time to get moving.

I start uphill and Scott follows me. Not far from where we were sitting, we see black dirt kicked up in the snow. These are hoofprints, the mere sight of which feels as if I just lifted a clump of duff and uncovered a mushroom: Yes! It's a game trail, and not the subtle, summery kind we were finding two weeks ago. Many hooves have trodden along this path, and recently.

I can't tell which way the animals were headed, so we

follow the tracks downhill for a while, then turn around and walk uphill. Through the trees ahead, I see a flat clearing that I recognize as an old, overgrown logging road. Trees and shrubs have reclaimed too much of it for a vehicle to drive on, but it would make a convenient path for large animals. As I get closer, I see what I think are more prints on the road, so I tell Scott to lag back a little farther while I go check it out. He stops behind a boulder and watches me walk ahead.

Just as I reach the road, I look to my right and see what looks like an elk walking toward me. I crouch down behind a scrawny snowbrush shrub, my hands shaking with excitement. Ever-so-slowly, I raise my binoculars to my face to get a better look at the animal.

Sure enough, it's a cow elk, probably seventy-five yards away from me. She's alert, with her head raised and ears pointed forward. She slows, then stops and stares in my direction. I freeze, knowing that because elk are social, there could be others here, too. If she spooks, she will alarm the rest of the herd and they will all clear out. After what feels like minutes, but is probably just seconds, she turns and trots away. She is leery but not panicked. If she had identified me as human—by smell, for example—she would have sprinted away. Instead, she turns to face downhill. She looks back at me again, hesitating. Then she jogs into the trees and out of sight.

I twist my torso around, look at Scott and raise my finger to my lips. He nods and takes a step back, tucking himself behind the boulder. I turn again to the little road that stretches in front of me. I sit down on the ground, to give my legs a break, and take a deep breath.

There's a good chance that other elk, possibly even a bull, are in the vicinity. If so, it seems most likely that they're downhill, where the cow ran. From where I sit now, a thicket of small trees blocks my view. I need to find a way to peek down there. But if I walk down the open road, the elk below might see me. I clutch my gun with both hands and look uphill, thinking. I need to hatch a plan. Perhaps I could creep through the trees on the other side of the road. Branches and shadows would hide me while I peeped downhill, past the road, in the direction the cow headed.

Then I look back down the road and gasp. A bull elk—I can see his antlers with my naked eyes—is walking toward me, following in the footsteps of the cow. Is this a mirage? Am I so desperate that I'm imagining wildlife now? I shoulder my rifle and peer through the magnifying scope. Adrenaline gushes back into my veins. He is about seventy-five yards away, and looks more relaxed than the cow did, plodding rhythmically toward me with his head hanging low.

I remind myself: Stay still. He could notice me at any moment and disappear faster than I could blink. How has he come this close without seeing me? Now he's fifty yards away.

And then, without warning, time slows down. What starts as just the *idea* of calm, the abstract goal of it, somehow builds, like a fluorescent bulb that warms up and then illuminates everything around it. I watch through my scope as he continues to walk toward me, never missing a step. I try to sit deep in the ground, place my feet flat and anchor my elbows on my bent knees, so my rifle is as still as possible. I rest my thumb on the safety, just in case I get a clear shot.

This doesn't seem likely, though. The elk is facing me head-on, and the heart shot—the only fatal place I know to aim—would require him to turn broadside. But he is still walking toward me. The least I can do is be ready, just in case.

Now the elk is impossibly close, maybe twenty-five yards away, and still plodding. Wait, now he slows down. Does he see me? No, he's turning to face downhill. Maybe he's looking for the cow.

I center what I think is his heart in my crosshairs.

If I am ever going to get a shot, this is it.

I slide the safety forward and pull the trigger.

He charges forward at the bang of my gun, downhill and out of sight. I turn to Scott and give him a thumbs-up. I can hear the elk—my elk—thrashing through the trees and brush. He stumbles. A pause. Is that it? No. A loud, terrible wheezing sound, almost like a donkey braying, emerges from the woods. Later, Scott tells me that it reminded him of the sounds his grandfather made as he was dying of lung cancer and emphysema.

I look back at Scott and motion for him to come toward me. He walks up and together we wait.

Even when hit in the heart, an animal rarely drops dead immediately. Instead, a surge of adrenaline propels the animal to run; the body gives itself what could be its last shot at life. Every hunter I've known has advised waiting at least ten minutes—some say thirty minutes or even more—before tracking it. The idea is that an animal who knows it is being chased will produce even more adrenaline and keep running instead of lying down to die.

While we wait, I alternate between feeling triumphant—*It was a good shot,* I tell Scott, *I'm sure I got it*—and wondering if maybe I somehow missed. And anyway, do I really want the elk to be dead?

I can't help but think of Nathan, wondering if my killing of this elk will bring even a fraction of the sadness and suffering to the herd that my brother's death caused in my family. I wince at the thought. Then, as always happens during intense moments since Nathan's death, I wonder what he would say if I could call and regale him with the latest stunner in the story of my life.

Desperate for something to do while the minutes tick by, I tie some flagging around my snowbrush shrub. If we get turned around while tracking the animal, the flagging will remind us where I sat when I fired the shot.

When twenty minutes is up, I bound over to where the elk last stood. I bend down and examine the melting snow for blood. There's nothing, just some strands of light fur scattered around. Dread creeps in. What if I missed the elk altogether, or just nicked some fur off his back? Maybe that wheezing sound wasn't the elk's final breaths at all, but some sort of able-bodied warning to the others. What if I did hit the elk but it wasn't fatal and we end up tracking him for hours, only to have to shoot him again to put him out of his misery? Or worse, what if he's wounded but we never find him? Together Scott and I follow the elk's tracks downhill. The vegetation is dense and I focus on the ground, examining the snow and dirt for hoofprints and, hopefully, drops of blood.

Scott interrupts my thoughts. "There he is."

"Where?"

"By that tree."

It takes me a second to see the elk, even after Scott points to him. He lies on his back, as if he slipped down the hill until a stout tree stopped him. He's just a dozen or so yards from where I shot him. Branches are entangled in his legs. His eyes are open. He looks enormous.

I approach him slowly, with my gun loaded and pointed at him. I've been warned that a wounded animal lying still might not be dead. One book advises to take a branch and poke the animal's eye; only an animal who's truly dead will remain still. But I can't bring myself to poke him—it feels disrespectful. And he could be alive.

"He's dead," Scott says. "Really dead."

Nervously, I lower my gun and slide my hand onto the elk's fur, which is thick, coarse and slightly oily. I'm still dizzy from the adrenaline, and amazed that all of this is really happening. Already, my feelings are not as pure as what I've come to expect when bird hunting. Guilt has a bigger, prompter presence this time around. So does awe. What have I just done?

I wrap my hand around one of his antlers. I count the points—four on each side, which indicates that he is fully grown but still young, maybe two and a half years old. Soft ears, filled with wild tufts of blond fur, flop beside each antler. Again, I am shocked by his size. He is more horse than deer. His long, slender legs are covered in short, dark fur. His furry face falls somewhere between the narrow deer and the bulbous moose, an endearing, unfamiliar middle form.

I'm disappointed in myself that in all my research, I never thought to memorize a hunting prayer. Almost every hunting culture has some traditional words that a hunter recites over her prey. I stroke the long, dark mane that covers his neck and I fumble for something to say to this majestic animal. When nothing else comes to mind, I lean in and whisper, "Thank you. I'm sorry."

I shove my fluttering flock of emotions—awe, remorse, guilt, giddiness, gratitude—down into my gut before one flies loose and destroys my composure. Big-game hunting is an exercise in compartmentalization. I will release these feelings later, one at a time, and roll over each one in my mind, to savor it and try to understand it. Right now, there is work to be done. I'm nervous about accomplishing all of it here, in the middle of nowhere, with no seasoned adviser. What if I can't gut and quarter this animal? What if we can't get it back to our car?

Scott takes my picture with the bull, then we try to drag him a few feet, away from the tree and into a more open patch of ground. I grab hold of his antlers while Scott grabs the hind legs. We both heave with all our might but the beast moves maybe an inch. We try a few more times before giving in. The best we can do is to rotate him so that his belly faces downhill. Using some rope from my backpack, we tie his antlers to one tree and a hind leg to another, to prevent him from sliding downhill while I'm below. I get out my elk tag and cut notches in the date and month, then tie it securely around one of the antlers. Scott snaps another photo.

These photos will never be posted on a brag board, or even on my Facebook page. I remember too well the discomfort of looking at similar pictures without any context, and I don't want to put others through that. Nor do I want to demean my own elk, even in the eyes of non-hunters. I will share the snapshots with a few friends who hunt, and some understanding family members. Mostly, the photos are for me, a physical reminder of an experience so intense and so unique that I will sometimes reminisce about it in disbelief, as if it had been a dream.

I stand downhill from the elk, near his belly. I get out my knives and a pair of latex gloves. I have never watched a large animal being field dressed in person. I've watched YouTube videos and studied diagrams in books. I have one book in my pack—*Making the Most of Your Deer*—that explains how to field dress a closely related species.

"Do you want me to get out the book?" Scott asks, watching me hesitate.

"No, I think I know what to do... First I'm going to cut open the abdomen."

"Don't you have to tie off the anus and penis first?"

I hesitate. I think I can slice open the abdomen—or most of it—before resorting to my most dreaded step in field dressing.

"I'll work my way up to it," I say.

I squat down and feel around for the animal's sternum, then drag my knife along it, lightly. I go back and trace the same path, again and again, until the skin suddenly bursts open to reveal wet, pink viscera. I move the knife down and extend the cut, first a light trace through the outer layer of skin, then another layer, then another, then zip—the incision glides open. On the next stretch of cutting, I poke the knife in too far; the skin unzips and the tip of my knife punctures something below. A bilious bubble oozes out of it.

"Uh-oh, I think I nicked the intestines."

"Does it smell bad?"

"Um, I think so." I can smell something, and it's not pleasant. But it's not overwhelmingly awful, which is how I've heard hunters describe the scent of a pierced intestine.

I lean back and wipe my hair from my eyes.

"Oh no." I jump up. "I think I just got blood on my face. Do I have blood on my face?"

Scott inspects my forehead.

"I don't see any."

"Oh, thank God." In half an hour I will be drenched in blood and laughing at my earlier squeamishness.

I lengthen the incision until it stretches the whole length of the abdomen. I have literally reached the genitals; now I have no choice but to deal with them. I cut off two eighteen-inch lengths of nylon cord. Then I grab the penis in one hand, pull it away from the elk's body, and tie one length of cord around it, tight, into a bow.

"That wasn't so bad," I say cheerfully.

"For you," Scott says, looking away.

The anus is worse, but not by much. I'm used to cutting the elk by now. I make a deep incision all the way around, then reach in and pull the whole organ out. I tie the second cord around it, pull it tight and knot it again.

Then I walk back around the hind legs and return to the abdomen. I extend the incision as far as I can at both ends of the animal. The belly is split wide open now, exposing the giant, four-lobed stomach—it looks like an exercise ball, fully inflated despite my having nicked it with my knife earlier. Below, there's a never-ending pile of dark, coiled intestines. The problem is, all of these organs are still firmly inside the elk. I stand up and turn to Scott.

"I thought once I got him open, all of this stuff was supposed to just fall out."

He gets the book out of my pack and starts flipping through it for advice.

"Okay, it says you might have to reach up in there and disconnect any tissue that's holding everything in place."

The anus? The penis? That, it turns out, was the easy stuff. I spend the next hour with my right arm fully submerged in the elk carcass, breaking strands of... what? I guess the term is "connective tissue." I have to take breaks every few minutes to step away from the elk and cool off my arm, breathe some fresh air. Heat wafts out of his body. So does the wet, metallic smell of blood.

"This is a well-made elk," I say, plunging my arm in yet again.

After a while, though, I start to get comfortable with it. I know when I'm feeling an organ—maybe a kidney? the liver?—and when I'm touching a string that needs to be ripped. I think back to field dressing my first pheasant, and how simple that was by comparison. Three years ago, I couldn't have imagined reaching my whole arm, up to my shoulder, into the cavity of an elk. Yet here I am doing it, without much of a fuss.

With each torn piece of tissue, the intestines and stomach bulge farther out of the abdominal cavity. Eventually, they sag downhill in a pile that looks larger than the body they fell from. Next, I pull out the wide, floppy mass of a diaphragm. A gush of blood follows.

I shove the offal a little farther downhill. We will leave it here, and assume that coyotes, vultures and other animals will feast on it. Andy has told me that when he or a family member shoots a deer on public land near his parents' house, they hear coyotes later that night, yipping with pleasure at their discovery of the gut pile. We humans are, after all, just one fraction of the food chain.

I reach up into the elk's body cavity again, this time with a knife, to sever the windpipe and pull out the heart and lungs. I can barely reach that far up in the animal. I manage to cut the windpipe, but I can't get my hand in far enough to grip it and pull it out. I ask Scott to do it, then explain what he should feel for, what he should grab.

He kneels on the ground below the elk and reaches up, into it.

"Whoa." He retracts his arm. "I just felt the heart."

"That's okay. Keep reaching past it."

"No, it's like it had this energy. I mean, it's this elk's *heart*."

Scott's words hang in the air for a moment, yet another reminder of what I have done by taking this animal's life. The heart, the motor of life for this giant animal. Perhaps it does have an energy of its own.

Scott reaches back in, up to his shoulder, and pulls. Out come the final organs.

Next we cut the elk in pieces, to transport it back to our car. We saw off the bottom half of each leg. Then we sever the front half of the animal from the back, just above the pelvis. This is our first glimpse of how thick the muscle is—several inches of meat surround the spine—and of how much food we will reap from this animal. Not that we are in the mood for eating.

"I think I'm a vegetarian now," Scott will tell me as we pack the animal out of the woods.

Along the spine, we separate the animal again—two hindquarters, one half of a rib cage attached to one shoulder and foreleg, then another half of the rib cage attached to its own shoulder and foreleg. The head and neck are a fifth piece, just as heavy.

Together, we hoist one piece of the animal into the air and then Scott holds it in place while I open a canvas drawstring bag that I purchased at an outdoors store before deer season. Scott lowers it in. We repeat this for each quarter, and each time the process grows more difficult.

"How are we going to get these pieces out of here?" I ask Scott. "We're exhausted and we haven't even started yet."

"We'll just do it," he says, shrugging.

We strap one of the canvas-covered quarters onto Scott's backpack, and then I struggle to help lift it onto his shoulders. Our hike back to the car isn't long by elk-hunting standards—a mile and a half, perhaps—but it's over steep terrain, littered with trees, rocks and downed wood. I worry that Scott will injure himself, hiking with so much weight on his back.

I take a second quarter and wrap it in a tarp, then sling a nylon strap around it. I will try dragging it, like a makeshift sled. The trek is difficult. My tarp doesn't slide particularly well over the uneven ground. I have to crouch down and shove it over bumps and logs. Other times I stand downhill and heave like I'm in a tug-of-war match. Both of us have to take breaks every few minutes, to catch our breath and regain some muscle strength. It takes us two hours to reach the car.

On the next trip, I drag the elk head using the same tarp-and-strap setup. Scott hoists another quarter onto his backpack. During our breaks, we talk about how physically demanding big-game hunting is.

"I never thought of hunting as an extreme sport before," Scott says.

We get back to the elk for the third and final excursion. We place the last quarter—the largest one, a front shoulder with leg and ribs attached—on the tarp, and each of us holds one strand of the nylon strap. The sun is setting, and there is no way we'll get back to the car before dark. We both pull on our headlamps before we begin the haul.

"Just one more trip through Fortitude Valley," Scott grunts.

I smile. And then I stop. Stunned. Make no mistake, we are standing deep in Fortitude Valley. I am drenched in blood and sweat. My legs and arms are so fatigued that they are sore to the touch. I'm so thirsty that my tongue sticks to the roof of my mouth. We drank the last swig of water from our packs hours ago.

Yes, I'm in Fortitude Valley and yet, somehow, I'm loving it. Even though I'm still reeling from the gutting process, I'm already looking forward to all of the meat we will reap from this elk. I'm proud of myself and proud of Scott. I had doubts the whole way, but in the end I did it. We did it. It's pitch black outside when we get back to the car. Grunting, we heave three of the quarters into the Rocket Box on top of our Subaru. We set the last quarter and the head in the back of the station wagon and close the door. Eleven hours have passed since I pulled the trigger.

★ ★ ★

That night, I toss and turn, replaying the day that just happened. I worry that the meat hasn't cooled properly and we'll find it spoiled the next morning. This becomes the latest worst-case scenario to avoid.

We awaken early and drive home, where Andy and Jessie (who have returned to Bend following graduate school) come over to help us butcher. Most books and experienced hunters recommend hanging the quarters for a week or more. The idea is that this gives some of the connective tissues time to break down and tenderize the meat. But the weather has warmed up, and we don't have a place to hang the meat at a safe temperature. Tough meat seems a small price to pay to keep this animal from going to waste.

We carry the four quarters into the backyard and hang them, using nylon ratchet straps, from a pergola that stands over our grill. Andy is gentle in his critique of my field-dressing and quartering methods. He says that we got all the meat, which is the most important thing. But we could have separated the quarters more easily by knowing which joints to break. We should have skinned the elk in the field, to make it lighter. We could have left ribs and more of each leg behind, to pack it out more easily. Instead, we skin each quarter as it hangs. With one hand, I stretch the hide down and away from the leg. With the other, I use the knife to sever any thin, stringy connections. The inside of the skin is gelatinous and slippery-smooth. I shake salt over the inside of each piece of hide, then roll it for later tanning.

Next, we take down the skinned quarters and set them on tables we improvised by laying sheets of plywood across sawhorses. Andy hands me a sharp knife and explains that my next job is to separate the muscles. The highlight of my entire education was the dissection unit in seventh-grade biology, so butchering is right up my alley. It's fascinating to take apart an animal and see the glorious intricacies of life's own design. Scott and I focus on this part of the butchering while Andy oversees us and slices the detached muscles into steaks or stew meat or roasts. Jessie packages the cuts using a vacuum sealer. Into buckets go the scraps, which we'll grind into burger meat the next day.

As the sun goes down, I finish one hindquarter and move on to the right front, which is where the bullet entered the elk. The trauma has changed the physical makeup of the meat: Clotted blood forms a dark, gelatinous coating over it. I have to discard layers of it to unearth any normal-textured meat that remains.

Jessie, who has shot and butchered two deer of her own, will later tell me that she views this as a critical part of hunting. Seeing the bullet trauma up close helps her understand the consequences of pulling the trigger. Earlier, as the butchering lagged on, I found myself forgetting, for long stretches, that we were picking apart a real life. The elk had started to look like meat—regular old sides of beef that hang in butcher shops or fancy steak houses. But now, as I poke the formations of congealed blood, this trauma brings me back to the stark truth of what I have done, of the cost of all this meat.

As we butcher, Andy grills one of the fillets so we can try it. It's delicious and as tender as store-bought filet mignon. Elk meat tastes a lot like grass-fed beef, or a strong-tasting version of regular grain-fed beef. It's much leaner than beef, so burgers don't hold together well on the grill. Soon I will find myself adding a spoonful of olive oil to recipes such as meat loaf.

The day after the butchering is done, I buy a chest freezer. We tuck it in a corner of our basement and fill it with the wrapped meat of my elk. Because I didn't take the animal to a professional butcher, it was never weighed and I don't know exactly how much meat we got out of it—but it's in the hundreds of pounds, for sure.

Food isn't the only thing this elk has given us. I freeze three pieces of hide, and try my hand at tanning the fourth, for a furry wall hanging or, at the very least, some fly-tying materials for Scott. I scrape the flesh off the inside, then stretch it and dry it in the basement. In the summer, I will soften it with a store-bought tanning kit.

Elk have two rounded molars in their upper jaw that are sometimes called ivories. Old-timers carry these teeth around in their pockets for good luck. So I carefully cut these teeth out of the gums, then dry them and brush off the remaining tissue. These tokens of my hunt are too precious to carry around in a pocket, so I tuck them in my jewelry box.

I take the head to a taxidermist, for what's known as a European mount—no fur or glass eyes, just a cleaned white skull and antlers. On the way to pick it up, I worry that it will look creepy: the hollowed-out areas where eyeballs used to sit, the broken edges where nose cartilage once met bone. But when I see it, I get to relive the satisfaction and thrill of the kill all over again. We hang it carefully over our mantel. To me, it looks elegant. It is art by nature's incomparable design, a token of a rite of passage that still feels so mystical that without such physical reminders, I might doubt it ever happened. When I look at it, I am reminded of all the feelings summoned during the hunt: intense pride, satisfaction, exhaustion, awe, gratitude and, yes, guilt.

Guilt is an unavoidable part of hunting. Nearly every hunter I've ever met has admitted to feeling guilty about killing animals. Since I shot the elk, this guilt has also started to feel appropriate, even necessary.

What got me interested in hunting in the first place was the archetype of a hunter who respects her prey. But at times, I fell into my old habit of wondering if that "respect" wasn't just a convenient form of self-deceit, like claiming that a prenuptial agreement is a symbol of love.

The idea of respecting something and also eating it is a tricky one indeed, and it has confounded humans for a long time. In his paean to deer, *Heart and Blood*, Richard Nelson describes his love for the species and acknowledges that as a hunter, his actions might seem to conflict with this love. Then he adds: "If this seems contradictory, then the whole living process is a contradiction. We love apart from ourselves that which we also kill to sustain us: great trees become our houses and furniture, flowering plants become our vegetables and fruit, fellow creatures become our food and clothing."

A local Indian woman once explained to me how her tribe, the Umatilla, reconciles this apparent paradox. They view eating an animal not as a heartless act of cruelty, but as a display of gratitude and respect. They believe that many animals already inhabited the earth when man first arrived. The Creator called all of these creatures together and told them to prepare for a new being, man. The other animals would have to take care of man, who would first appear as an infant, the Creator said. Salmon was the first animal to step forward and offer to help nourish man. Deer was the second. It's no surprise, then, that salmon and deer are such important animals. If one year went by and no humans ate salmon or deer, the animals would be deeply offended. Both would feel as if they had lost their great importance.

Regardless of what you think of this creation story, the last part rings true. The more we eat something, the more it becomes an integral part of our culture and the more we will fight to keep it around. One can't imagine humans letting cattle go extinct. Or potatoes.

According to Erich Fromm, hunters have always had respect—perhaps even love—for their prey. There is, he writes, "no evidence for the assumption that primitive hunters were motivated by sadistic or destructive impulses. On the contrary, there is some evidence to show that they had an affectionate feeling for the killed animals and possibly a feeling of guilt for the kill. Among Paleolithic hunters, the bear was often addressed as 'grandfather' or was looked upon as the mythical ancestor of man."

In addition to the guilt over killing such a large, beautiful animal, as time goes on I begin to feel guilty about how quickly I encountered it and what a close, easy shot I had. My elk-hunting experience was, as a friend points out, "the aesthetic ideal." It was on public land, in a place I had come to know and love during deer season. It was also very, very lucky. Faced with waiting another year to bag a large animal, I had started making backup plans. I contacted landowners who had crop-protection preference tags that extended the season and briefly considered hiring a guide to show me the ropes in an area that was unknown to me but familiar to him. I am grateful that I didn't have to fall back on these other options.

Most of all, though—more than the meat or the ivories or the mount or the photos—I came away from the elk hunt relishing my memory of the experience. It marks the beginning of something that I hope will be part of the rest of my life. The elk has lifted me over some imaginary threshold, though I still consider myself a novice hunter. Perhaps I always will. The more I learn about hunting, the more I realize I still don't know. I understand why some hunters become obsessed; there is so much to learn from each individual animal. Every fall, a hunter gets to see a new group of deer behaving slightly differently from the previous ones. She gets to see if the ground where she found last year's buck is still attractive to a new generation of animals. Just as Heraclitus said that no man steps into the same river twice, it is also true that no hunter visits the same land twice.

It's a pleasant surprise how many friends are eager to try the elk meat. We invite friends and family over for elk chili with adobo. We give away packages of meat to friends, who report back with recipes and stories of how their own special meals were received.

Twice a week, Scott and I eat elk for supper. With so much in our own freezer, I try to avoid buying meat. Every month or so I buy a whole chicken, which I roast and then stretch into other meals such as enchiladas, stir-fry and chicken soup. The rest of our dinners are meatless. I notice that our weekly grocery bills are more than twenty dollars lower than usual.

That winter, Scott and I face another circle-of-life moment as we talk, yet again, about starting a family. The subject arises as it always does: suddenly. Only this time, there is no long discussion, no back-and-forth hypothetical debate. After years of talking about it, we simply decide to ditch the birth control and see what happens.

At times, I struggle to shoo away nagging questions. (Will I be able to get pregnant? What if we have a baby who has major health problems? What if she or he dies suddenly, like Audrey? Or like Nathan?) In May, I discover that I am pregnant and due in January.

I had expected pregnancy to be riddled with fear and anxiety. But the instant I see those two pink lines, my only feelings are shock and joy. Scott takes pictures of the three of us—him, me, Sylvia—to remember the moment. Weeks tick by and my heart remains wide open. Yes, my future could hold unfathomable sadness. I could lose this baby—to miscarriage, stillbirth or death—at any moment. I do think about this. But somehow, for some reason, I don't dwell on it. I vow to deal with life as it unfolds. Each day, hope surprises me by trumping fear.

My overarching wish for this child is that he or she will grow up strong, curious and brave. Tucked underneath is the smaller hope that someday we will hunt together, absorbing the world around us side by side. Hunting has allowed me to explore some of my greatest questions and fears. Ultimately, it helped lead me to the decision to bear this child.

Hunting can be fun and it is certainly physically challenging, but I no longer consider it a sport. It is life and death. It is a forced reckoning of the questions that hide in the corners of every day: What is this place where I live and what did it used to be? How do I fit into the natural order of things? What am I capable of? What is the right thing to do? Hunting is history. It is human.

There are so many things that we stand to lose, as a society, if hunters go extinct. Of course there's the money from hunting and fishing licenses. If environmentalists won't or can't hunt or fish, then it's time to revamp our conservation funding model, which is still based almost entirely on license fees. Hikers and bird-watchers are literally not paying their share.

Hunting license fees pay for the majority of wildlife conservation programs in the United States, including those that protect non-hunted animals. Hunters who pursue migratory waterfowl in any state, for example, must pay for a federal license each year—called a "duck stamp"—to fund wetland conservation. In 2010, my stamp cost $15.00, and the U.S. Fish and Wildlife Service says that $14.70 of that went to purchase or lease wetlands for the National Wildlife Refuge System. That same year, I paid $58.00 for a general Oregon fishing and hunting license and an additional $20.00 for the right to hunt upland birds and waterfowl.

Hikers and bird-watchers, who did not pay these taxes, benefited from the ones I did. My license fees helped pay for biologists to study animals and restore important habitat. Some of my money even helped non-hunted animals like songbirds and pygmy rabbits (tiny rabbits that burrow in sagebrush country). As hunting and fishing have declined nationwide, conservation projects have multiplied and become more expensive. To fill the funding gap, states have jacked up the costs of hunting and fishing licenses. In 1982, for example, 16.7 million Americans spent a collective $259 million on hunting licenses and fees. Averaged out, that's about $15.50 apiece. In 2003, 14.7 million hunters spent a total of $679.8 million, or about $46.12 each. Costs more than tripled in twenty-one years, much faster than the rate of inflation.

Even with all of these official costs aside, hunting is not a cheap hobby. At almost every turn, I have discovered another piece of equipment I needed to buy. A chest strap to keep my binoculars handy but also out of the way. A vest to carry ammunition and birds. Warmer, sturdier pants. Waterproof boots. A gun case. A lock for my gun case. Once, I added up the costs associated with shooting one duck—hunting licenses, gas and ammunition—and discovered that its meat cost me more than twenty dollars a pound.

It's no surprise, then, that the wealth of the average hunter is also on the rise. Private land that was once open to hunting has been cordoned off as public opinion has turned against the sport. The rising price of agricultural land means that hunters whose extended family members once owned farms and ranches are now looking to hunt on public land. (If they're not retiring altogether.) Public lands face more and more hunting restrictions. As wealthy hunters turn to pricey private reserves, the sport faces a possible future as a pastime of only the very rich. This has long been the case in land-poor Europe, and it's one of the key differences that set our culture apart.

As American hunting declines, the social losses far eclipse the financial ones. At its best, hunting balances short-term use with long-term preservation, and therefore provides us with a model for how to approach environmental protection. Our lives have become more computerized and climate-controlled, yet we must not forget that we are still animals, still dependent on clean water, fresh air and a functioning ecosystem. Hunting and angling teach us to understand the earth in a way that more passive activities such as hiking and nature-watching cannot. We must not lose our fluency with the natural world—because if we do, we will lose our greatest reason for protecting it.

But how we hunt matters. Too many hunters rely on vehicles and high-powered firearms to compensate for a lack of hard-earned knowledge and familiarity with the land. Hunting is difficult work. We must be in shape and willing to hike away from roads to find wildlife. We must take the long view and come down on the right side of important wildlife issues: That means agreeing to a ban on lead ammunition, no longer baiting wildlife, and acknowledging that predators play an important role in any functioning ecosystem. Nationally, hunters are more likely to vote for property rights than for habitat protection, even as rampant development displaces game species faster than any non-human predator. Clearly, forging a connection to a place is not the same as protecting that place. If it were, all hunters would already be the staunch environmentalists that they ought to be.

It is hard for me to picture a time when I won't want to hunt anymore. Yet I promise myself that if I ever get to the point where I kill an animal and don't feel in awe of it, I don't feel twinges of guilt upon reflection or I don't feel grateful for its life, then I will stop. Perhaps the biggest threat to hunting is not gun control advocates or even environmental destruction but those hunters who refuse to give the practice the respect it deserves, who treat it as no more sacred than NASCAR or *Monday Night Football.* Hunters who show more respect for guns, the tools they use, than the lives they take.

Hunters can and should be powerful advocates for the species they pursue. A family's favorite hunting spot, year after year, should become an irreplaceable heirloom that they fight to protect. Good hunters should be intimately familiar with their prey. They should know what the animal eats, how it responds to different weather, what constitutes a particularly small or large specimen.

Ernest Hemingway used to worry that if American men stopped hunting, they would cease to be men. It's a chauvinistic attitude, but I agree with a version of it. So much of American history—of human history, really—can be boiled down to the battle between man and nature.

In 1960, when our nation first considered designating wilderness areas, Wallace Stegner wrote a letter in support of the *idea* of wilderness, which, he argued, is itself an American resource: "We need wilderness preserved—as much of it as is still left, and as many kinds—because it was the challenge against which our character as a people was formed." The same argument should be extended to hunting. It connects us to our past because we engage in the same pursuit our ancestors did for thousands of years. After a few seasons of hunting, I can't help but feel new respect for what my ancestors went through. Their survival depended on their hunting abilities, and they hunted without the help of optical scopes, waterproof fabrics—many of them without guns. To paraphrase Hemingway: If humans stop hunting, we could lose some of our humanity.

My baby, this new life, is a constant reminder that amazing things can happen any day. Some of these wonders are tragic, like the death of my brother. Some are joyous, like welcoming a new family member into the world. And some, like shooting the elk, are a little bit of both.

Guilt resurfaces with each elk dinner. Also with each meal, my relationship with the animal extends and deepens. I didn't spend much time tracking him, or observing him in the wild. But now we share a new ritual. I walk downstairs to our dirt-floored basement, open the freezer and sift through packages of his meat. I select one and carry it upstairs to the kitchen. While it defrosts on the counter, I flip through cookbooks or browse websites for recipes. Most are simple—grilled steaks or burgers, spaghetti with meatballs, stir-fry. Every couple of weeks, I transform some of the elk into a special meal, such as a classic bourguignonne recipe capped with buttery, homemade crust and baked for a decadent potpie.

Scott and I sit down at the built-in dining nook in our kitchen, with the rich smell of elk meat wafting from our plates. Unlike most of the meat that I've eaten in my thirty-one years, I am fully conscious of what this meal is and what it once was. That fatal shot created between us a bond that will last longer than the animal's earthly life. Its flesh nourishes me, and the new life growing inside me.

In return, I bear a responsibility to the species this meat belonged to, and to the land that nurtured it.

Before we lift our forks, we raise our glasses and make a simple toast, loaded with thanks: *To our elk.*

And then we eat.

ACKNOWLEDGMENTS

These pages reflect the efforts of many people, because this is my first book and because I had to learn to hunt before I could write it. Jessie Fischer, Andy Fischer, Charles Eisendrath, Gary Lewis, Russ Seaton, Del Jeske, E. V. Smith, Jack Jones, Marc Thalacker, Hank Fischer, Carol Fischer, Kit Fischer, James Johnston and many others welcomed me into the world of hunters and made me proud to call myself one.

My talented friends—Jill McGivering, Jessie Fischer, Andy Fischer, Kayley Mendenhall, Betsy Querna Cliff, Patrick Cliff and Lauren Dake—improved the early drafts. Shana Drehs helped hone my original proposal and nudge it toward publication.

I had great luck in finding two publishing professionals who understood my vision of the book and shepherded it into existence. My agent, Daniel Greenberg, has been a wise and levelheaded adviser. And my editor, Emily Griffin, gave each draft the kind of sharp, thorough read that

I'd heard no longer existed. I shudder to imagine what this book would have been without her. Also at Grand Central Publishing, I give thanks to production editors Leah Tracosas and Tareth Mitch, copyeditor Laura Jorstad and publicist extraordinaire Erica Gelbard.

The Institutes for Journalism and Natural Resources planted the seeds for this project in my head. Many of the thoughts around which I built these chapters arose in casual conversations, so I'm grateful to Frank Edward Allen, Jack Ward Thomas, Andy Buchsbaum, Roland Kalama and Nina Raff for saying just the right things to spur my curiosity and imagination. Adam Short, Susan Parrish, Barbara Smuts, Sarah Buss, Sally Schmall, Gregory McClarren and Durlin Hickok recommended books and articles that were pivotal to my research. The Knight-Wallace Journalism Fellowship at the University of Michigan and my home newspaper, *The Bulletin*, supported this project from the start.

It would have been understandable for my parents, Mel and Dee Raff, and my sister, Gretchen Raff—peaceful city folk, all three—to retch when I told them I was learning to hunt, loving it and finally writing a book extolling it. Instead, they were a source of enthusiasm and encouragement. Scott McCaulou was at my side every step: talking through sticky ethical quandaries, proposing new titles, packing out the heaviest elk quarters and all the while insisting he was comfortable walking beside me while I carried a loaded gun. I couldn't have asked for a better partner—in fishing, hunting, writing and life.

This is my own story in my own words, but many other lives are intertwined with mine. To all the people woven throughout this book: Thank you.

NOTES

Chapter 1: *Going West*

14 *[Lava Butte] erupted seven thousand years ago and covered nine square miles with black, porous rock:* "Lava Butte Vicinity, Oregon," last modified April 15, 2008, http://vulcan.wr.usgs.gov/Volcanoes/LavaButte/Locale/framework.html.

14 *NASA actually trained astronauts for the moon landing on these desolate lava beds:* Judy Jewell and W. C. McRae, *Moon Handbooks: Oregon* (Jackson, TN: Avalon Travel, 2010), 497.

16 *Most come from California:* The 2006 American Community Survey found that in 2005, 56,379 people migrated from California to Oregon, the sixth-largest interstate flow in the United States that year. The report is available at www.census.gov/acs/www, accessed July 24, 2011.

24 David James Duncan, *The River Why* (San Francisco: Sierra Club Books, 1983).

Chapter 2: *Pulling the Trigger*

33 *Back in the 1960s and 1970s, more than 10 percent of the city's twelve thousand residents worked full-time at the town's two lumber mills:* Population figures (11,936 in 1960 and 13,710 in 1970) provided by the City of Bend. Historic mill employment estimate (twelve

hundred full-time employees at peak) from the Deschutes County Historical Society.

37 *Larix occidentalis:* Latin name and details provided by the USDA Forest Service, available at www.fs.fed.us/database/feis/plants/tree/larocc/all.html, accessed July 24, 2011.

40 *Bambi* DVD. Directed by David Hand et al., 1942 (Los Angeles, CA: Walt Disney Video, 2005).

40 William Golding, *Lord of the Flies* (London: Faber and Faber, 1954).

42 *As I drive back to Bend, where hunting season is about as noticeable as National Lead Poisoning Prevention Week:* www.cdc.gov/nceh/lead/nlppw.htm.

44 Douglas Brinkley, *The Wilderness Warrior: Theodore Roosevelt and the Crusade for America* (New York: HarperCollins, 2009), 5.

44 *In 1887, more than a decade before taking over the Oval Office:* Ibid., 202.

44–45 *By the time he left office in 1909, Roosevelt had created:* Ibid., 19, 818–30.

45 *"the greatest good to the greatest number for the longest time":* Gifford Pinchot, *The Fight for Conservation* (New York: Doubleday, 1910), 48.

45 *Roosevelt was condemned by some:* Brinkley, *Wilderness Warrior,* 702.

45 *In 1950, two-thirds of Americans lived in cities or suburbs:* From "Urban and Rural Population: 1900 to 1990," prepared by the U.S. Census Bureau (October 1995), available at www.census.gov/population/censusdata/urpop0090.txt, accessed July 29, 2011.

45 *Now more than four out of five Americans are packed into 366 metropolitan areas:* From census data cited in A. G. Sulzberger, "Rural Legislators' Power Ebbs as Populations Shift," *New York Times,* June 2, 2011.

45 *Our image of environmentalism began to shift in the 1960s:* Carolyn Merchant, *The Columbia Guide to American Environmental History* (New York: Columbia University Press, 2002), 177–82.

45 Richard White, "Are You an Environmentalist, or Do You Work for a Living?" in *Uncommon Ground: Rethinking the Human Place in Nature*, ed. William Cronon (New York: W. W. Norton, 1995), 172.

45 Felix Salten, *Bambi: A Life in the Woods* (New York: Simon and Schuster, 1929).

46 *the name "Bambi" has become "virtually synonymous with 'deer'":* Matt Cartmill, *A View to a Death in the Morning: Hunting and Nature Through History* (Cambridge, MA: Harvard University Press, 1996), 66.

48 *The age of the average hunter is rising, too... Surveys have found that just one in four children raised by hunting parents will learn to hunt:* Data from the 2006 National Survey of Fishing, Hunting and Wildlife Associated Recreation, available at http://wsfrprograms.fws.gov, and from a telephone interview with Steven Williams, president of the Wildlife Management Institute, in September 2007, partially reported in Lily Raff, "Recent Surveys Show a Steady Decline in the Number of People Who Hunt and Fish. So, Are Hunters and Anglers... Endangered Species?" *The (Bend) Bulletin*, September 9, 2007, F1.

48 *nationwide, hunting has been on a steady decline:* Since the U.S. Fish and Wildlife Service began tracking hunting licenses in the 1950s, license sales peaked in 1982, when about 16.7 million Americans paid for the right to hunt. In 2006, about 12.5 million people hunted, a 25 percent drop in hunters even as the national population grew more than 30 percent. Sales figures for 1982 found in the U.S. Fish and Wildlife Service National Hunting License Report, published December 2, 2004. Figure for 2006 from the National Survey of Fishing, Hunting and Wildlife-Associated Recreation. Both reports available at http://wsfrprograms.fws.gov.

49 Michael Pollan, *The Omnivore's Dilemma: A Natural History of Four Meals* (New York: Penguin, 2007), 281.

Chapter 3: *Gun-Shy*

57 *The Last Shot*, DVD, directed by Alan Madison (Chatham, NY: Alan Madison Productions, 2001).

58 *Cheney was carrying a diminutive 28-gauge shotgun during his quail-hunting accident. But if he had instead used a burly 10-gauge (typically reserved for downing geese), his friend probably would not have survived:* Extrapolated from Paul Farhi, "Grace Under Fire: Since Dick Cheney Shot Him, Harry Whittington's Aim Has Been to Move On," *Washington Post*, October 14, 2010.

62 *"Yet cars kill way more people every year than guns do":* This quotation is backed up by 2009 U.S. data, which shows that motor vehicles caused 35,900 deaths and firearms caused 13,872 (excluding suicides). *Injury Facts* (Itasca, IL: National Safety Council, 2011), 94, 143.

66 *In 2009, for example, 138 children ages eighteen and younger:* Numbers of children, ages zero to nineteen, killed by accidental gunshot (138), accidental drowning (1,056) and automobile accidents (6,683) from ibid., 32.

66 *the astonishing number of guns that we, as a nation, own—roughly 250 million:* A 1994 survey published in 1997 by the National Institute of Justice as *Guns in America: A National Survey on Private Ownership and Use of Firearms* estimated that Americans owned 192 million firearms. Other organizations, including the National Rifle Association and the American Firearms Institute, estimate that we own 250 to 280 million guns as a nation.

66 *Nearly three out of five gun-caused deaths:* Gun-related suicides, murders and accidental deaths from National Safety Council, *Injury Facts*, 143.

66 *The lifetime odds that you will be murdered by firearm:* Odds of dying from murder by firearm (1 in 306), accidental gunshot (1 in 6,309), cancer (1 in 7), heart disease (1 in 6) and any cause (1 in 1) from National Safety Council, *Injury Facts*, 37.

66–67 *Forty percent of U.S. households contain at least one gun. One in four adults owns one or more guns. Yet only 11 percent of firearm-owning households say they hunt. The rest keep their weapons primarily for self-defense:* From National Institute of Justice, *Guns in America: A National Survey on Private Ownership and Use of Firearms* (1997).

68 *Pulp Fiction*, directed by Quentin Tarantino (Los Angeles: A Band Apart, 1994).

Chapter 4: *Pull*

87 *Trapshooting was first developed in England in the 1700s:* D. H. Eaton, *Trapshooting: The Patriotic Sport* (Cincinnati: Sportsmen's Review Publishing, 1921), 1–4.

87 *Today trapshooters take aim at standardized four- and five-sixteenth-inch disks:* General information from the Trapshooting Hall of Fame in Vandalia, OH: www.traphof.org.

Chapter 5: *Guts*

91 *This image was appropriated back in 1949, when a Pablo Picasso lithograph of a dove was selected as the emblem for that year's World Peace Council meeting in Paris:* Ina Cole, "Pablo Picasso: The Development of a Peace Symbol," *Art Times* May–June 2010, www.arttimesjournal.com/art/reviews/May_June_10_Ina_Cole/Pablo_Picasso_Ina_Cole.html, accessed July 21, 2011.

92–93 Mourning dove Latin name, migration pattern, diet, habitat and fertility information as well as mortality statistics from D. B. Marshall, M. G. Hunter and A. L. Contreras, eds., *Birds of Oregon: A General Reference* (Corvallis: Oregon State University Press, 2003), 304–5.

93 *threatened or endangered species . . . fall under the purview of the U.S. Fish and Wildlife Service or the National Oceanographic and Atmospheric Administration:* "Summary of the Endangered Species Act," last modified March 2, 2011, www.epa.gov/lawsregs/laws/esa.html.

93 *The U.S. Fish and Wildlife Service began tracking hunting licenses:* Sales figures from 1982 found in the U.S. Fish and Wildlife Service National Hunting License Report, published December 2, 2004. Figure from 2006 found in 2006 National Survey of Fishing, Hunting and Wildlife-Associated Recreation. Both reports available at http://wsfrprograms.fws.gov.

94 *Doves are adaptable eaters, munching on a wide variety of seeds, grasses and forbs:* Marshall, Hunter and Contreras, *Birds of Oregon*, 304–5.

94 *A flock of doves is sometimes called a dule or a dole:* "Animal Congregations, or What Do You Call a Group of...?" last modified September 29, 2006, www.npwrc.usgs.gov/about/faqs/animals/names.htm.

99 *More than 96 percent of Americans:* Vegetarian estimates derived from a series of surveys by the Vegetarian Resource Group, found at www.vrg.org/nutshell/faq.htm#poll; population from U.S. Census Bureau.

99 *As a nation, we raise and slaughter nearly ten billion land animals:* Henning Steinfeld, Pierre Gerber, Tom Wassenaar, Vincent Castel, Mauricio Rosales and Cees de Haan, *Livestock's Long Shadow: Environmental Issues and Options*, prepared by the Food and Agriculture Organization of the United Nations (Rome, November 29, 2006).

99–100 *That portions out to about two-thirds of a pound of meat per person per day, or 241 pounds a year... It's also more than twice the international average:* "Livestock and Poultry: World Markets and Trade," prepared by the U.S. Department of Agriculture (Washington, DC, April 2011).

100 *The average American today eats eighty pounds more meat per year than in 1942:* Per-capita meat consumption in 1942 was 161 pounds, according to Roger Horowitz, *Putting Meat on the American Table: Taste, Technology, Transformation* (Baltimore: Johns Hopkins University Press, 2006), 16.

100 *Add up the weight of all land animals on the planet... and domestic livestock accounts for one out of every five pounds*: Steinfeld et al., *Livestock's Long Shadow*, xxiii.

100 *Thirty percent of the earth's land surface is now used to raise meat, either for grazing or growing grain for feed:* Ibid.

100 *three species go extinct from our planet every hour on average:* Ahmed Djoghlaf, "Statement from Executive Secretary of the United Nations Convention on Biological Diversity" (speech, New York City, May 22, 2007).

100 *The meat industry is... responsible for 18 percent of all greenhouse gas emissions, which is more than transportation:* Steinfeld et al., *Livestock's Long Shadow*, xxi.

100 *The production of merely 3.8 ounces of beef... releases about as much carbon dioxide into the atmosphere as a midsize sedan emits by driving eighteen miles:* Calculated from a 2007 study by the National Institute of Livestock and Grassland Science in Japan, cited in Mark Bittman, "Rethinking the Meat Guzzler," *New York Times*, January 27, 2008.

100 Jonathan Safran Foer, *Eating Animals* (New York: Little, Brown, 2009), 143.

Chapter 6: *First Kill*

104 Ring-necked pheasant Latin name, origins and population decline information from Harry Nehls, *Familiar Birds of the Northwest* (Portland, OR: Portland Audubon Society, 1981), 60. And from Marshall, Hunter and Contreras, *Birds of Oregon*, 174–75.

114 Pollan, *Omnivore's Dilemma*, 353.

114–15 Herman Melville, *Moby-Dick: Or, The Whale* (New York: Modern Library, 2000), 776.

115–16 Garrison Keillor, Lake Wobegon Days segment of *A Prairie Home Companion*, American Public Media, March 22, 2008.

116 *Ironically, this has coincided with a sharp rise in meat consumption:* Horowitz, *Putting Meat on the American Table*, 16.

116 *there are now at least 4,385 [farmers' markets] in the United States. But only about 3 percent of farmers' market vendors sell meat:* National Farmers Market Survey, published in May 2006 by the U.S. Department of Agriculture's Agriculture Marketing Service, 2, 27.

117 *hunters play an important role in population control:* From multiple sources including telephone interview with Steven Williams, president of the Wildlife Management Institute, in September 2007, and Frank Miniter, *The Politically Incorrect Guide to Hunting* (Washington, DC: Regnery, 2007), 105–16.

118–20 Interview with Greg Cazemier in Bend, OR, September 9, 2010.

119 Mule deer study results from Richard Cockle, "Study Shows Surprising Rate of Mule Deer Poaching," *The Oregonian*, November 15, 2010.

Chapter 7: *Off the Mark*

128 Eastern cottontail Latin name, habitat description, diet and hunting regulations provided by the Oregon Department of Fish and Wildlife. Species information may be found at www.dfw.state.or.us/species/docs/rabbit.pdf.

130 Pollan, *The Omnivore's Dilemma*, 23.

130 Roy Wall, *Fish and Game Cookery* (New York: M. S. Mill, 1945), 126.

132–33 NRA background information from Robert J. Spitzer, *The Politics of Gun Control* (Washington, DC: CQ Press, 2004), 75–83.

133 *On its website, the NRA proudly calls itself the "largest pro-hunting organization in the world.":* www.nra.org.

133 *a semi-automatic Glock 19 pistol—which can shoot more than one round per second:* William Hermann, "Gabrielle Giffords Shooting: Pistol Used Is a Mainstay for Law Officers," *Arizona Republic*, January 10, 2011.

133 *The NRA has about 4.3 million members:* From www.nraila.org/Issues/Faq, accessed July 29, 2011.

135 *Nationwide, three thousand tons of lead are shot into the environment by hunters every year, another eighty thousand tons are released at shooting ranges and four thousand tons are lost in ponds and streams as fishing lures and sinkers:* From "Petition to the Environmental Protection Agency to Ban Lead Shot, Bullets, and Fishing Sinkers Under the Toxic Substances Control Act," submitted by the Center for Biological Diversity, American Bird Conservancy, Association of Avian Veterinarians, Project Gutpile and Public Employees for Environmental Responsibility on August 3, 2010, available at www.biologicaldiversity.org/news/press_releases/2010/lead-08-03-2010.html, accessed July 29, 2011.

135 California condor information from the American Bird Conservancy, available at www.abcbirds.org, and from Katharine Mieszkowski, "Condors vs. the NRA," Salon.com (September 22, 2007), available at www.salon.com/news/feature/2007/09/22/condors, accessed July 29, 2011.

136 *the president announces that one of the organization's top priorities is to oppose Endangered Species Act protection for gray wolves:* M. David Allen, "Standing Up for Elk Country," *Bugle* 28, no. 2 (March–April 2011), 9.

136 *But elk numbers didn't drop nearly as much as some biologists had feared. In fact, they noticed changes that helped the entire ecosystem:* Sandi Doughton, "Can Wolves Restore an Ecosystem?" *Seattle Times*, January 26, 2009.

137 *At the same time that he denounces wolves, the group's president pledges that the organization will "become more engaged in the core issues of our time that threaten our hunting heritage and that of our children":* Allen, "Standing Up for Elk Country," 9.

138 *the National Wildlife Federation, which was founded by a hunter back in 1936:* Available at www.nwf.org/About/History-and-Heritage.aspx, accessed July 24, 2011.

Chapter 8: *Wild Tastes*

141–44 I took creative license with several details of the goose's life. Facts including Latin name, population statistics, social and mating habits, size and migratory pattern from Marshall, Hunter and Contreras, *Birds of Oregon*, 76–78.

144–47 I took creative license with precise details of the chicken's life, but based it on factual accounts of factory farms, beak cutting, commercial habits for light and feeding, injury rate, crating technique, slaughter method and life span from multiple sources including Pollan, *The Omnivore's Dilemma*; Foer, *Eating Animals*; the Farm Sanctuary (www.farmsanctuary.org), and Karl Weber, ed., *Food, Inc: A Participant Guide: How Industrial Food Is Making Us Sicker, Fatter and Poorer—and What You Can Do About It* (New York: Public Affairs, 2009), 62.

149 Tom Standage, *An Edible History of Humanity* (New York: Walker and Co., 2009), 4.

149 *Today more than nine out of ten land animals killed for food in the United States are broilers. Fast-food chain KFC alone buys nearly one billion per year:* Weber, *Food, Inc.*, 62.

149 *more than 250 million chicks are destroyed each year, most of them layers who . . . turned out to be males:* Foer, *Eating Animals*, 48.

149 Pollan, *The Omnivore's Dilemma*, 333.

150–51 *A 2009 study found that more than 40 percent of all food produced in the United States is thrown away instead of eaten:* Kevin D. Hall, J. Guo, M. Dore, C. C. Chow, "The Progressive Increase of Food Waste in America and Its Environmental Impact," *PLoS ONE* 4, no. 11 (2009), e7940.

154–55 *Only a few known species of mushrooms are, if ingested, capable of killing an otherwise healthy adult. Most poisonous mushrooms cause run-of-the-mill food-poisoning symptoms: diarrhea and vomiting*: David Arora, *Mushrooms Demystified* (Berkeley, CA: Ten Speed Press, 1986), 893–96.

156–57 *The capped mushrooms that we're used to—the kind you could slice and sprinkle on a pizza—are actually reproductive structures:* Fungus life cycle and mushroom species information, including Latin names, from ibid., 4–6, 662, 191.

157 *"a provocative compromise between 'red hots'and dirty socks.":* Ibid., 191.

160 Lily Raff, "A Taste of Tradition," *The (Bend) Bulletin*, August 17, 2008, F1.

161 Horowitz, *Putting Meat on the American Table*, 44–70.

161–62 Hal Herzog, *Some We Love, Some We Hate, Some We Eat: Why It's So Hard to Think Straight About Animals* (New York: HarperCollins, 2010), 183.

Chapter 9: *Good Dog, Bad Wolf*

169–70 *Wolves and dogs are the same species, and their genomes are almost impossible to distinguish:* Barbara Smuts, "Behavior of Domestic Dogs," *The Encyclopedia of Animal Behavior*, ed. Michael D. Breed and Janice Moore (New York: Elsevier Academic Press, 2010), 562–67.

170 *Other than humans, wolves were once the mammal with the most varied geographic distribution on earth:* From http://nationalzoo.si.edu/Animals/NorthAmerica/Facts/fact-graywolf.cfm, accessed July 29, 2011.

170 *In one particularly odd theory, wolf pups were stolen from their dens at one or two days old, and lactating women nursed them from their own breasts:* Mark Derr, *Dog's Best Friend: Annals of the Dog-Human Relationship* (Chicago: University of Chicago Press, 2004), 21.

170 *each theory rests on some level of consent from the wolves themselves:* Smuts, "Behavior of Domestic Dogs," 562–64.

170 *some scientists believe that Siberian huskies stem from a population of semi-domesticated wolves who flocked to nomadic tribes of people when hunting became difficult during harsh winters:* Ray Coppinger and Lorna Coppinger, *Dogs: A New Understanding of Canine Origin, Behavior, and Evolution* (New York: Scribner, 2001).

170–71 *Humans and dogs have been hunting together for as long as forty thousand years, according to some estimates:* From Herzog, *Some We Love, Some We Hate, Some We Eat*, 105.

171 Modern breed descriptions from the American Kennel Club, available at www.akc.org/breeds/index.cfm.

172 *In 2009, wolf hunts open in two Western states:* Montana and Idaho opened wolf hunts in September 2009, following the Obama administration's decision four months earlier to remove gray wolves from the list of endangered species. Sources include Montana Fish, Wildlife and Parks (http://fwp.mt.gov) and Idaho Fish and Game (http://fishandgame.idaho.gov).

173 *Non-profits in Montana and New Mexico, for example, actually bring wolves into school classrooms:* Sources include www.wildspiritwolf sanctuary.org/ed_presentations.php and www.wildsentry.org.

174 *In a recent study, well-regulated hunts were found to help migratory animals adapt more quickly to habitat changes:* Todd Brinkman, Terry Chapin, Gary Kofinas and David K. Person, "Linking Hunter Knowledge with Forest Change to Understand Changing Deer Harvest Opportunities in Intensively Logged Landscapes," *Ecology and Society* 14, no. 1 (2009).

175–76 Chukar information, including Latin name, habitat, physical description and diet, from Marshall, Hunter and Contreras, *Birds of Oregon*, 171–73.

180 John Cage, *Silence: Lectures and Writings* (Middletown, CT: Wesleyan University Press, 1973), 93.

182 *Daniel Lieberman, an evolutionary biologist at Harvard, believes that the uniquely human capacity for long-distance running... is a vestige of our ancient method of "persistence" hunting, or chasing a wild animal to exhaustion and eventually death:* Charles Bethea, "Fair Chase," *Outside*, May 2011.

182 *"Buck," the word for a male deer, for example, is slang for one dollar:* Richard Nelson, *Heart and Blood: Living with Deer in America* (New York: Random House, 1997), 101.

183 Wall, *Fish and Game Cookery*, 90.

184 *A study in 2009 found that the horns of Canadian bighorn sheep have shrunk because of hunting:* Chris Darimont, "Human Predators Outpace Other Agents of Trait Change in the Wild," *Proceedings of National Academy of Science*, January 12, 2009.

184 Chris Darimont quoted in Anne Minard, "Hunters Speeding Up Evolution of Trophy Prey," *National Geographic News*, January 12, 2009.

Chapter 10: *Friends for Dinner*

190–91 Salmon life cycle information from Jason Cooper, *Life Cycle of a Pacific Salmon* (Vero Beach, FL: Rourke Publishing, 2003).

193 Erich Fromm, *The Anatomy of Human Destructiveness* (New York: Holt, Rinehart and Winston, 1973), 135.

196–97 Full Boone & Crockett definition of fair chase, as well as 2005 statement against canned hunts, available at www.boone-crockett.org.

197 *One short-lived but highly publicized ranch tried allowing clients to operate a gun over the Internet and, with the click of a mouse, take the life of a captive hog.:* Zachary M. Seward, "Internet Hunting Has Got to Stop—If It Ever Starts," *Wall Street Journal,* August 10, 2007, A1.

199 *In 2010, the Whitetail Pro Series began hosting deer-hunting tournaments in which contestants stalk deer, zero in on them using digital scopes equipped with memory cards and then fire blank shells:* James Card, "A Kind of Hunt That Even Deer Can Get Behind," *New York Times*, October 16, 2010.

200 *And in England, since fox hunting with dogs was outlawed in 2004, dedicated hunters on horseback have eliminated the fox altogether and taken to pursuing a designated human:* Frances Stead Stellers, "Coakham Hunt's Greatest Game Pits Bloodhounds Against Man," *Washington Post,* January 12, 2011.

200 José Ortega y Gasset, *Meditations on Hunting* (Belgrade, MT: Wilderness Adventures Press, 1995), 103–4.

201 *Each day in the United States, six thousand acres of open space, including working farms and forestland, are developed:* Statistics provided by

the USDA Forest Service, available at www.fs.fed.us/projects/four-threats/facts/open-space.shtml.

202 Herzog, *Some We Love, Some We Hate, Some We Eat*, 176.

203 *While the exact number of vegetarians is difficult to discern, most surveys and polls produce estimates of seven to eleven million. That means we have about as many vegetarians as residents of the state of North Carolina:* Vegetarian estimates derived from a series of surveys by the Vegetarian Resource Group, found at www.vrg.org/nutshell/faq.htm#poll; population from U.S. Census Bureau.

203 *one survey found that 60 percent of the people who identified themselves as vegetarians admitted they had consumed meat in the last twenty-four hours:* Herzog, *Some We Love, Some We Hate, Some We Eat*, 195.

203–4 Pollan, *The Omnivore's Dilemma*, 326.

204–5 Foer, *Eating Animals*, 13–14.

205 *in the United States, vegetarians are outnumbered, three to one, by ex-vegetarians:* Herzog, *Some We Love, Some We Hate, Some We Eat*, 200.

205 Peter Singer and Tom Regan, eds., *Animal Rights and Human Obligations* (London: Prentice Hall, 1989).

205–6 Cora Diamond, "Eating Meat and Eating People," *The Realistic Spirit: Wittgenstein, Philosophy, and the Mind* (Cambridge, MA: MIT Press, 1995), 322.

207 Pollan, *The Omnivore's Dilemma*, 333.

208–9 Foer, *Eating Animals*, 198.

Chapter 11: *Year of Death*

229 *Gunshot is the most common method of suicide in the United States. In fact, gun-inflicted suicides outnumber gun-related homicides and accidental deaths combined:* National Safety Council Injury Facts 2011 Edition (Itasca, IL: National Safety Council, 2011), 164–65, 47.

Chapter 12: *Killing Bambi, Reviving Artemis*

231 Lily Raff, "Tribes and the River," *The (Bend) Bulletin*, January 25, 2005, A1.

236 *Several states . . . have amended their own constitutions to protect citizens' right to hunt:* By the end of 2010, thirteen states had amended their constitutions to guarantee the right to hunt, according to the National Conference of State Legislators in a report by Douglas Shinkle prepared November 2010, www.ncsl.org/default.aspx?tabid=21237, accessed July 29, 2011.

237–38 *One in ten American hunters is female—our gender's highest participation rate in history—and we are the only demographic of hunters currently on the rise:* From the 2006 National Survey of Fishing, Hunting and Wildlife-Associated Recreation, available at http://wsfrprograms.fws.gov.

238–39 Interview with Jessie Fischer in Bend, Oregon, March 2011.

239–41 Description of Artemis paraphrased from Robert Graves, *Greek Myths* (London: Penguin Books, 1981), 33–34.

240 *Some versions of mythology claim that Artemis killed Orion:* Cartmill, *A View to a Death*, 33.

Chapter 13: *Deer Diary*

245–46 White-tailed deer information including Latin name, related species, habitat, diet and population control, and mating habits from Nelson, *Heart and Blood.*

246 *"Everyone in North America who lives each day on agricultural foods . . .":* Ibid., 310–11.

247 *"looked as if mice were running up and down inside her esophagus . . .":* Ibid., 67.

257 Gary Lewis, *Deer Hunting: Tactics for Today's Big Game Hunter* (Bend, OR: Gary Lewis Outdoors, 2003).

Chapter 14: *Big Game*

269–70 General information about elk gleaned from multiple sources including Jay Houston, *Ultimate Elk Hunting: Strategies, Techniques & Methods* (Minneapolis: Creative Publishing International, 2008), and the Oregon Department of Fish and Wildlife.

289–90 Nelson, *Heart and Blood*, 71.

290 *A local Indian woman once explained*: Interview with a source who wishes to remain anonymous, in Richland, Washington, May 2006.

290–91 Fromm, *The Anatomy of Human Destructiveness*, 133.

294 *the U.S. Fish and Wildlife Service says that $14.70 of that went to purchase or lease wetlands for the National Wildlife Refuge System*: According to the service, "The 98 cents of every dollar generated [from duck stamps goes to] purchase or lease wetland habitat for the National Wildlife Refuge System." Information about the federal duck stamp program is available at www.fws.gov/duckstamps/Info/Stamps/stampinfo.htm.

294 *My license fees helped pay for biologists to study animals and restore important habitat. Some of my money even goes to help non-hunted animals like songbirds and pygmy rabbits*: From "ODFW 2009–2011 Fee Increase Fact Sheet," prepared by the Oregon Department of Fish and Wildlife, available at www.dfw.state.or.us/agency/budget/docs/2009/budget_fact_sheet.pdf, accessed July 29, 2011; and from "On the Ground: The Oregon Conservation Strategy at Work," prepared by the Oregon Department of Fish and Wildlife in November 2006, available at www.dfw.state.or.us/conservationstrategy/news/2006/Nov2006.asp, accessed July 29, 2011.

294 *In 1982, for example, 16.7 million Americans spent a collective $259 million on hunting licenses and fees. Averaged out, that's about $15.50 apiece. In 2003, 14.7 million hunters spent a total of $679.8 million, or about $46.12 each. Costs more than tripled in twenty-one years, much faster than the rate of inflation:* Figures obtained from the U.S. Fish and Wildlife Service National Hunting License Report, published December 2, 2004, and available at http://wsfrprograms.fws.gov/Subpages/LicenseInfo/Hunting.htm. Inflation calculated by West Egg at www.westegg.com/inflation.

295 *the wealth of the average hunter is also on the rise:* 2006 National Survey of Fishing, Hunting and Wildlife-Associated Recreation, available at http://wsfrprograms.fws.gov.

297 *Ernest Hemingway used to worry that if American men stopped hunting, they would cease to be men:* Hemingway's views about hunting and masculinity are efficiently illustrated in his short story "The Short Happy Life of Francis Macomber." Paul D. Staudohar, ed., *Hunting's Best Short Stories* (Chicago: Chicago Review Press, 2000), 41–74.

297 Wallace Stegner, "Coda: Wilderness Letter," in *The Sound of Mountain Water: The Changing American West* (New York: Penguin, 1980), 147.

ABOUT THE AUTHOR

Lily Raff McCaulou was born and raised in Takoma Park, Maryland. She graduated from Wesleyan University and worked in the independent film industry in New York City before becoming a journalist. She writes an award-winning newspaper column in Bend, Oregon. This is her first book.

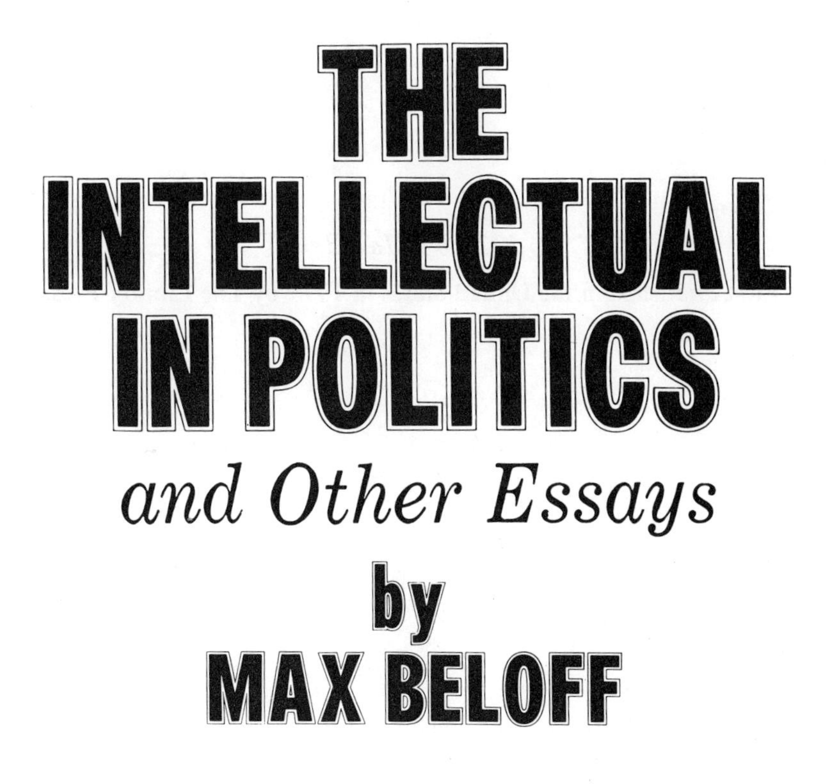

THE LIBRARY PRESS
New York
1971

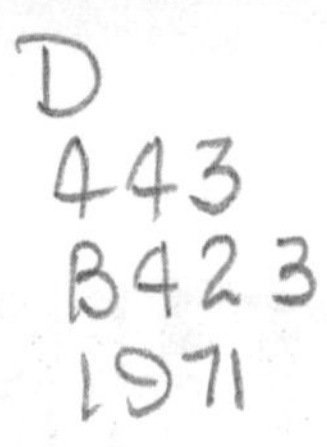

First Published in the United States in 1971 by The Library Press

Library of Congress Catalog Card Number 78–141854
International Standard Book Number 0–912050–02–0
Printed in the United States of America

FOR DENIS BROGAN

Contents

Acknowledgements

I am indebted to the editors of journals where many of these pieces first appeared and to other holders of copyright for permission to reproduce the following:

Nottingham University for 'Prime Minister and President'; to the Council of Christians and Jews for 'The Challenge of Barbarism'; to *Government and Opposition* for 'England in Transition' (April 1966); to *The Times* for 'Soviet Studies – and Russian Reality' (22 September 1965), 'The Planners' Place in Foreign Policy' (18 January 1963), 'Defining the Limits of Official Responsibility' (11 September 1967), 'Examining the Working of Whitehall' (19 June 1963), 'Making the Civil Service More Outward Looking' (12 November 1968), 'Apartheid Policy under New Leadership' (22 September 1966), 'Isolation of South African Universities' (4 October 1966); to *Public Administration* for 'The Plowden Report on the Foreign and Commonwealth Services' (1964); to *Encounter* for 'Comes the Devolution?' (April 1969), 'Before the Flood' (January 1968), 'A Neglected Historical Novel' (March 1970), 'Rootless Cosmopolitans?' (November 1969), 'English as a "Second Language"' (July 1964); to the Jewish Historical Society of England for 'Lucien Wolf and the Anglo-Russian Entente 1907–14'; to Chatto & Windus for 'The Sixth of February' (first published in *St Antony's Papers,* vol. 5, 1959); to the Presses Universitaires de France for 'The Anglo-French Union Project of June 1940' first published in *Mélanges Pierre Renouvin;* to Messrs Hamish Hamilton for 'The Special Relationship: An Anglo-American Myth' first published in *A Century of Conflict,* edited by Martin Gilbert (1967); to the *Journal of International Affairs* and Columbia University Press for 'Reflections on Intervention' (vol. XXII, no. 2, 1968); to the *Jewish Chronicle* for 'Zionism as Nationalism' (27 August 1965) and

'The Israel-Haters' (21 November 1968); to the South African Institute of Race Relations for 'The Crisis in Parliamentary Democracy'; to the India International Centre and the *Journal of Education and Psychology* (Vallabh Vidyanagar, July 1967) for 'Indian Universities and their Problems; Some First Impressions'; to *The Listener* for 'Imperial Sunset' (16 January 1964). 'The Intellectual in Politics' and 'The Machinery of Government' have not been published before.

All these pieces are reprinted as they first appeared with the addition and deletion of a few footnotes and the imposition of stylistic consistency.

I am indebted to Mr John MacCallum Scott for suggesting that I make this collection, and to Mr Melvin J. Lasky for his help and encouragement.

My secretary Mrs Ann Strong has been of the greatest help in the preparation of this volume, and I am most thankful to her, as also to Mrs Margaret Croft for compiling the index and to Miss J. F. Maitland-Jones and Mr Michael Beloff for reading the proofs.

December 1969. MAX BELOFF
All Souls College, Oxford

Introduction

Any decision to publish a volume of lectures and essays, most of which have already been in print, demands some explanation. One part of the defence is easy enough; a number of the pieces here, including some of the most important ones, have appeared abroad or if in this country in pamphlet series not likely to reach more than a small number of the readers who might be interested in them; the other part, that directed to the question of their intrinsic value and inter-connection, is more difficult.

A glance at the table of contents will show that the essays cover what appear to be rather diverse topics treated in very different ways; some historical set-pieces alongside rather immediate reactions to particular events or publications. It is indeed a much more heterogeneous collection than my earlier one which was almost exclusively concerned with problems of international relations.[1] But although there are one or two cases where the decisive reason for inclusion was a wish to make a particular piece of historical research available, the guiding principle has been to bring together out of a much larger number of occasional writings those which illustrate my own intellectual interests over the last few years. If international relations do not bulk very large, it is because I summed up my approach to the current phase in world politics in my Edward Beatty memorial lectures at McGill University in 1967[2] and in my short book, *The Future of British Foreign Policy*.[3]

Although, as with many occasional pieces of writing, most would never have been written had not a precise invitation or occasion presented itself, none of them fall outside my own sphere

[1] *The Great Powers* (London, 1959).
[2] Published under the title, *The Balance of Power* (Montreal and London, 1968).
[3] London, 1969.

of academic work; and the decision to reprint them is in part due to my wish to make yet one more gesture of protest against what I believe to be a deplorable consequence of specialization in the case of those whose work is ultimately concerned with the interpretation of the contemporary world and even with practical issues of policy. The distinctions made between the work of the historian and that of the political scientist, or between domestic and foreign concerns, while appropriate in undergraduate studies where depth rather than breadth is called for – an unpopular view, I know – seem untenable when one's concern is with the actual world of men and ideas and not, as so often today, with questions of method.

For most of the relevant period, my interests have in fact been directed towards two very closely connected fields of inquiry; the way in which Britain has come to decline from being the centre of an imperial system to being primarily a European power of the middle rank, and the way in which its governmental and political institutions have responded to these and other changes in the present century. The principal outcome of the first investigation is intended to be a multi-volume study of which the first instalment has actually appeared.[4] Out of the second I hope to produce a book on contemporary British government, and its problems.[5]

My own view remains more akin to the nineteenth-century students of the *res publica,* many of them men of affairs rather than pure academics, than to that of today's political scientists with their addiction to 'quantification' and the gadgetry that goes with it.[6] To attempt to understand the institutions of any country without some attempt to steep oneself in its history and culture seems to me a task foredoomed to failure. More can be gleaned from apprehension of the physical environment, from painting and from the novel than from answers to questionnaires, depth interviews, content analysis and all the other tools of the 'behavioural scientists'.

Again, while I sometimes go so far as to doubt, despite the contrary example of my own teacher Sir Denis Brogan, whether it is ever possible fully to understand the workings of a country other than one's own, to study only one country, albeit one's own, seems

[4] *Imperial Sunset,* vol. 1: *Britain's Liberal Empire, 1897-1921* (London, 1969).

[5] This will form part of a series of books on the governmental systems of different countries which I am at present editing for the publishers of the present volume.

[6] I have said a little more about this in a broadcast 'The Americanization of British Intellectual Life' reprinted in Derwent May (ed.) *Good Talk: an anthology from BBC radio* (London 1968).

in itself an important limitation upon one's chances of arriving at the truth, or even of asking the right questions. On the whole, I think it has been fortunate for me that from the age of twenty-five to the age of fifty, my own academic interests were almost wholly directed away from this country, to the United States, France and Russia. For it is this experience that has enabled me, I hope, not to take Britain for granted, but to look at it both lovingly and critically but from the outside. The fact too, that although a Londoner by birth and upbringing, my parentage was Russo-Jewish, and that although a British subject by birth I would not think of myself as an Englishman, has I think been an advantage rather than the reverse. Perhaps only those who are British by accident rather than ancestry fully appreciate the exceptional merits of British ways of life and British institutions[7] and fully mourn their contemporary retreat. I do not therefore regard what I have included here on Russian or Jewish themes as irrelevant to my main purpose.

Since I have been working on Britain not just as a single country but as the centre of an imperial system I have spent some of this time getting acquainted with some of the areas where Britain once exercised sovereignty or as in parts of Latin America, an informal Empire hardly less important. Not all my journeying has produced material that I want to reproduce here, but both India and South Africa are represented; and my Canadian experience is rarely out of mind when I emphasize the linguistic aspect of politics, as I do in one essay here, and *ad nauseam* to my pupils.

Friends of mine would know that one growing concern of mine over the last few years is under represented in this volume; though the essence of my position is contained in my lecture 'The Challenge of Barbarism'. I have indeed felt impelled to enter into controversy in what appears to me to be the most crucial aspect of Britain's affairs today, educational policy. It is certainly my view that the inroads of egalitarianism into the educational field – by which I mean subjecting the requirements of science and learning and intellectual discipline to vague and only half-formulated social objectives – is the greatest threat that Britain faces in maintaining its position among the great nations of the world. Whether through

[7] It is perhaps for this reason that though the holder of a chair founded in memory of Mr Gladstone and although a member of the Liberal Party, I still cherish a cult for Disraeli that goes back to my schooldays.

the destruction of the grammar schools in favour of so-called 'comprehensive education', or in the attack on independent schools, or above all in the ever increasing and utterly destructive intrusions of the State into the internal affairs of the universities, politicians, alas of all parties, seem bent on wrecking what most knowledgeable people outside Britain regard as one of its most precious possessions, its system of secondary and higher education. If I have not reprinted any of my contributions to the educational debate, it is not that I have changed my views but that the situation has been changing so rapidly (for the worse) that anything one writes is liable to seem outdated a few years or months later.

One final piece of self-justification seems in order. The sources of these pieces range from learned journals to newspapers; I hope that the former group avoid obscurity and the latter, vulgarity. I do not believe that in matters which relate to important national and international questions one should ever use the kind of jargon which makes it certain that only other users of the same jargon would dream of reading what one writes; on the other hand, I do not see that one need, in writing for the press, adopt some different and more 'popular' approach. The point of writing is to make one's knowledge and ideas accessible to others who may find what one has to say useful; much of current political science writing ignores this salient fact, and will perish where Bagehot, Bryce and Bodley survive.

I THE INTELLECTUAL IN POLITICS

I The Intellectual in Politics[1]

A memorial lecture should have reference to the man one is commemorating and to problems that interested him or that are related to his own career. I never knew Ramsay Muir though I was once taken as a schoolboy to hear him speak at the National Liberal Club – I cannot remember, alas, on what subject. In order to prepare this lecture I have been reading what he wrote about himself and what others wrote about him, and with growing interest. For in Ramsay Muir you have an example of someone who was clearly an intellectual – an historian and scholar – but who was driven by conscience, I think, rather than ambition to seek for an active role in the politics of his time. Briefly an MP, several times a candidate for Parliament, the founder of this Liberal Summer School and the performer of many services to the Liberal Party in the long winter of its decline, Ramsay Muir might be regarded as the type-figure of the twentieth-century intellectual who takes his political commitment as implying not merely the duty of commenting on current political events but also that of going down to the grass-roots, and performing even the most tedious of political chores without which no party can hope to survive, let alone flourish.

I happen personally to find him a most sympathetic figure and partly no doubt for the robust character of his liberalism; his belief in the constitutional and parliamentary system of this country; his unwillingness to condemn the British record overseas, his pride in aspects of the British achievement and his ability to distinguish on what might be called the progressive side of British politics what was truly Liberal from that which clearly was not. I suspect that much of what passes as Liberalism in some quarters today – a heady mixture of socialist economics, anarchist politics and a style

[1] The Ramsay Muir Memorial Lecture, 19 July, 1969.

of life which denies the kind of serious commitment to study which was part of Ramsay Muir's Liberalism – would have struck him as very remote from the core of the Liberal faith, as likely to repel rather than attract the good solid citizens of Rochdale and Scarborough among whom he preached the Liberal creed. I do not think he ever contemplated that people calling themselves Liberals – whatever excuses might be made for youth and inexperience – could denounce parliamentary government, and try to substitute for political campaigning the technique of the demonstration, the sit-down and so forth.

For one result of the fact that Ramsay Muir came into politics at a relatively mature age was that he had both a good deal of practical experience, particularly in relation to university administration and finance, and of course a deep and thorough knowledge of the political tradition out of which liberalism grew. He could place himself and his work historically; and to be able to do this is the beginning of political wisdom.

But all this, though relevant to the man, is not the essential question which I wish to discuss in this lecture. We must ask both what political life gains or loses when it recruits intellectuals and what they gain or lose from the experience of immersion in it.

Ramsay Muir himself realized that his change of vocation was not all to the good. As he says in the unfinished autobiography written towards the end of his life: 'Perhaps my work as a historian had been spoilt because I was too much of a politician. Later my work as a politician was to be spoilt because I was too much of a historian to be completely immersed in the party game.'[2]

The same point is made in an appreciation of him by Ernest Barker, a man who cherished many of the same ideals but never let himself stray outside the strict confines of academe.

> He was [he writes of Ramsay Muir] in two worlds, or – more exactly – he stood between two worlds. That was the difficulty of his position. Perhaps he was too much of the professor, and spoke too much *ex cathedra* to suit the role of politician. I say 'perhaps' for I have not sufficient knowledge to judge if my guess is true. I can say with more assurance that he was too immersed in politics – too steeped in political feeling – to achieve the cool

[2] S. Hodgson (ed.) *Ramsay Muir, an autobiography and some essays* (London, 1943), p. 109.

and calm profundity of the highest historical scholarship.[3]

Barker goes on to say that had Ramsay Muir got further in politics and then after retirement been granted time in which to meditate on his experience as well as to study further 'he might have achieved an historical work of a memorable quality. The fates were not so kind. He was destined to live in the ebb of his party's fortunes; and he laboured like a man to give it way and course against the ebb.'[4] And to us who have inherited the party and the label which he strove to preserve the tribute is sufficient.

Fortunately for our more important question – the use which politics can make of intellectuals – the history of English politics and of English Liberalism furnishes us with other examples. For instance, as is clear from David Mathew's recent biography, a very different figure and a great historian, Lord Acton, was also lured by the appeal of politics and the desire to act as well as reflect, and he too had a brief experience as an MP, and even as a junior member of a government as a Lord-in-Waiting. As he had by then reached the ripe age of fifty-eight, he had in a sense been a Lord-in-Waiting for a very long time. What he had hoped for earlier, the Berlin embassy, had never been offered him. Close though his links were with Gladstone it is hard to see that on any important issue his political influence was of real importance.[5]

Among the colleagues of Gladstone who objected to the idea of giving Acton cabinet office was John Morley, and one of Morley's reasons may well have been his jealousy of Acton as a competitor with himself for the confidence and friendship of Gladstone. For Morley, as Mr D. A. Hamer so well brings out in his interesting and important study, affords the best of illustrations of the central dilemma of someone whose essential avenue into politics is the avenue of the intellect; a fondness for general ideas and the desire for opportunities to make them effective in legislation. Morley in the course of a long career made an attempt at utilizing all three of the techniques that are open to someone whose primary purposes are of this kind. He tried to do it through his writings, notably in his editorship of the *Fortnightly Review,* he tried to do so by finding a patron among active politicians through whom to work,

[3] *Ramsay Muir, an autobiography and some essays,* p. 159

[4] *Ramsay Muir, an autobiography and some essays,* p. 160

[5] See for a recent account of Lord Acton's relations with Gladstone, David Mathew, *Lord Acton and his Times* (London, 1968).

first Joseph Chamberlain and then Gladstone; and finally he entered politics himself and held high office. Not only did he play this role of an intellectual in politics but he played it consciously, reflecting on it a good deal, and especially upon the problem that above all others is liable to arise, the degree to which an intellectual is bound to dilute his principles in order to make possible the combination of forces upon which successful political action depends.[6]

Morley's decision to enter the political arena would seem to have been very much a personal one; belief in the life of action as a remedy against the introspectiveness and with it depression that went in his view with the purely literary life. It was a form of private therapy. Since, however, politics is a matter of taking decisions, and since Morley found taking decisions difficult and indeed often agonizing, it was not even a very successful piece of therapy. The case may not be unique.

Morley's career as depicted by Mr Hamer illustrates several of the inevitable dilemmas of the intellectual whose interests lie in public affairs. On the one hand, Morley attached great importance to the opinion-moulding function and to the power of those who were in a position to launch new ideas into the arena of public debate – a position perhaps easier to maintain when the suffrage was still limited and the world of politics a relatively narrow one. On the other hand, he was anxious not to be thought a pure intellectual, even a doctrinaire; and at the India Office he welcomed any indication that his merits as a practical man of affairs were achieving recognition. Again, his relationship with his political patrons, Chamberlain and Gladstone, was illustrative of what is almost bound to happen unless the intellectual is indifferent as to whether his advice is taken, or the politician lacking in ideas of his own and ready to accept whatever his intellectual mentors recommend. Morley and Chamberlain found themselves drifting apart even before the great divide over Ireland because of different views on the proper tactics for the party. In Gladstone's case, Morley was more prepared to sink his individuality in his leader's and may by his near adulation have supplied something that the older man needed even more than the benefit of his advice.

It would seem appropriate to consider this subject as we have been doing from the point of view of three prominent intellectuals

[6] D. A. Hamer, *John Morley, Liberal Intellectual in Politics* (Oxford, 1968).

connected with the Liberal Party in this country. But it would be insular to imagine that the problem is peculiar either to this country or to Liberals. What can be said is that it takes on a particular aspect in democratic countries – the intellectual advisers of princes and dictators present a problem of a different kind. It is also the case that it will more often than not be with reforming parties and administrations that intellectuals will wish to cast in their lot, so that in a very broad and general sense this is an issue that concerns the Left in politics rather than the Right.

Much illumination can be got from American experience in the present century, both because the presidential system clarifies the locus of power and because there is so much intimate material about its working. An intellectual who has ideas about the general course that politics should take must either become President like Theodore Roosevelt or Woodrow Wilson or attach himself to an existing or likely President as did so many young hopefuls to Franklin Roosevelt or John Kennedy. The former experience is not often open nor without dangers of its own. Theodore Roosevelt survived partly because he was so many things as well as an intellectual, and partly because in what he wanted to do he was to some extent swimming with the tide. Woodrow Wilson, a natural solitary and doctrinaire, in the end lost touch with the world of practice and was broken by the experience. But to seek influence as a Presidential adviser has its pitfalls too.

Since no two Presidencies are alike, generalizations are difficult. The very personal and unstructured way in which Franklin Roosevelt used the members of his brains trust was obviously different from the way in which John Kennedy brought individual intellectuals into the White House itself or even into the Departments. But there are some constants which derive both from the nature of the Presidential office and from the characteristics of intellectuals themselves.

In the end, the power of decision is wholly the President's. The intellectual has no leverage upon him, unlike, say, a Congressman who may have political strength with which to bargain. The intellectual can give his advice; he has no means of knowing how seriously the President is going to take it, or what other advice on the same topic is being sought at the same time. A President may indeed, as Franklin Roosevelt did, make the balancing of conflicting advice, individually obtained, his main technique in the field of decision-making. It is not to be wondered at that some intellectuals who were close to Roosevelt in the early days did not stay the course and even ended as opponents.

The tragic end of the Kennedy administration does not permit one to learn whether things might have followed the same course; some of the Kennedy team stayed on with his successor, of whom not all stayed very long; others departed to find another standard-bearer and were to find the rivalry between Robert Kennedy and Eugene McCarthy further strain on their loyalties. The problem of allegiance to the living and the dead is one which intellectuals find harder to solve than politicians though it is not unknown among the latter, as witness in our own history, the Canningites, the Peelites and the Gaitskellites – the last of these groups being, it is true, made up of both sorts of men.

Another kind of problem is set by the nature of the relationship in the American climate between the President and the intellectuals upon whom he draws for advice. In a sense the latter must play the role of courtiers and defer to the President's position however much he may appear to treat them on an equal footing. In the case of a President who is particularly vain, no flattery may suffice; but even so eminent an intellectual as Justice Frankfurter found it necessary to deal with President Roosevelt in terms that might seem to err a little on that side.[7]

Most important of all is the degree of risk to their own intellectual integrity and to the respect of their intellectual peers that is incurred by those who cast in their lot with a particular President. While they serve him, they are bound to accept his verdicts on policy and to carry out his wishes as executives, speech-writers, negotiators or in whatever spheres of activity he may find for them. If the policies pursued are unsuccessful or turn out unpopular it is no use saying that they were opposed or doubtful from the beginning; if so, why did they continue to serve? Nor does it help to say that they made an honest error and that they now share the objections to the policy in question. To revise an intellectual position is one thing; to say one was mistaken about something that may have cost treasure and lives is quite another matter.

So far, I have been using the word intellectual without defining

[7] The relationship between the two men can be followed in the volume *Roosevelt and Frankfurter, their correspondence 1928-1945,* ed. M. Freedman (Boston, London, 1968).

it, and consequently in what I take to be its broadest significance; that is someone whose concern is with general ideas rather than the possessor of particular knowledge or techniques; and it is in this sense that the contacts between intellectuals and rulers have been most persistent ever since Plato acted as adviser to Dionysius of Syracuse and Aristotle was engaged as a tutor for Alexander of Macedon. For what the Greeks meant by philosopher is closer to our intellectual than to the philosophers of today who belong rather to the class of expert.

Nevertheless the modern differentiation between the several branches of intellectual endeavour has had an important effect upon the role of the intellectual in politics because some of the new specialized fields – economics, political science, sociology and so on – would claim to have a particular relation to political action, so that the intellectual's role is partly at any rate that of a purveyor of particular techniques not simply of general ideas or principles. Indeed if we may return to Ramsay Muir for a moment, his own role in the 1920s represents something of a bridge between the old intellectual – the Lord Actons, the John Morleys – and the new expert. Ramsay Muir, an historian and man of wide culture, belongs to the older group; but many of those who were associated with him in the Liberal Summer School or in the work that went into the Liberal 'Yellow Book' were professional economists claiming recognition in the light of their particular competence in that field.

It is obvious that in activities of this kind where a climate of opinion is being created or a general programme of action drawn up, such expertise is not only welcome but indispensable. Even if the social sciences do not yet possess, and may never possess, the credentials of the physical sciences there are broad areas of agreement on analysis and policy which it would be absurd to ignore. Indeed much of the task of the intellectual in the broader sense of the term must be to make accessible the insights of the specialist to those unfamiliar with his language or techniques. But when it is a question of doing more than this, of playing an active role in the framing of policy or the sponsoring of legislation, there are dangers in the employment of academic experts, in this case not so much to them as to the government or individual statesman they serve.

Two separate though not unconnected lines of reasoning explain why this should be the case. In the first place, as has been said, the social sciences are themselves in a very imperfect state. It is one thing to say that some particular approach to an economic problem is sufficiently attractive to be worth pursuing at a theoretical level; it is quite another to say that it is well enough established for national fiscal or other policies to be based upon the assumption that it is immediately applicable in practice. And if this is true in so relatively well established a discipline as economics, how much more true this must be of such uncertain and fragmentary fields as educational theory, criminology and so on, where the experts themselves are by no means at one on the fundamental assumptions and theories upon which their prescriptions rest.

In the second place – and this aspect is perhaps the more serious of the two – the field of practical politics is one in which it is most improbable that any theory can be fully applied. The inertia of human societies, the strength of vested interests will always make it more likely than not that the actual policy will be a compromise between what the intellectual proposes and what the practical man is prepared to treat as possible. The intellectual is again faced with the choice either of destroying his influence for good by speaking out, or of compromising his own intellectual integrity by pretending that what is being done is consistent with his own beliefs, and will bring about the result which the policy in its original undiluted form might have been expected to do. If he does the latter – and he usually does the latter – then he damages the standing not only of himself but of his own branch of social science.

A particular twist to this argument occurs when there is a transference of the approach and techniques elaborated in one field of activity to another and quite different one. The most prestigious word in contemporary social science is 'quantification'. It is not all that new of course as an idea. When I was an undergraduate – a long time ago – my tutor told me that Sir William Petty was the first modern man because when other people said; 'let us chase out the Papists and take away their lands!' he said, 'first, let us find out how much land the Papists have'. Still, numeracy is even more fashionable today.

Operational research using mathematical techniques was responsible for important improvements in our war-making capacity during the Hitler war. But the kind of problem that was usefully handled in this way was a limited and concrete one about the most profitable use of scarce resources in shipping, manpower or whatever it might be. It was not at all clear that the same techniques or indeed any mathematical techniques could deal with the much vaster problems that arise in the conduct of war as a whole, however helpful they might be to those who had to frame budgets or determine peacetime priorities in expenditure. When therefore the Americans under President Kennedy reshaped the Department of Defence to give the principal weight to intellectuals of this kind – to the exponents of particular analytical techniques – they took a major risk. Whatever view one may take of the morality of the Vietnam War – and I personally do not subscribe to the facile denigration of our Allies or to the high esteem in which their enemies and ours are held in some quarters – it remains a fact that in terms of the original objectives of the United States, this is a war that has probably been lost, and lost in a costly and terrible fashion. It is also a war entered into on the advice of intellectuals, and fought according to the dictates of a particular intellectual school whose monument is their record of false predictions.

The depredations of intellectuals in British government in recent years have been even more obvious but not so lethal: the failure of the prices and incomes policy and the blatant follies of the selective employment tax and so forth. The intellectual servants of Kennedy and Johnson bear grimmer responsibilities than that.

In talking like this one is always likely to be misunderstood and I must therefore make my own position clear. I do not deny that defence is necessary nor that some intellectuals may not have a useful, indeed an indispensable, role to play within the defence community. In some fields – disarmament, arms control, and peace-keeping – their contribution may be indispensable and far more likely to bring about results than the emotional and unsystematic rhetoric which all too often surrounds these crucial topics. Actually, as a founder-member of the Institute for Strategic Studies and as a member of the Foreign Office's advisory panel on disarmament questions. I would be badly placed to argue the opposite.

What does worry me is the intellectual arrogance which assumes that all these great questions which involve the most delicate of assumptions about the political and human emotions of vast communities, and upon the successful resolution of which human survival itself may depend, are purely technical ones, that it is all a matter of correctly programming the computer. What worries me only a little less is the assumption that intellectuals are entitled to carry with them into political office – or administrative office – the prestige they enjoy not simply in virtue of whatever their own personal intellectual achievements may have been but in virtue of the contributions of the academic community as a whole. I do not mean to say that this blurring of the categories of thought and consequent liability to error is unique to social scientists. Natural scientists, who enjoy much greater prestige in modern industrial societies, are just as often victims of the same illusion. Knowledge of how a hydrogen bomb is made gives no one a special claim to tell us how we can best make certain that it is never used. The record is unfavourable to the view that natural scientists are particularly good at social or political diagnosis; indeed why should one expect them to be good at it, since so much depends upon a depth of specialized knowledge which their own successful devotion to their own subject will hardly have allowed them the time to acquire. Politics should welcome the scientist as scientist, but beware of him if he offers his credentials as an intellectual.

Having in this way removed some possible misunderstandings about the negative aspects of my argument, it is time perhaps to move towards a more positive set of conclusions. What conclusions one actually comes to on this topic will of course ultimately depend on one's ideas about politics and particularly about democratic politics in this country.

My own view would be that we must regard politics as being in the nature of a continuing dialogue in which the elected leaders of the people – the professional politicians – supported and assisted by fully trained professional administrators and experts in the civil service, confront as the other partner, society in all its complex and ever-changing diversity. Society in its evolution throws up problems and presents government with demands which it is the politicians' ultimate business to try to answer.

But in order to meet the demands it is necessary to understand them and to weigh them one against the other when they conflict. Because society is multifarious its demands will rarely be sufficiently clear-cut for the politician to know exactly what is being asked of him and what he is expected to do about it. Indeed the only unambiguous demands will be those emanating from highly organized groups, and as such they should be suspect. The primary task of the intellectual is, through his own sensitiveness to social change and to developing sensibilities, to make himself the voice of the inchoate and confused and complex aspirations of the society of which he forms part. He must be poet and prophet rather than legislator. If he can contribute to bringing some order out of the chaos or impress some order upon it so that the politician looking for a cause can find it, he will have performed his most important task, perhaps his only really important task.

It will be held by some that this is too easy and too modest a role. My reply would be that, on the contrary, it is extremely difficult, far more difficult today than when Morley could set the tone of political debate by editing a single journal of opinion, and therefore by no means a modest ambition. Most of the great things that have been done in politics have been made possible because some general idea has made an irresistible impact upon the public mind, so that all that was left to do was to translate it into legislative or administrative action as the case might be. The ideas themselves as ultimately formulated on the hustings may seem absurdly simple: free trade; the national minimum; full employment. But behind each phrase there is a world of ideas some of which at least are highly sophisticated.

It is also difficult, because it demands above all things independence; not the echoing of what appears to be the public demand for this or that, but the capacity to penetrate beneath the surface and to insist upon the recognition of things that most practical men would rather forget. It is the duty of a political leader to seek popularity; under a democracy there is no other way for him to get the power which he needs if he is to be any use. On the other hand, while the intellectual should not actively seek unpopularity it is probably healthy that he should be rather widely hated. It is probably a sign that he is on to something useful from the point of view of the community at large and threatening to some vested interest.

Independence does not of course mean saying the opposite of what everyone else believes; there is no virtue in paradox for its own sake. But it does mean complete freedom to follow the argument where it leads, and to change one's mind if what one finds is a dead end. That is why it is better that intellectuals should proclaim their views as publicly as possible, and not just whisper them in some politician's ear. It is only through public debate that their ideas can be tested, and that they can contribute to the consolidation of public opinion around one major issue or another. This does not necessarily mean that they should avoid commitment to a party; for in a democracy parties are the natural vehicle for the transmission of ideas and their translation into practice. But if the price of party allegiance is a denial of what they believe to be true then they are better off outside.

It may be that for some intellectuals there comes a point when what they have to offer is not an intellectual service at all but a directly political one, that they are not going to have any more original perceptions but may have a talent for translating their own or other people's perceptions into practice. In that case, they should not hesitate to change roles. It is probably desirable for the general health of the body politic that a proportion of those entering public life should be intellectuals. A party which recruited no intellectuals to its body of parliamentarians and prospective ministers would be a poor party.

But it is equally true that there is a limit to the proportion of intellectuals that a party can safely tolerate in its upper strata if it is to be a governing party in the full sense of the term. The history of the British Labour Party is very revealing in this respect. The Labour Party, which became the alternative party of government after the first world war, was from the beginning an uneasy amalgam of practical politicians mainly nurtured in the trade-union movement and intellectuals. The latter were drawn from two sources: the Fabians, who had been increasingly linked with the Labour Party since before the war, and the newer intake of renegade Liberals whose conduct was largely determined by their attitudes on foreign policy which were of a pacifist and Little England slant.

The amalgam was an uneasy one from the first. The diaries of Beatrice Webb show the poor esteem in which that quintessential intellectual held most of the members of the first two Labour governments. Some of her scorn was justified; not all of it. And few Labour ministers proved more ill-equipped for office than that other intellectual, Beatrice Webb's husband. Fortunately for the Labour Party and for the country, the third Labour Government was preceded by a period of coalition under the stress of war in which the fitness for office of future Labour ministers could be tested. The 1945 Labour Government was on the whole dominated by practical politicians, and among its leading figures only

Aneurin Bevan, and in a very different way, Sir Stafford Cripps, might be thought of as some kind of intellectual.

What has happened since then to the Labour Party is a matter of familiar record and well-worn commentary. As a result of new attitudes in the trade-union movement and of wider educational opportunities for boys and girls of working-class origin, the Labour ranks in Parliament and the Labour leadership became increasingly weighted by intellectuals or would-be intellectuals. The Wilson government differed from the Attlee government from the very beginning in the greater role of intellectuals, and as the years passed the trade-union element has been progressively extruded from positions of importance, so that by now we have a government drawn to a wholly unprecedented degree from the party's intellectual element. Mr Wilson and his colleagues would make a splendid senior common room; one could visualize most interesting seminars; but as a government, it clearly leaves much to be desired.

I do not mean to imply that intellectuals are uniquely unfitted to play a dominant role in government. For rather different reasons, businessmen and trade-unionists who have reached the top of their respective ladders of promotion have also often failed when called to high political office, and the same can be said of some civil servants. Government is itself a skill, and the politician's trade like others must mainly be learned in practice and by early and prolonged apprenticeship. The blend between command and persuasion that is the most important element in statesmanship is difficult to acquire for those coming from walks of life where the relationship between the two is quite different.

Most intellectuals, therefore, who seek to serve the public weal must do so in my view outside political life itself or only on its margin. But the role of external oracle which is what I am proposing as the most suitable one carries with it dangers and temptations of its own. To fix one's eyes upon the next practical step only would be to duplicate the work more properly and probably better done by civil servants. To look at the far horizons and declaim about utopia is probably unhelpful and almost certainly risible. It is the middle distance that is all-important – the creation of opinion on subjects of vital concern but not yet fully ripe for positive legislation.

Quite apart from the actual intellectual problems which activity of this kind necessarily involves, there are other disincentives to take into account. As I have said, one must not expect popularity – at least not in one's own generation. With posterity, the case may be different. It is not the exponents of the values of early industrialism in this country that we now honour, but those who gave early warning of the human costs that industrialism involves. Something of the same kind of honour will I suspect (if humanity survives) be accorded to those who are at long last calling attention to the rate at which man is disfiguring and destroying the natural environment. Posterity will be more likely to accord us bad marks for our rate of destruction of arable soil and pure water than good marks for our so-called rates of growth. But posterity can neither feed one's family nor get one onto the honours list.

The second danger is one that I have already touched upon, the danger to independence arising out of the intellectual's often necessary relationship to government as such. I do not refer to the queasy consciences of those who feel that scientists who lend their skills to the national defence are in some way a discredit to the academic community. Indeed there is something to my mind rather nauseating about highly-paid and privileged people like university professors in the United States decrying the society at whose hands they accept these privileges, and denying the legitimacy of protecting it against its enemies. The loss of nerve in American academe is one of the saddest of contemporary phenomena.

What concerns me is the wider problem that arises from the fact that so much modern intellectual activity is seen to involve research, that research requires money and access to information and that the Government is the great provider of both. If one really wants to know about some important aspect of contemporary affairs, it is obvious that the best thing to do is to hope for appointment to the relevant Royal Commission or committee of inquiry. But on such a body, the desire to be practical, the wish to avoid controversy with one's colleagues, perhaps even terms of reference as wickedly loaded in advance as say those of the Public Schools Commission, all push one in the direction of the comfortable consensus. It is a commonplace that the only profit from such officially-sponsored inquiries is to be found in their minority reports.

The effect of immersion in the official atmosphere with its built-in and of course reasonable and even necessary restraints is likely to be a long-lasting one, and to mark an individual even after he has resumed his role as an independent intellectual. I can think of one very eminent historian for whose earlier writings I have an admiration that falls little short of idolatry, who, having been once employed as an official historian, continued to respect its limitations to the extent of refusing to make use of information acquired outside his official functions where it related to matters which at that time were in the restricted period as far as the rules governing the archives were concerned. In my view the duty of an intellectual is to observe the laws, including of course the official secrets acts, but to regard it as his duty also to pursue the truth by all legitimate means without respect for the governing conventions.

Independence is not only the thing that an intellectual most needs to cultivate if his services to the community are to be of any importance; it is also a very difficult thing to attain and particularly at present. Obviously compared with the intellectuals of Czechoslovakia or Russia or many other countries, the position of British intellectuals is a highly enviable one. But the degree of conformity – especially of conformity in dissent – is rather high as compared with some previous periods in British history. We pride ourselves on our new-won freedoms and on the fact that writers of fiction can refer at will to bodily functions previously described only in medical textbooks or in *graffiti*. But where political criticism is concerned this is a mealy-mouthed generation indeed. Neither our institutions, nor those who man them, are confronted with their failures or failings. Indeed there was much more vocal criticism of British institutions and British public men at a time when Britain was a dominant world power than there is in this period of decline and uncertainty.

I think therefore that even within the limitations I have described there is a role for intellectuals in politics and above all in the Liberal Party. As both Jo Grimond and Jeremy Thorpe have shown, the bridgehead that the Liberal Party holds in Parliament means that topics are raised and aired which would otherwise be buried for good by the common consent of the two front benches. The important thing for intellectuals in the Party is to take advantage of the platform it commands to make their own contribution to public debate. It is different from the role of the political leadership and it works to a different time-scale. But it is a role. It was perhaps the role that Ramsay Muir whose memory we are gathered to honour played best. Apathy in the stalls and rotten tomatoes from the gallery must not divert us from it. With proper modesty and a proper awareness of his own limitations the intellectual in politics – some current examples notwithstanding – need not be a figure of fun.

2 *Prime Minister and President*[1]

I should like to begin by expressing my appreciation of the honour conferred upon me in asking me to give this year's Hugh Gaitskell memorial lecture. I can offer few credentials. I knew Hugh Gaitskell only very slightly; I do not share the political beliefs, still less the party allegiance, that were the mainspring of his public life. I do not even attach the priority which he did to the science of economics as the sovereign guide to the solution of all our problems. Nevertheless it seems to me proper that one should honour the memory of someone so out of the common run of politicians as to believe that the views one holds should determine one's course of action rather than that the limitations of action should determine what beliefs it is proper to express. As a man of courage in a profession in which other qualities are far more widely distributed Hugh Gaitskell commanded respect while he lived and deserves to be remembered now.

The subject which has been assigned to me by the organizers of this lecture – the problems of the modern Executive as exemplified by the offices British Prime Minister and of President of the United States – would seem particularly appropriate for this occasion since it is reasonable to believe that had he lived, Hugh Gaitskell would now be occupying the first of these offices, and occupying it with distinction.

No fewer than three of the contributors to the memorial volume edited by Mr W. T. Rodgers make allusion to Gaitskell's qualifications for the highest post.[2] Mr Douglas Jay who saw him at close quarters during the 1945 government goes so far as to say

[1] The Hugh Gaitskell Memorial Lecture, 21 January 1966.
[2] W. T. Rodgers (ed.), *For Hugh Gaitskell 1906-1963* (London, 1964).

that 'he was probably better fitted to be Prime Minister than any other party leader of this century.' Earl Attlee with the advantage of having held the position himself writes with characteristic reserve: 'I have very little doubt that had he survived he would have made a very good Prime Minister.' Such categorical statements tell us little about the man or the office. Mr Roy Jenkins, the present Home Secretary and the biographer of Asquith is more helpful: 'Gaitskell,' he says, left to English politics, 'the promise of being a great Prime Minister, not because he would necessarily have avoided mistakes, but because he would have influenced the whole Government with a sense of loyalty and purpose, and made men of widely differing gifts and character proud to serve under him.'

It would seem that if Mr Jenkins is right then we have at the outset one way of setting off the Premiership against the Presidency. Mr Jenkins sees British government in traditional terms as the product of team-work in which the test of captaincy is not only to be measured by the number of runs the captain makes himself but also by his success in maintaining the harmonious co-operation of his team, and in imbuing it with a common loyalty. And of course in this case, the captain is also the sole selector.

Few people assessing the kind of President an American public figure might have made, or discussing the reputation of actual American Presidents, would give anything like the same importance to the qualities that Mr Jenkins singles out. While the American President shares the role of selector with the British Prime Minister, and has indeed a much wider range of posts to fill – perhaps seven times as many at the policy-making level – he is not selecting a team, but a series of individuals whose relationship with himself is clearly one of subordination and from whom no corporate spirit is normally expected nor indeed could be. The White House, as Sir Denis Brogan has observed, is easier to understand by analogy with the court of an absolutist monarch, than by analogy with other democratic centres of power. What matters to courtiers is the right of access.

The historians of the two countries reinforce this view in their treatment of the past; when one thinks of the great constructive periods in the history of British government in modern times, Gladstone's first cabinet or the Campbell-Bannerman-Asquith administration, one thinks not only of the Prime Minister but also of his colleagues in the principal offices. Even in war-time when the inclination to personal rule is necessarily greater, one still cannot envisage Lloyd George's role apart from that of the war cabinet and of some of the principal ministers outside it, nor even Churchill's apart from the contributions on the home front of half-a-dozen leading figures.

In American history on the other hand, the great Presidents stand out like isolated peaks in a plain, not just as the highest summits in a range. Who remembers more than one or two members of Lincoln's cabinet, or any of Wilson's or even of Franklin Roosevelt's?

Nor, I think, is the historian's vision here deceptive. For it would go along with neither the constitutional powers of the President, nor with the American political style, nor with the personalities who are likely to reach the summit under American conditions, to rely upon collective wisdom or collective endeavour. Indeed, President Eisenhower, who in a sense tried to introduce something of the British version of cabinet government, is almost universally considered to have failed in the principal *governmental* role of an American President – he has other non-governmental roles: namely to act as the prime mover in all the major aspects of policy and administration.

So far from striving to create a spirit of common loyalty, many of the most able and successful American Presidents have quite deliberately fragmented the authority confided to their subordinates and placed their full trust in none of them. There is more than one way of reaching a President's natural objective, which is to preserve a free hand for as long as possible and to retain in one's own keeping the right to decide on all important matters. The naturally reserved and secretive Wilson stands at the opposite pole from the apparently extrovert and garrulous Franklin Roosevelt. Yet both contrived to leave their subordinates (and later historians) guessing as to their real intentions; indeed the dividing of authority and the provoking of conflicts within the administration itself seems to have been developed by Roosevelt into almost a philosophy of government. Some profess to believe that the current British division of authority – otherwise hard to justify – between the Chancellor and the First Secretary has been a device for leaving the decision-making power in the hands of the Prime

Minister. But there is no public evidence in fact which would entitle us to pronounce upon this claim.

It would be absurd to suggest that a sense of loyalty and commitment has not been present within the institutions of American government even when these have not been fortified by the pressures of war. But these seem to have been to a large extent the product of self-identification not with a President, not certainly with a party, but with a cause. Why else should every reforming President have found it necessary to coin a slogan to suggest his purpose: 'The Square Deal'; 'The New Freedom'; 'The New Deal'; 'The Fair Deal'; 'The New Frontier'. Of course, there have been individuals to whom the cause and the individual were indistinguishable; both the Roosevelts and even Wilson attracted men to themselves whose sense of personal involvement with their fortunes and later reputation is an essential clue to understanding American politics. People have sometimes been taken by surprise by the direction of President Johnson's interests since he took office, because they have overlooked the fact that he entered federal politics as a Roosevelt 'New Dealer'. We have ample testimony also to the attraction that the late President Kennedy exercised over a large number of men, particularly of his own generation. Nevertheless in the case of most Presidents one suspects that it is the cause that is loved rather than the man; one feels that the idea of the 'Great Society' may appeal to many Americans not temperamentally attuned to its embodiment in President Johnson.

So far I have accepted Mr Jenkins' criterion of Prime Ministerial excellence at its face value. But I have done this in order to bring out certain inherent differences between the two offices as sharply as possible, not because I believe that the British Premiership always complies with some preordained text-book model of cabinet government. Indeed what the two offices have in common above all else is their undoubted susceptibility to the accidents of personality. Unlike lesser offices where the role of the incumbent is laid down by rules or binding conventions, the greater offices constantly remodel themselves to suit the individual's habit of work.

It may also be the case that the increasing variety and complexity of the task itself forces upon successive holders of the office some modification of their inheritance and that we ought perhaps to revise our view of the Premiership as well as of the Presidency.

Our difficulty here is that our information about the inner working of British institutions in the present or the immediate past is so much scantier than that available to us for the United States. We are indeed reasonably well informed about Churchill's methods of conducting business; and he took care that we should be. We shall no doubt know even more when the authorized biography and its supplementary volumes of documents reach the war-time years. But the illumination that beats upon Churchill's war-time premiership only serves to increase one's awareness of the darkness on either side. The reticence imposed by the Official Secrets Act, the weaknesses of British political journalism by comparison with the best American journalism, and the general reticence that is normally the hall-mark of British public life make all attempts to form a clear idea of how central government actually operates, and of how the Prime Minister in particular allots his time and energies, exceedingly difficult to bring to any but the most anodyne of conclusions.

My own belief until Mr Wilson's Premiership began was that the general trend was towards the transformation of the office into something nearer that of the American President in the sense of a progressive increase in the power and authority of the Prime Minister as against those of even his most senior colleagues. If one accepted a view of this kind, it was not difficult to find reasons for believing that it was true The increasingly important role of external affairs to which it was inevitable that a contemporary Prime Minister would direct much of his attention, the interaction of these with domestic affairs and of every branch of domestic affairs with almost every other, owing to the high degree of competition for scarce resources, with the consequence that the coordinating function of the Prime Minister and his own choice or priorities were increasingly decisive, and lastly the growth of television and the consequent focusing of the electorate upon the political fortunes of the single individual with the easiest and most constant access to this medium – all these seemed powerful reasons for believing that government by Prime Minister rather than cabinet government was the key to the British system.

It is possible however that the peculiar ascendancy of Mr Harold Macmillan made these arguments seem rather more conclusive than in fact they are. Mr Macmillan allowed Lord Salisbury to doom himself and the Tory Right to an impotence from which even Mr Ian Smith has so far failed to rescue them. He got rid later on of half his cabinet with much less effort than it seems to have cost Mr Wilson to make a couple of relatively minor changes. In world politics Macmillan appeared as an interlocutor of Kennedy and Khrushchev with no less right to speak for his country than they for theirs. He overrode or circumvented the natural and ingrained hostility of much of his party to the idea of Britain entering Europe. But does this necessarily mean that similar feats are open to all his successors?

It may be argued that Sir Alec Douglas-Home's Premiership was too contested in its origins and too short in its duration for the test to be a fair one. And on the face of it, it may be said that Mr Wilson for all his lenience to the weaker brethren in his team, would seem in other respects – as the opinion polls indicate – to have achieved a very impressive degree of authority both in the country at large and within his own party. If Mr Macmillan made the Conservatives swallow the Common Market and bow before the wind of change, Mr Wilson has made the Labour Party swallow the even larger pills of the independent British deterrent and of unconditional support for American policy in Vietnam. It could be argued that in the case of Rhodesia, he has even managed to revive for his own benefit the original key-function of the Prime Minister, that of the first confidential adviser of the sovereign. People can make easy jokes about it; 'the poor man's Melbourne' or 'the poor man's Disraeli', but it does seem to me the case that the Rhodesian affair has seen the importation of the Crown into a highly controversial political issue for which it would be hard to find adequate precedent in the recent past, and that the Prime Minister's role in this development has undoubtedly been of advantage to himself, if perhaps more dubiously to the advantage of the Crown.

Nevertheless one knows too little to be certain of the precise standing of the Prime Minister in the present cabinet – although one might add that there has never been one in which the fact that the power of advising upon a dissolution is the Prime Minister's own and not shared with the Cabinet has been more important.

In a very remarkable paper read last September to the annual meeting of the American Political Science Association, Professor Richard Neustadt pointed out that although the British system is not becoming presidentialized in that the President has duties and functions which cannot be filled except through his initiative and action, while the British system can operate in a looser or more collective fashion, an activist Prime Minister can make a lot of difference; and in Professor Neustadt's view, Mr Wilson clearly belongs to this category. 'In externals Number 10 looks no more like the White House under Wilson than it did a year ago. But in essence, Mr Wilson comes as close to being "President" as the conventions of his system allow. He evidently knows it and likes it. So I take it did Macmillan.'

But if you accept Professor Neustadt's general view that we have to look at the instruments by which government is carried on as well as at the attitudes of rulers, then I think that there is some evidence that a process of 'presidentialization' which could have been observed under Macmillan has been somewhat checked under both his successors.

Here the differences between the two offices appear at their most striking: 'Presidents', to quote Professor Neustadt again, 'with their twenty-odd high powered assistants, and with a thousand civil servants in their executive office – Prime Ministers with but four such assistants in their Private Office (three of them on detail from departments) and a handful more in the Cabinet Office which by definition is not "theirs" alone.'

But Professor Neustadt gives us a clue as to what might happen when he points out elsewhere that there are signs that the White House staff is under Johnson becoming something like a Cabinet Office in its duties and functioning. Could the Cabinet Office then become a little more like the White House staff? The problem of who serves the Prime Minister is in any event a real and important one. Even a Prime Minister – even a President – is human. He has only so many hours a day for affairs of State and many of them, far too many, are spent on the move. For this reason, it does not merely matter what kind of a man he is, or what view he takes of his job; it is also important to know what institutional aids he possesses, and in particular what advice he can draw on other than that of departmental or other Ministers. In other words is the Prime Minister an institution as well as a person? Under Mr Macmillan, we were beginning to get hints that the Cabinet Office was ceasing to fulfil the purely coordinating function that had been its role since it was first devised by Maurice Hankey, and that it showed some signs of being something closer to a Prime Minister's department. Certainly Sir Norman Brook (as he then was) was publicly described as the Prime Minister's principal official adviser. But if this was becoming true of the Secretary of the Cabinet then, it would hardly seem to be true now. If Sir Burke Trend (as some suggest) has followed in Lord Normanbrook's footsteps, he has done so more discreetly, nor of course does he hold the same powerful combination of offices. It may be that there are one or two acquisitions to the Cabinet Office who play some direct role as advisers to the Prime Minister but there is no sign of a Prime Minister's department or, to use the American terminology, Executive Office developing. To take the two most important of the new fields of modern government: there is no British equivalent to the Council of Economic Advisers or to the President's Special Assistant for Science and Technology. These functions are departmental in Britain, Presidential in the USA. Still less does the very small staff at Downing Street itself appear to be a factor of political importance; less than it was perhaps in Mr Macmillan's day. There is no Downing Street aspect of the present governmental arrangements for some future British Schlesinger – if one can conceive of such a thing – to chronicle.

I do not think that this point can be emphasized strongly enough. If the Prime Minister has no information other than what comes to him through ministers and from their departments then he can only act in the traditional way with their consent and through their agency. We shall get nothing from him like President Kennedy's direct confrontation with United States Steel.

It may be that in the interest generated by the intensely personal view that we have of American Presidential government, we fail to realize the institutional importance of the differences between the two systems. On the one hand, the President cannot exert the same direct command over every branch of executive government that a Prime Minister can if his cabinet is well in hand; indeed much Congressional ingenuity in setting up independent agencies has gone to make sure that he cannot. The importance of this for economic affairs is brilliantly brought out in Mr Andrew Shonfield's *Modern Capitalism*. On the other hand, the American President has in the Executive Office at one level, and in the White House at a more intimate one, machinery that can be used to shape and promote policies of his own, which may not necessarily correspond to departmental wishes or advice. Indeed, the President can be seen not as the apex of the Executive branch of government but as a separate arm, partly superior to but partly competing with the regular administration as well as with the legislature which (unlike the Prime Minister) he cannot normally hope to dominate unquestioned.

Indeed, to return to Professor Neustadt, the burden of his paper would seem to be not so much that the Prime Minister is getting more like the usual view of what the President's role must be but rather that the usual view is partly illusory, and the President is much more like the Prime Minister and shares more of his limitations than orthodox students are inclined to admit: 'British government may not be presidential but our government is more prime ministerial than we incline to think.'

It is true, says Professor Neustadt, that superficially, the American President can command where the Prime Minister must negotiate, or try to by-pass possible objections, until it is too late to challenge his policies. Mr Macmillan's handling of the Common Market and of the Nassau agreement can be used to illustrate both methods. The question, says Professor Neustadt, is not whether the President has to negotiate, and therefore to accept the reality of other people's independent judgement, but who these other people are.

In the British system it is still the Cabinet that counts. Professor Neustadt detects an additional obstacle in the possibility of civil servants standing in a Prime Minister's way, but as he points out this is only because they stand behind their ministers who speak for them. It is therefore itself an outcome of the cabinet system. Now clearly the American Cabinet, not being one in our sense, can normally be ignored or over-ruled. It is not office but persons who count: 'an informal shifting aggregation of key individuals, the influentials at both ends of Pennsylvania Avenue.' They may be Cabinet members, individuals on the White House staff or leaders of Congress. And this may even at moments be institutionalized. Indeed, sometimes no President could act alone.

It is clear that this must be so in respect of so awesome a decision as that invoked in a nuclear confrontation; and we have in the Cuba crisis of October 1962 a well-documented instance of how the Presidential system can operate under such conditions. I follow now the story as told by Mr Theodore Sorensen in his admirable *Kennedy*.

The crucial event was the discovery by air reconnaissance of the existence of Russian missile sites in Cuba. The President was told of this at 9 a.m. on October 16. At 11.45 a.m. the President called a meeting in the Cabinet room (but not of course of the Cabinet itself as a British Prime Minister would). To quote Mr Sorensen: 'Those summoned to that session at the personal direction of the President, or taking part in the daily meetings that then followed, were the principal members of what would later be called the Executive Committee of the National Security Council, some fourteen or fifteen men who had little in common except for the President's desire for their judgement.'

They included: from the State Department, the Secretary, Mr Rusk, the Under-Secretary, Mr Ball, and three or four experts on Russia and Latin America; from the Defence Department, the Secretary, Mr McNamara, and his two principal civilian subordinates as well as the newly appointed chairman of the Joint Chiefs of Staff, General Maxwell Taylor; the Director of the Central Intelligence Agency; the Secretary of the Treasury, Mr Dillon; from the White House staff, Mr Sorensen himself and Mr McGeorge Bundy. At some meetings, the Vice-President attended and the President's Appointment's Secretary, Mr O'Donnell. Occasionally the group was reinforced by the presence of Mr Adlai Stevenson, a cabinet member in virtue of his post as American representative at the United Nations, and Messrs Dean Acheson and Robert Lovett. One might note that Messrs Acheson and Lovett were at the time private citizens.

The preparations for American action and the basic decisions as to the course to be followed were hammered out in this group. Only on October 22, when all was prepared, was the circle widened. The three living ex-Presidents were consulted by telephone, and a meeting was called of the full National Security Council which retrospectively endorsed the procedure that had been followed by (as we have seen) giving the inner group the status of an Executive Committee of the Council itself.

At 4 p.m. on that day, the President at last met his Cabinet, but not for consultation. To quote Mr Sorensen: the President 'briefly explained what he was doing and promptly adjourned the meeting. His presentation was tense and unsmiling. There were no questions and no discussions.'

At 5 p.m. the President met the leaders of Congress which evoked, so Mr Sorensen tells us, 'the only sour note of the day.' Since the Congressmen 'had been plucked from campaign tours and vacation spots all over the country some by jet fighters and trainers,' their critical mood is perhaps understandable. Some of them, notably the Democratic Senators Russell and Fulbright, argued that the proposed blockade would be too slow in its effect and urged an actual invasion of Cuba. Charles Halleck, the Republican leader in the House of Representatives 'said he would support the President but wanted the record to show that he had been informed at the last minute, not consulted.'

The President was however adamant, we are told. He was acting, he claimed, by Executive Order, Presidential proclamation and the inherent powers of the Chief Executive, not under any mandate from Congress. He had in fact rejected suggestions for reconvening Congress, then in a pre-election recess, or for asking it for a formal declaration of war. Its leaders had only been summoned when everything had been decided. The President later said to Sorensen, the latter tells us, 'my feeling is that if they had gone through the five-day period we had gone through – in looking at the various alternatives, advantages and disadvantages – they would have come out the same way we did.'

Now if one looks at crisis handling under the British system – take Rhodesia for instance, though not a crisis on the same scale – one can see how different the picture is. If there is an inner group, it will consist of ministers, and its composition will be fixed largely by their departmental responsibilities. Only a little variation is possible.

The Opposition leaders may be informed, but otherwise Parliament will be asked to handle the matter like any other piece of controversial policy. In other words, the Prime Minister cannot create a new instrument of government by personal fiat.

Now one can see that the nature of the Cuban crisis did make the handling of it different in some respects from other exercises of Presidential power in that the constitutional need for Congressional action was absent. Unless the blockade failed, Congress did not need to act. This is different from the case of Vietnam where Congress is needed to vote men and money, and where therefore such cavalier treatment would be impossible. It is obviously even more the case that in regard to domestic policy, where almost all Presidential initiatives require support in the form of money, legislation or both, Congress's cooperation is of the essence.

The difference is that which arises from both the constitutional difference – the separation of powers – and the difference in the nature of the parties. On the Prime Minister's side all the principal leaders will be in the Cabinet, and so already involved in the decisions. Whether the Opposition's assent is required or not, will depend on the size of the majority, the solidarity of the government party, and on whether or not national unity is desirable or obtainable or both. I say national unity where an American would say 'bi-partisanship', not so much because I reject the thesis that we live under something called a two-party system, but because in most matters, and in nearly all domestic ones bi-partisanship is neither sought nor obtainable in this country; the presumption is that the Opposition will oppose. An American President will hope and usually need the maximum of support in Congress from both parties. It is only when we want to do something whose success depends on all the country wanting it – as in war or other conflicts – that we rightly talk of national unity, not of bi-partisanship.

What this means is that Congressmen and even more Senators have not got views that can be predicted on the basis of 'for or against' the Government, but are themselves independent centres of power as towards whom the President is in a bargaining not a command position.

May I quote for the last time from the invaluable Professor Neustadt?

> It is men of Congress more than departmental men who regularly get from Pennsylvania Avenue the treatment given Cabinet ministers from Downing Street. Power in the Senate is particularly courted. A Lyndon Johnson when he served there, or even a Vandenberg in Truman's time, or nowadays an Anderson, a Russell, even Mansfield, even Fulbright (all Democrats be it noted) – to say nothing of Dirksen – are accorded many of the same attentions which a Wilson has to offer a George Brown.

The conventions of 'bi-partisanship' in foreign relations,

established under Truman and sustained by Eisenhower, have been extended under Kennedy and Johnson to broad sectors of the home front, civil rights especially. These were never so much a matter of engaging oppositionists in White House undertakings as of linking to the White House men from either party who had influence to spare. Mutuality of deference between Presidents and leaders of congressional opinion, rather than between formal party leaderships, has always been of the essence of 'bi-partisanship' in practice. And men who really lead opinion on the Hill gain privileged access to executive decisions as their customary share of 'mutual deference.' 'Congress' may not participate in such decisions, but these men often do.

An example given by Professor Neustadt is Senator Dirksen's influence in the framing of the Civil Rights acts.

Now it is clear that the Prime Minister is not in this position with regard to Parliament; he may of course be in a bargaining position with regard to other individuals with an independent power base; on domestic issues, with regard to powerful trade-union leaders for instance; though a Labour Prime Minister has the alternative of muzzling them through Cabinet office as Mr Wilson has done in the case of Mr Cousins. But he needs their active support much less, and will normally give them little share in making decisions as a reward. The role of 'Neddy' as originally conceived may have borne some resemblance to this kind of thing in relation to both sides of industry; but the British tendency to bring things within the departmental hierarchy and to work through the Cabinet has been powerful here too.

With regard to the execution of policy rather than its formulation, there are weaknesses in the President's position not so much compared with that of the Prime Minister himself as with the Cabinet as a whole. The first is the one that I have already referred to; the diffusion of administrative responsibilities through agencies like the Independent Regulatory Commissions or the Federal Reserve System that are not directly subject to Presidential control, and where his powers of nomination are restricted by set terms of office. This administrative device sometimes goes here by the name of 'taking something out of politics' and we do make use of it. But I suspect it is not congenial to the style of British politics and one can already see the difficulties ahead in the case of the Prices and Incomes Commission, where an attempt has been made to separate the implementation of policy from its formulation, in a field where the implementation of principles is all important.

In other spheres the device may work, because for a long time the thing in question seems to be non-political; but then it becomes political and so governmental. We have seen this in the breakdown of the original idea of the Burnham treatment of teachers' pay. We are seeing it in the growing impatience of Parliament, and perhaps of Whitehall too, with the universities' conception of the financial implications ol academic freedom. Now after a century of general acceptance, we are finding a growth of mutterings in the Labour Party against the non-political criteria that guide the Civil Service Commission; we see it in the current attempt from Whitehall to dictate a single pattern of secondary education for the entire country and population.

The American tendency is towards the diffusion of power, ours towards its concentration. One has only to look at the weakness of our courts as compared with American courts, as soon as the powers of the State are raised.

All this strengthens the British Executive and so, if he is properly head of it, of the Prime Minister.

The second relative weakness becomes apparent when one probes one of the similarities between the two offices already referred to – their gain from the advance of the mass media. Both here have command of something to an extent shared by no one else. They make news, as well as being its source. But the American President can only use this power in an indirect fashion; not to demand a mandate which he is being denied by other elements in the governing institutions of the country. Not only has he no power to choose his moment for elections, but he cannot make them when they come a plebiscite in favour of his policies. People are not elected because they support him; indeed he may run behind not in front of powerful local figures from his own party. This is not accidental but characteristic; it can be found in State and even Municipal Government as well – look at Mr John Lindsay and the Democrats who surround him.

Even when he seems to have a mandate, as Roosevelt did in 1936, he cannot be sure it will be interpreted in this way. He can be defied as Roosevelt was over the Supreme Court. On the other hand a British Prime Minister is through the operation of the normal parliamentary system both endowed with a continuing licence for action – including legislative and financial action – and in a position, if challenged, to ask for its confirmation or renewal.

This of course is only an example of the difference made to the two offices by the difference in the party systems of the two countries. The British Prime Minister is normally the leader of a national party with a permanent and coherent organization in the country; his personal control of the machine is greater in the case of a Conservative than of a Labour Prime Minister, but in both cases it is stronger than anything a President has at his disposal.

On one last point there is an advantage on the Prime Minister's side. On the whole he gains rather than loses by being only Head of Government and not Head of State – particularly since as we have seen he can annex not only the prerogatives of the Crown but even its personal influence. What the Prime Minister does not have to perform are the tasks of a formal kind which arise for the President because he is Head of State as well as Head of the Executive. These are both domestic and foreign. If he neglects the one it may rebound on him politically; if he neglects the other, it may cause trouble abroad. And the latter – the international aspects – have become much more demanding with the multiplication of foreign countries and foreign heads of State for whom the call at the White House is some equivalent of a coronation.

As I said, what a man can do depends as much on the time and energy at his disposal as on his formal powers. The American system is very wasteful of the former, our own more economical which is one of the many clear advantages of constitutional monarchy over American-style presidential republics.

To sum up, I do not believe that one can really answer the question as to the extent to which the two offices differ or as to their relative strengths, nor even as to whether they are converging or not, though I am still inclined to the view that in British Government as in so much else in this country, the pull of the American pattern is very strong. The study of formal documents does not help much. The only true understanding, as always in politics, derives from seeing how actual things are actually done; and this kind of historical knowledge is always somewhat out of date – more so for Britain than for the United States. We need more information and since we have no Sorensens and Schlesingers in Downing Street we must rely on the Prime Ministers themselves.

No peace-time Prime Minister since Asquith has written memoirs, except for Lord Avon and his are coloured by the drama attached to the concluding episode of his Premiership and reveal little of his working methods. Mr Macmillan is now writing his memoirs; if well done they will be worth any amount of theory. The difficulty is of course that men who do things care more for what they do than for how they do it. No authority on public administration has yet occupied the Premiership. A pity, at least for the profession of political science.

3 *The Challenge of Barbarism*[1]

This annual lecture was founded in order to commemorate the work of a leading figure in Anglo-Jewry who devoted much time and energy to promoting mutual understanding between his own community and the larger Christian community which provides its environment. His abiding ideal was tolerance; and it is around the theme of tolerance that successive speakers from both communities have grouped their remarks. I do not underestimate the importance of this theme nor the value of the work done to further it by the Council of Christians and Jews under whose auspices we meet. Some might feel indeed that both Christians and Jews sometimes find it so hard to practise tolerance within the limits of their own communion that they may have little energy over for making sure that they practise it outside. And speaking as a Jew, I am sometimes gratified that at least my ecclesiastical authorities cannot, like the mediaeval Christian churchmen, hand over their heretics for punishment by the secular arm.

Nor indeed can one altogether be happy about the situation in this respect in the land of Israel – although it is not hard to understand what lies behind it. For tolerance does not seem to profit when religious organizations are in a position to exercise a measure of political power, or to control aspects of national life not strictly relevant to the free practice of their own creed. Perhaps one is being over sensitive; perhaps one has over the centuries become so accustomed to the role of the Jew as a victim of persecution that to become aware of Jewish persecutors, even mild ones, seems altogether unnatural and even abhorrent. If religious tolerance is to be upheld, there would seem to be as good reasons as ever there were to forbid access to the seats of political power by servants of any organized religion.

[1] The Robert Waley Cohen Memorial Lecture, 17 November 1965.

But although I feel obliged to begin by saying this, it is not of tolerance in this restricted sense that I wish to talk today. On the whole and with all respect to those who find even this measure of tolerance an uphill struggle, it does not seem to me to be by any means the most difficult of the tasks that confront us in this country and at this time. Previous lecturers in this series have more than once brought out the fact that there lies in the teachings of both our religions an important common core of belief from which the lesson of tolerance can without difficulty be derived. And this is, or should be, made easier by the fact that this teaching is so interpreted by most, if not all, contemporary leaders of religion. It should be made still easier by the fact that we are talking in this country of what are essentially two minorities. For whatever the statistical strength of our religious denominations may be reckoned at, the fact is that we live in a society in which the secular predominates, and in which the mutual attitudes towards each other of two groups of believers is not any longer itself an issue of the first importance.

Nor, of course, are we confronted here with the alternative of a closely thought-out system of secular or humanist philosophy; for many of those with whom political power now lies in modern industrial societies, Plato and John Stuart Mill mean as little as the teachings of the scriptures, and speak in an equally alien language. Where there is tolerance, it is largely the tolerance of indifference; as soon as it is challenged, the fragility of its foundations is demonstrated all too quickly. In other parts of the world there are important areas in which tolerance receives not even a theoretical approval and in which persecution is the norm rather than the exception, and violence endemic rather than sporadic.

Nevertheless it does not become Europeans to hold up their hands in horror at the lengths to which religious, political or racial intolerance has gone in newly independent countries on other continents. Nothing that has so far been done in the name of ideology outside Europe has even approached the events in Europe to which Dr Abba Eban devoted his own Waley Cohen Memorial Lecture four years ago.[2] I do not propose to harrow your feelings by entering once more into the horrors of the 'final solution'. Nor is it surprising that many people should feel that simply to rehearse

[2] Abba Eban, *The Final Solution: reflections on the tragedy of European Jewry* (Council of Christians and Jews).

them is of no advantage. What does seem to me worth saying once more is that what is surprising is that there should be so much resistance to the idea that understanding the causes of these events is an essential prerequisite of any hope that their repetition may be avoided.

It was as long ago as April 1962 that Mr David Astor proposed the setting up of an institute which should use the resources of the social sciences to study and examine the causes of such outbreaks of mass violence and in particular of large-scale massacres of which those perpetrated by the Nazis would clearly come first.[3] Since then much hard work has been put in on this proposal and the ideas for research put forward by Mr Astor have been turned into something closer to an academic blueprint. The difficulties we have found in seeking to go ahead – difficulties that now happily appear to be on the verge of being surmounted – have been financial. It is not that individual or corporate generosity are lacking but rather that the academic community itself has been slow in giving studies of this kind a high priority among many competing claims. A similar problem has faced those of us concerned with safeguarding the future of the great collection of records on the Nazi era in the Wiener Library and with assisting in developing out of it an Institute of Contemporary History.

It is a fact that a great deal of research into the problems of modern societies does go on in the universities of the western world; if the amount of it in this country is much less than what is being done in the United States, it is certainly not negligible. But it is hard to believe that much of this research can or should take precedence over trying to find out the possible explanation of events which have in themselves resulted in a massive retrogression in our civilization, and which, if repeated with the newly available techniques, might end up by annihilating humanity altogether.

I do not profess to know whether such research as has been proposed would either answer our questions about the past, or in fact provide protection against the recurrence of such events. It may be that the human mind cannot turn in upon itself sufficiently to disentangle the motives that lie behind human actions. But even

[3] Mr David Astor (who was chairman for Dr Eban's lecture already referred to) gave his address at a meeting in commemoration of the Warsaw Ghetto uprising. Its text was printed in *Encounter*, August 1963.

if the chances of advancing our knowledge were very slender indeed, the gamble would surely be worth while. We do not provide ships with lifebelts because we expect shipwreck: nor do we do away with them on the ground that shipwrecks are infrequent. The chances of finding anything useful to do on the Moon or Mars when we get there are very poor indeed; yet money is lavished on space research.

One is forced to one of two conclusions. Either our academic communities (including the great foundations) are themselves so appalled by the abysses of human depravity into which it would be necessary to peer, that they too would prefer to forget that these things happened and consign them to oblivion; or they are simply frivolous, preferring certain returns on easy and fashionable inquiries to a systematic and difficult confrontation with the greatest problem of all. Perhaps both conclusions are true. If so, we have here a field for moral suasion where Christians and Jews could perhaps find room for fruitful collaboration.

But I do not wish this evening either to appeal to your emotions or to make demands upon your pockets; my purpose is rather to suggest that the study of tolerance, and how to achieve it, is not something which can stand by itself but must be approached by asking broader questions about the societies in which we live. No creed and no anti-creed has an unblemished record; the reasons why heretics are burned in one century and not in another, why the materialists of today in the Soviet Union find it as difficult not to persecute believers as believers once found it hard to tolerate unbelievers or each other, are surely not to be found in the nature of the theological or philosophical positions that people hold. Nor of course is belief or the lack of it the only object of persecution, particularly in our own times.

Such facts are generally recognized, but the right conclusions are not always drawn from them. For it is too often assumed that because tolerance is a virtue in the thinking of people like us – or we would not be here this evening to commemorate someone who believed it was – it is a virtue which can be practised to the limit. Individuals can perhaps be tolerant to a very high degree, but even the most tolerant individual will rule out some forms of human behaviour as intolerable, and hesitate about the degree of freedom that ought to be accorded to expressions of the kind of belief that might justify such behaviour. It is not surprising that (with memories of the 'final solution' so vivid) we in the Jewish community should often urge stringent controls over the dissemination of racialist views. Some of the same people who take the lead in this find it hard to see any reason why other countries should regard the propagation of communist doctrines as calling for restriction, because of the overt consequences that have elsewhere followed from the imposition of communist rule.

Yet both those who would outlaw fascist beliefs and those who would outlaw communist ones would find it very hard to forgive people who persecute others on the ground that the doctrines they preached imperilled immortal souls. Some of those who say 'tolerance – but' are simply making choices between what they regard as tolerable because it does not affect them and what they regard as intolerable because it does.

These dilemmas have often been explored, and the lines of argument are familiar. What is important is not so much what the limits of an individual's tolerance may be – that is largely a matter of emotional and intellectual self-discipline – but what society's limits ought to be. No society can or should tolerate everything; all of us can think of forms of behaviour, and at their extremities even of forms of expression, which we would expect society to prevent, by coercion if necessary. And here again the argument is a perpetual one between political philosophers and between advocates of different schools of jurisprudence.

It is perpetual not only because almost all philosophical arguments are essentially timeless, nor because they cannot, unlike scientific arguments, be put very easily to empirical tests. It is perpetual because social evolution itself continually presents new situations in which it is not always easy to see how established rules apply. Just as societies are very different so too are the dangers they run; it is the essence of political genius to see what these are and to make provision against them. The difficulty is that the conventional wisdom by which statesmen and other public figures normally live is itself the product of the successful handling of the problems of a preceding era. They spend most of their energies perfecting solutions to old questions, and often miss newer and more important challenges to their ingenuity.

Sometimes, these failures are due to a complacent assumption that we are 'different', that 'it can't happen here'. It is not surprising in the light of general human experience that large-scale immigration where the immigrants are clearly marked out as different through physical appearance, speech and domestic and social attitudes and habits should place something of a strain on the host community. It would have been true anywhere; it was bound to be so in Britain. What is disturbing is not the emergence of a problem, or even the failures to deal with it on the practical levels of employment, housing, schooling and so forth, but the revelation of the abyss of prejudice that lies hidden behind a conventionally stable and seemingly firmly rooted social order. I am thinking for instance of the letters by people who still call themselves Liberals in the *Liberal News* in the summer following the publication of the report of a Party committee on the subject. What is liberalism worth if the reasoned attitude it enjoins can be swept aside so easily once prejudice is awakened?

We have in Britain been spared the violence that has marked the recent phase of race relations in the great cities of the United States. But it is clear from these experiences, and from those with our own 'mods' and 'rockers' on a thankfully much more restricted scale, that our industrial societies give no guarantee against endemic violence any more than the much and foolishly lauded hierarchical society of the Middle Ages avoided the bloody horrors of the jacqueries or the crusades. Even if we disregard the 'bomb' and all it signifies, it is clear that the crust of civilization on which we rest is a very thin one indeed. It is natural for modern man to expect to progress; sometimes one feels one must run as hard as one can so as to stay in the same place.

Let us then look merely in the crudest schematic outline at the course of human events in the western world in the past two centuries in the light of the famous revolutionary triad: liberty; equality; fraternity. It is a commonplace that from the eighteenth until well into the nineteenth century, it was 'liberty' that made the running. People who wanted to change things wanted above all to make room for creative, and indeed acquisitive, energies and for new ideas in many fields that had hitherto been cramped or confined by established hierarchies and ancient tyrannies. In so doing (as we now look back on them) they showed themselves astonishingly blind to the demands of 'equality', in the sense of their neglect of the impact of this release of energy on human beings who were the victims rather than the beneficiaries of the evolutionary process.

Even now it is difficult to imagine the state of mind that permitted people, most of whom believed in a just God, to exploit in so inhuman a fashion the working classes – men, women and children – of the industrial revolution, or the slaves of the American cotton kingdom. We look with horror too at the denial even of the simplest right – the right to survive – that was often the fruit of European and North American expansion. It needs a tremendous effort to believe not only that these things ever took place, but that sane and intelligent men had no difficulty in providing a rationale for what they did which satisfied their own consciences and those of most of their contemporaries. The mental world of early capitalism is as remote from us as the mental world of the Puritan witchhunters, or the mediaeval inquisitors.

Not only of course do we look down upon these people and even feel some vicarious remorse – for we are, after all, their heirs – but we do what we can to make good their failings and crimes. Instead of 'liberty', which has only an apologetic following among modern intellectuals, we exalt 'equality'.

Now of course much that is done in the name of 'equality' is good, in itself; just as much of the clearing of the ground in the name of liberty was worth while – the Bastille deserved destruction. We are now attempting to construct a welfare state on the basis of equality of access to at least the necessities of a decent life: food, housing, medical care. Because the indignities inflicted on the working class in the early stages of the industrial revolution were possible in large part because of their own lack of organization and consequent disfranchisement, we accept and applaud the privileges of trade unions and the advent to power of parties claiming to represent the working class as such.

Just as we repent of capitalism and legislate at home in the name of 'equality', so we repent of imperialism and welcome the advent to total independence of any community, however immature and unfitted for it, provided that it can claim to speak in the name of the equality of peoples. Indeed, we go further since we excuse, in the case of such countries, behaviour both internal and international that we would feel obliged to condemn in the case of European states.

Indeed, because we have feelings of guilt about the earlier neglect of the demands of 'equality' by the protagonists of 'liberty', we are now prepared to accept infringement of 'liberty' itself when these take place in the name of 'equality' – whether it be violations of individual rights by powerful trade unions or the harrying of minorities by the dominant elements in newly enfranchized states.

But what, one may ask, will the historian think, looking back on the mid-twentieth century? What values are we neglecting, as obviously as the ruling classes of the early capitalist era neglected values that are quite clear to us? In a sense, we cannot know the answer to this question. For it is part of the case that I am putting to you that in these questions hindsight is the only guide. But we can speculate.

It would be temptingly symmetrical to complete the triad and say that if the eighteenth century neglected 'liberty' and the nineteenth century neglected 'equality' so the twentieth century shows a blindness to the demands of 'fraternity'. And if by 'fraternity' we mean an extension of human sympathy without limits of race, creed, age-group, or class, it would not be difficult to put up a powerful case for this view. Future generations may find it difficult not only to understand the major horrors of our age to which I have already alluded, but other phenomena in respect of which the demands of common humanity have been lacking. We have, for instance, obliterated in warfare most of the distinctions between the combatant and the non-combatant which previous generations have believed to be an essential mark of civilization. Both sides in the war of ideologies have committed crimes against humanity; which crimes one is indignant about usually depends upon which side one is on. We do not in fact, although we do in theory, make as much of the sufferings inflicted upon men of coloured races as we do when the victims are white; the sufferings inflicted by the coloured on the coloured interest no one but the victims.

In our domestic societies, we permit certain groups – the old, the immigrants – to get much less than a fair share of rising standards of living, while we lavish purchasing power upon heedless adolescents. Our sense of solidarity as a national community is weak when sections of it still allow themselves to pursue limited interests with an overt disregard for the conflicting interests of others or of the nation as a whole. While decrying individual acquisitiveness, we assent to acquisitiveness on the part of powerful groups and are unconcerned about the corresponding sacrifices demanded from others or from the general community. People who call each other 'brother' may not in fact practise fraternity.

Finally, we shall surely readily come to regard the international economic arrangements of our world with the same shocked surprise with which we contemplate the internal economic arrangements of early industrial societies. We shall come to regard the existence of food surpluses alongside great areas of the world which are at or over the starvation limit with as much horror as that with which we now recall the nineteenth-century contrasts within single societies, and find it difficult to understand the archaic trading and financial arrangements that can lead to surpluses accumulating in a world of want. I suspect too – though here I am on more delicate ground – that humanity may also come to look back in amazement at the insistence of men (in the name of religion) on placing obstacles in the way of controlling the excessive growth of population that is at the moment perhaps the principal cause of misery over much of the world. In any event, a case can be made out for saying that our descendants, if we have any, will regard a lack of the spirit of fraternity as an obvious feature of mid-twentieth-century existence, and whether they condemn this in the name of the common fatherhood of God, or in the name of some secular philosophy, they will certainly find it hard to comprehend.

But I am not certain that this is in fact the most profitable way in which to look at the weaknesses of our society. We can explain some of the lack of tolerance that we find in terms of an insufficient sense of fraternity, but that is by no means the whole story. What is perhaps more disturbing, and indeed more sinister, is the lack of any general understanding of where the idea of tolerance fits in to the general picture people should have of a civilized life. What we are witnessing today in many forms is not so much an attack upon particular ideas, or the institutions or groups that uphold them, but a general unconcern with certain attributes that are essential to civilization as such. It is this unconcern that I call the challenge of barbarism – and it is to the further exposition of this idea that I wish to devote the rest of my time this evening. Even so, I can only indicate what I have in mind and give some very tentative examples.

It would seem clear that civilization itself is the product of a perpetual and fruitful tension between tradition expressed in established hierarchies, institutions and codes of conduct on the one hand, and the spirit of innovation on the other. If innovation is inhibited by an intolerance of new ideas or by too great a resistence to their implementation then stagnation ensues; if on the other hand, innovation is too rapid, if existing patterns of conduct are subjected to change at too rapid a rate, then one risks a general breakdown in existing value-systems and a rapid alternation between absolutism and anarchy. Such reflections are common to most people who have thought at all about what can be learned from human history. But the utility and applicability of the lessons to be derived from such reflections is of necessity very limited. We can see in retrospect that change was too slow at one point and too precipitate at another, but we conclude that this was so from observing the actual consequences of what was done or left undone. We have no means of devising measuring-rods to enable us to say what is happening in this respect to our own society, where judgements are apt to be wholly subjective and reflect for the most part the individual's sense of whether he or his like are gaining or losing most by the current rate of change. If one were to ask people the question: do you think things in Britain today are moving (a) just fast enough, (b) too slowly or (c) too fast, one could probably forecast the trend of people's replies according to criteria of occupation, age, political allegiance and so on.

Where barbarism enters from one angle is where the question itself is dismissed as meaningless, where it is suggested that change is possible at any speed and without any limit in accordance with the unmediated effects of technological advance, where the whole idea of a creative tension is ignored in favour of a purely mechanical conception of progress, and where all existing institutions are regarded as expendable if they impede innovation. Such shallow thinking is evinced in our own day by the clamour for giving 'science' – that is, the natural sciences – a more and more important place in our educational system. It is not that science is not itself one of the major ennobling occupations of which man is capable – ranking only a little below the creative arts – nor that science has not got many further ameliorations of our material lot to offer. What is wrong and barbarous is the view that science as such provides an adequate preparation for dealing with human problems, including the problems that science itself creates. There is no evidence that physicists or chemists have a sensitivity greater than that of any moderately literate citizen to the human problems we have been discussing, and other such basic issues, and some evidence that their sensitivity is rather less. It is a matter of common observation that can be derived from the study of the signatories to almost any political manifesto of an extremist nature, that scientists are among the most gullible members of western society in everything outside the range of their own competence. The false prestige of men of science in matters outside their specialized range is a symptom of barbarism. But this would not matter so much were it not for the fact that the influence of the scientific approach is so often to minimize the real difficulties in the way of keeping our civilization afloat at all. Civilization is a fragile plant; the more complex, the more fragile. All mechanistic analogies miss the point completely.

Attitudes of this kind are often displayed by those who profess an undivided allegiance to 'science'. Our educational institutions come under attack because, so it is claimed, an insufficient place is given in them to the claims of 'science', meaning by this the natural sciences. And it is argued that if the 'scientific spirit' were more widespread most of our human and social problems would disappear. Such naïveté is, I hardly need say, rarely shown by scientists themselves, or at any rate rarely by those who stand in the first rank of their profession. They are well aware of the close relationship between the human environment and the progress of science itself, and still more of that between scientific discovery and the applications of which it may be capable on the one hand, and the general social framework on the other. But they do not imagine that scientists themselves are particularly fitted to solve the social and political problems involved, or that their training often gives them any special qualifications for doing so.

In any healthy civilization the relations between those who are extending the boundaries of knowledge and the remainder of the community are of course very important. And no one would claim that the relationship is at present a wholly satisfactory one anywhere in the western world. But it will not necessarily be improved by insisting that the natural sciences monopolize all the best brains in the younger generation. Least of all should major educational changes be justified on the assumption that certain subjects are more 'practical' than others. No doubt there are many fields of practice that are undermanned, and a wise society will take appropriate measures of a non-coercive kind to fill the gaps. But many of those who raise the cry of 'more science' have not really thought the matter out. In their use of 'science', it is lowered from being one of the main vehicles through which we learn to understand the external world and to alter it for human purposes. It becomes instead a kind of magic able to deal at will with any human situation. The debasement of science into magic seems one clear indication of a relapse into barbarism.

As a footnote to this one might add the deleterious effect that this conception of science has on the humanities themselves. It is increasingly asserted that the study of human affairs is only valid in so far as it can make use of the methods and language of the natural sciences. Up to a point, there is of course much of value to be gained from mathematical techniques: demography is an obvious example; economics, in some of its aspects, another. But the disciplines in which such techniques are appropriate all seem to lie on the borders between biology and humane studies. Elsewhere – as in the talk of 'quantification' in such subjects as political science or international relations – the imitation of the natural sciences merely darkens counsel. It is not that the subjects demand the techniques, but that it is thought that the techniques make the subject respectable. Again we are stumbling from science into magic; barbarism once more.

The general effect of this misuse of the idea of science is to minimize the depth, and to distort the nature, of the problems that we face. It is not just a question of marginal improvements to civilization that is at stake; whether our civilization is to survive at all is always an open question.

A quite different aspect of the challenge of barbarism arises directly from our preoccupation with equality. Because if it is important not to overlook the importance of continuity and hence of the established institutions within a civilization, it is even more important to emphasize that their value is only as a counterweight to its motive force, namely innovation. And innovation, though it may owe something to social organization – and rather a lot in the case of modern science and technology – is above all the result of the work and impact of individual human beings, of unusual gifts and qualities.

The most critical challenge of modern barbarism is the denial of the importance and sometimes even of the existence of individual excellence and of the dependence of the rest of humanity upon such excellence, whether of the preacher, the teacher, the artist, the scientist or in some practical sphere of life. Indeed, the most important of the classical arguments for tolerance is the fear that to insist upon conformity might rob one of the vital message of an innovator in thought whose capacity for innovation derives precisely from his uncommon qualities. Democracies, as Tocqueville noted, are in themselves essentially suspicious of anyone who stands out from the common ruck of mankind, and will generally tend towards enforcing a high degree of conformity. To some extent, this observation might seem belied by subsequent history and by the apparently wide range of ideas to which our modern democratic societies normally give hospitality. But if one looks at the matter more closely one is led to believe that what they tolerate so readily are ideas that do not worry them unduly.

For one very important idea which gets less and less of a hearing is precisely the idea of different degrees of excellence as a factor of differentiation between human beings. Once again the educational scene offers a good vantage ground for observation; a society's educational philosophy and practice are as good an indication of its basic philosophical allegiance as one can easily find. We have progressed in this country in a very short time from the view that equality demanded that differential access to education should no longer be affected by social position or parental income to the view that differential access is wrong and unnecessary in itself, that 'equality' demands that the clever and the stupid, the industrious and the idle should be educated in the same schools and as far as possible in the same classrooms. Such arguments as can be put forward in favour of this approach are entirely concerned with doing justice to the less able child, with preventing him from having a feeling that he is suffering from discrimination. No one looks at it from the point of view of the more able child who may be held back and frustrated by being kept at the pace of slower brains, or diverted from cultural pursuits by the ubiquity of 'pop-culture' so-called in his environment. And it is of course easier to take this line in view of the fact that clever children and their parents are in a minority, and that the really exceptional children upon whose talents society's hope of innovation rests are even rarer.

Were such positions only adopted by the less well-educated sections of the population and motivated by mere envy, they would be easier to understand and condone. Envy, which used to be condemned by moralists, has now acquired considerable standing as the backbone of a fashionable political creed. But this is not altogether the case. On the contrary, parental ambition, that great motor of progress, has not yet been stamped out in the democracies, nor for that matter in the 'people's democracies', nor even as far as we can see in the communist fatherland itself. What is worse and more inexcusable is that the campaign against giving its proper place to excellence in education is led by intellectuals who have themselves very often achieved their positions of power or influence through the exploitation of educational advantages that they have enjoyed themselves, but are determined to deny to others. But this is only part and parcel of the treason of the intellectuals that is at the root of every society's decay. Not only are we told by the fashionable pundits that we need make no special provision for excellence where individuals are concerned, but that there is no hierarchy of values among the subjects of an educational curriculum any more than among the different forms that the arts may take. If one believes that the Beatles are no worse than Beethoven – or at least deserve equal attention from critics and Prime Ministers – then why not say that woodwork or folk-dancing are on a par with Latin grammar or calculus?

We have had a reversal of snobbery; dukes once looked down upon dustmen; dustmen would certainly now be expected to look down on dukes. By the principle of utility this can certainly be justified. But it becomes dangerous when the reversal is not one of snobbery but of intellectual values; when one has to plead not that the intelligent should refrain from turning up their noses at the less intelligent, but that society should give to the minority of the excellent that same toleration and scope that every other minority demands for itself.

Some may feel that I am concealing behind an abstract argument some deep bias of an ideological or even party kind. I do not think so. The tendencies I have been describing and deploring seem to me traceable in all parties and among the professional advisers in the Civil Service and elsewhere upon whom all parties depend. If the so-called conservatives seem to be pursuing some of the policies that follow from such tendencies, for instance in education, with less determination than their rivals, this is due rather to a greater degree of inertia on their part than to any conviction that the fashionable ideas are false in themselves.

My conviction, in brief, is that a combination of a mistaken view of the proper role of science together with an inflated egalitarianism are the principal forms that the challenge of barbarism takes today in the West, that we are in consequence dangerously near to living wholly on our cultural capital instead of adding to it, and dangerously close to accepting an obsolescence rate for our institutions – particularly but not by any means wholly in education – more rapid than society can stand.

To say this kind of thing is to be at once the target of abuse. And the current intellectual orthodoxies are no more tolerant than their predecessors. It has become in some quarters the fashion to deny even a hearing to one's opponents, and in the younger generation this tremendous confession of weakness is actually encouraged by those who should know better. Self-expression that consists in shouting people down is meaningless, and the sad experience of this country's universities with the so-called 'teach-ins' last summer is a portent worth noting. However, in a lecture dedicated to the idea of tolerance, there seems something appropriate in the lecturer concluding by demanding toleration at least from this audience for an expression of views about whose basic unpopularity at this juncture in our affairs he has no illusions at all.

4 England in Transition[1]

To write the history of one's own country in one's own time – Mr Taylor was a boy of eight in 1914 – demands a combination of qualities given to few historians; it is a measure of Mr Taylor's gifts that not only is the challenge superbly met, but that one can think of no one writing today who could conceivably have done the job better. Mr Taylor refers in his preface to having been 'slighted in his profession'; one can only comment that were it not for the fact that professors as such are not very highly regarded in this country – thank God! – it would indeed be a disgrace that Mr Taylor has never held a chair at his own university. As it is, he can afford to rest on his achievements while lesser men rest on their adventitious prizes.

Many possible ways of approaching the writing of contemporary history present themselves; one can accept the standpoint of one or other of the great bodies of opinion into which a modern democratic nation normally divides; one can attempt a 'middle of the road position' which normally means accepting that what was done, was done for the best, and that in the long run all men of good-will travel together; one can try to avoid the problem altogether by regarding one's task as that of a chronicler rather than that of an historian, accepting as important and worthy of record those things that bulk largest in the obvious sources; or one can try to impose upon the material the definite imprint of one's own personal philosophy, and make judgements that accord with its dictates.

People familiar with Mr Taylor's writings will expect him to have opted for the last of these and they will be right. One has only to take Mr Taylor's 'heroes', George V, Ramsay McDonald and Montgomery for instance, to see that his view of the immediate

[1] Review of A. J. P. Taylor's *English History 1914-1945* (Oxford, 1965).

past will hardly fit into the current mythology of the Labour Party; on the other hand his clear preference of Lloyd George to Asquith, his lack of respect for such figures as Balfour and Smuts, his highly critical treatment of Churchill's war-leadership, his less than adulation of Keynes – all these make it certain that no section of the 'Establishment' will now hurry to take him to its bosom.

Mr Taylor is equally oblivious of current intellectual fashions when it comes to deciding as to the proportions to be given to different elements in the period. Little is said about science except when defence is directly involved and not much about other branches of scholarship; the analysis of changes in British society as expressed in the categories of the sociologist does not interest him; the development of the administrative machine central and local is not a topic much to Mr Taylor's taste. For these reasons the book will not be used (like some other volumes in the series) as a work of reference in which one can be sure to find through the table of contents a brief but reliable account of whatever it is one wants to know about. On the other hand, it is (what most of the others are not) almost compulsive reading.

For in this sense, and in this sense only, Mr Taylor has in him the trace of the chronicler – he is telling a story in the most straightforward way; the hero happens to be a collective person – the people of England – not, be it noted, of the British Isles or the Commonwealth – and not an individual, but the method is the same, to concentrate on successive events in time and to bring in description and analysis only when required to explain what happened. It is the hardest kind of history to write well; and far the most rewarding to read.

But it is of course a story told from a highly individual viewpoint. Since all historical writing is of its nature subjective there is not reason to be put off by this fact. But there are two conditions to be met if danger is to be avoided; the point of view must be consistent in itself, and it must be overt. Mr Taylor is, I think, on the big things consistent. This does not prevent him from time to time from allowing his strong predilection for the paradox or the *bon mot* to carry him a little too far. And the danger here is that he may sometimes be unfair to a cause or to an individual.

Let me take one example of each. Mr Taylor maintains the thesis (if in a less provocative fashion) that he first advanced in his *Origins of the Second World War*, according to which Hitler did not actively contrive his successive aggressions as part of a master-plan, but took advantage of circumstances as they presented themselves in order to fulfil his general purpose of recovering and improving upon Germany's pre-1914 position in Europe. I have always held that Mr Taylor is perhaps nearer the truth here than some of his more vociferous critics.

From this standpoint Mr Taylor proceeds to criticize those who from the very beginning of the Nazi regime believed that it was inherently evil and dangerous and would need to be forcibly opposed, and those reporters on the German scene who gave this picture substance. In Mr Taylor's view, 'The Nazi dictatorship was no worse than that in some other countries, particularly that in Soviet Russia.' Now it is undoubtedly true that a case can be made for saying that the ordinary German did better under Hitler between 1933 and 1939 than the ordinary Russian was doing under Stalin. But that was not the whole of the matter: Mr Taylor himself goes on to say that what put Englishmen against the Nazi regime was its treatment of the Jews; but he adds 'here again, Jews were treated as badly in other countries, and often worse – in Poland, for example, with whom nevertheless, Great Britain remained on friendly terms'.

From these statements, Mr Taylor goes on to make two points; that there was more indignation in England about dictatorship in Germany than in Russia, because more enlightenment was expected of the Germans (and in this there may be an element of truth); and secondly, that the Jews of Germany got more sympathy than those of Eastern Europe, because more of them were wealthy or cultivated or had personal contacts in other western countries. And no doubt there is an element of truth here too. But the main point is that the 'alarmists' were right and Mr Taylor is clearly wrong.

No doubt anti-semitism was endemic in Eastern Europe and productive of much suffering. True, the nationalist anti-semitic parties of Poland and other countries might wish to impose limitations upon the professional and political rights of Jews; they might in extreme cases envisage some way of getting them to leave the country altogether; but no movement of this kind envisaged mass extermination of the Jews as the 'final solution', and no such genocide was in fact practised except under direct German command or inspiration. Nor indeed did the Nazi ideology limit the functions of mass extermination to getting rid of the Jews. Russians, Poles and Ukrainians went the same way, and would have gone to their deaths in millions likewise, but for the German military defeat. There was in their attitude to the human personality itself, to the very idea of the dignity of man, a real gulf between the adherents of the Nazi creed and the rest of Europe in the inter-war period and if some people divined this important and tragic fact despite the seeming 'moderation' of the early years of the Nazi dictatorship, they seem to merit praise rather than scorn.

The second example comes from the most original and interesting parts of Mr Taylor's account of the 1930s. It has always been difficult (owing partly to the problem of sources) to relate the growth of British armaments in the period to the content of British diplomacy. And this has made it difficult to make any but the most provisional judgements on the merits or demerits of 'appeasement'. Mr Taylor now suggests, and gives some evidence to support the view, that in fact the question itself is a false one in that once the basic decisions about re-arming had been taken, their implementation was a semi-automatic affair of the services themselves, based on rule-of-thumb assessments of the programmes of other powers, and was not clearly related to particular decisions in the field of foreign policy or estimates as to the occasion and nature of a future conflict. On the other hand the choice of the direction in which to re-arm had an inevitable effect upon policy since the concentration upon the bomber aircraft actually made Britain more rather than less isolationist.

One would of course require before accepting the whole of Mr Taylor's thesis a good deal more evidence than one has; how correct is he in his flat assertion that the change in policy signalized by the 1935 White Paper was the result of pressure from civil servants and service chiefs upon reluctant or indifferent ministers? Here, as so often, the absence of source-references (generally understandable in a book of this kind) becomes rather frustrating. But the point I want to make is that this is on Mr Taylor's own showing a field in which individual contributions positive and negative were exceptionally important. It is therefore significant that Mr Taylor should write of the replacement at the Air Ministry of Lord Londonderry by Lord Swinton that the latter was 'soon also to be as discredited as his predecessor'. Whatever may have been the public view at the time of Lord Swinton's record, it ought surely to be said that from the point of view of those intimately concerned (Sir Henry Tizard for instance) Lord Swinton was the one minister of the period with both the grasp and the energy necessary for the job.

To return to the more general issue, it could be argued that if Mr Taylor is on the whole consistent in his standpoint, that standpoint is not made sufficiently overt for the unwary not to be misled. And it may well be that for the undergraduates and sixth-formers who are inevitably going to form so large a proportion of Mr Taylor's readership the method of staking everything upon narrative has its disadvantages. After all a lot of people who should certainly have known better have taken his *Origins* as being an exculpation of the Nazi record. But for the reader who is prepared to take the new book as a whole and to consider its implications in full, there is little excuse for error.

Mr Taylor's attitude is a personal one but not an eccentric one; that is to say it is a recognizable variant of a general school of thinking about English society and politics and about historical development generally that has its roots in the radicalism of the nineteenth century. It involves maintaining a number of propositions which do not perhaps form a consistent political philosophy but which are certainly connected in the sense that they appeal to men of the same temperament.

In the first place, Mr Taylor, writing about a collectivist age, sharing its view that society has a duty to its less fortunate members, regarding therefore the development of the Welfare State as something to be rejoiced in, and the degree of all-party agreement attained on this subject as itself both inevitable and desirable – so that John Wheatley and Neville Chamberlain figure together in his improbable pantheon – remains at heart an individualist. Unlike much of the British Left whose hero (not Mr Taylor's) was Sir Stafford Cripps, Mr Taylor lacks both a puritan and a dogmatic streak. He believes that society and government should be judged by the extent to which they permit 'the pursuit of happiness', but that it is for individuals to judge in what their happiness consists. For this reason, Mr Taylor has no sympathy for the 'planners' and rejoices when he can find evidence of their discomfiture.

While the weight of 'informed' opinion was on the side of measures intended to revive the old export industries and the old industrial areas, Mr Taylor is on the side of those who voted 'with their feet' in favour of the newer industries, the new consumer-oriented society, the new suburbia. Looked at from the point of view of the statistician and the economist, Britain was a poorer country after the first world war than during it; but the people themselves felt better off, behaved as though they were better off and (except in the areas of continuing unemployment) were indeed better off than they had ever been before.

In the second world war, with unemployment at an end and a feeling of mutual responsibility transcending social divisions, there was (for all the suffering and anxiety) a further move towards the kind of life that most people wanted for themselves.

In other words one has here an approach to the history of England in the period that is a long way indeed from that of both the Fabians and the Marxists between whom the intellectual Left has mainly divided its allegiance. Without concealing from himself or his reader the fact that great disparities in wealth still existed or that important elements of privilege had not been eradicated, Mr Taylor uses as his measure of progress the simple test of whether more people were getting what they wanted, and finds it answered in the affirmative.

Similarly on the cultural side, Mr Taylor does not regard it as the duty of the State to decide for its citizens between push-pin and poetry. If people prefer watching football or going to the cinema to attending symphony concerts or highbrow drama, that is their look-out. Mr Taylor has little use for the school of thought represented by Lord Reith. Indeed in direct contrast to the puritan 'uplift' school, he is prepared to allot a considerable share in human happiness to the simplest pleasures of all – the development of efficient contraceptive devices seems to him to have done more for human happiness (and particularly for the happiness of the female half of the population) than most of the more publicized achievements of the age.

But to appreciate Mr Taylor's viewpoint on the social history of the period we must make allowances for quite another element in it, namely the relatively unimportant role that Mr Taylor assigns to people like himself – namely to the intellectuals. As I have said, intellectual history bulks less large in this volume than in most of its predecessors – though occasionally Mr Taylor makes an excellent point, as for instance when he calls attention to the fact that though educational developments have been studied in considerable detail, no one seems to know what it actually was that children were taught or how far schooling shaped their later attitudes.

For Mr Taylor, there are two principal components of English history in the period – there is the great mass of the people, earning its living, seeking its varied forms of personal satisfaction, and taking little part in making the great decisions except when deciding to follow (or more rarely not to follow) its accepted leaders. On the other side, there is the leadership making its efforts, often muddled and unsuccessful, to keep up with the changes in the national and international environment, seeking for the support it requires to keep going, competitive as to the relations between groups, parties and individuals, and yet basically a single entity – in which the 'outs' of the early part of the period – the trade-union and Labour leaders – are by the end of it as much part of the system as the plutocrats, once also a group of 'outs', had earlier become. It is perhaps the perception of the broader significance of this development (though he does not say so) that gives edge to Mr Taylor's silhouette of the Beaverbrook-Bevin struggle in the wartime Churchill cabinet. On the one side a Canadian-born press-baron, on the other side the product of the in-fighting of the Labour movement, and presiding over them a maverick aristocrat.

For those in between – the providers of ideas and the managers of society's institutions – Mr Taylor has little time to spare; those who read the *Daily Express* and those who are too grand to read at all triumph over those who read *The Times* or even a *Manchester Guardian,* already declining from its provincial vigour. The attitude is perhaps best exemplified in Mr Taylor's cavalier dismissal of the idea that this was a period in which civil servants were acquiring a new importance (except in relation to preparations for war) and his silence on such questions as the role of the Treasury under Warren Fisher or its contested relations with the Foreign Office in the 1930s.

In rather the same way both constitutional history and the history of political parties seem marginal to Mr Taylor's interests except to the extent to which successive general elections determine who shall hold office next. The idea of political parties as the connecting link between the rulers and the ruled does not seem greatly to interest Mr Taylor; nor indeed does his interest in the movement of population extend to considering its effects upon the electoral map. The society he pictures is a somewhat inarticulated one, not an active and throbbing democracy with its myriad of voluntary organizations; it is, as he sees it, highly individualistic. Whether this is closer to reality than the alternative and more common view is a problem that might certainly engage the social historian, but it does undoubtedly represent a thoroughly consistent and thought-out attitude to the period under discussion.

My own view is that Mr Taylor's best chapters are those on domestic history; partly no doubt because on international affairs he is bound to go over ground that has become very familiar to him, and therefore occasionally to take short cuts; he says for instance that the Nazi-Soviet pact contained 'secret clauses limiting the gains which the Germans could make in Poland'. But, of course, the secret protocol went much further than this, in arranging for a virtual partition of Eastern Europe; and this is of importance for interpreting later events. Again there are aspects of foreign affairs that Mr Taylor finds uncongenial and therefore allows himself to get slightly wrong. Thus although he rightly recommends in his bibliography – itself an invaluable survey – Mr Leonard Stein's book on the Balfour Declaration, he obscures one important point of controversy in relation to this document when he says that the Declaration 'recognized Palestine as a national home for the Jews'; what it did was to talk much more ambiguously of 'a national home . . . *in* Palestine'. Again while there were no doubt those Zionists who looked forward to some early form of statehood, it is not the case that in the early 1920s the Jewish immigrants 'made no secret' of their intention to turn the national home into a state. It took nearly two decades before Dr Weizmann accepted the need to cut the connection with Britain that had been the basis of his political strategy.

On the other hand, there is no doubt but that most people will turn to Mr Taylor's book precisely because of the space he devotes to foreign policy. In part this will be due to the fact that the only other comparable general history, Professor Charles Mowat's *Britain between the Wars,* is much stronger on domestic than on external issues. For this reason it is again essential to realize the point of view from which Mr Taylor writes; and here the problem is rather harder. Most historians of this subject are either traditionalist exponents of a power-oriented system, in which case they tend to accept the validity of the national interest as defined at the time of which they are writing, or they are pacifist-inclined or internationalist utopians extending a basic revulsion against war to an emphasis upon the real or imagined possibilities of transcending armed conflict by institutional means.

Mr Taylor belongs to neither camp. In accordance with his general radical standpoint he has no doubt but that peace is a more desirable state than war; on the other hand, he knows that in fact wars have played a decisive role in determining the fate of nations, and that the historian cannot avoid studying them as something more than mere aberrations. Unlike most left-wing historians, he is therefore strong on the campaigns themselves, and on the relevance to the changing balance between states of developments in weapons. He is thus untouched by the myths of the inter-war period – the chances allegedly missed in disarmament or 'collective security'.

From the point of view of Britain, Mr Taylor is faithful to the 'Little England' view that he expounded in his book *The Trouble-makers*. He does not respond to the mystique of empire; indeed one criticism might be that he underestimates the extent to which his 'England' is a fiction, the extent to which, that is, so many people in many strata of society had experiences or ties that made them feel personally affected by what went on in other parts of the Commonwealth or empire. As for Europe, he does not regard it as self-evident that England's interest resided in upholding the Versailles settlement despite the irrelevance or inexactitude of many of the criticisms levied at it. The shape of Europe would depend upon the degree of power that could be exercized by the different forces playing upon it, and he is sufficiently aware of the Commonwealth and American aspects of the outlook of the British ruling elite to know how their reluctance to become involved fortified the basic popular desire to avoid a repetition of the 'western front' of the first world war.

Seen in this perspective the conspiracy element that some historians have professed to find in 'appeasement' is reduced to its proper proportions, and its relationship to the main elements in the British political spectrum made clear at last: 'Appeasement', he writes, 'never sat comfortably on Tory shoulders. It was in spirit and origin a Left-wing cause and its leaders had a Nonconformist background.'

It is because of these basic insights that Mr Taylor can so successfully expound both the illusions that went into the claim that 'collective security' was an alternative to a national policy of resistance to the challengers of the *status quo* and the essential illogicality of the way in which British policy changed in the course of 1939. It was, as Mr Taylor shows, the House of Commons that 'forced war on a reluctant British government'. The British people, unconsulted, 'accepted the decision of parliament and government without complaint'.

On the other hand, once in, and still more once alone, the British people accepted unquestioningly the logic of this decision and turned victory over Hitler into a national objective. Given the nature of the enemy, Mr Taylor himself cannot forbear to cheer.

But, of course, the idea of defeating a Europe united under German domination through British might alone was an illusion – a Sorelian myth. In fact only new alliances could make meaningful the decision to fight on in isolation. And once the Russians and the Americans were in also, although victory was almost certain, its consequences were obscure, Britain in this respect was divided. For the man in the street, the Russians were the heroes, bearing on their shoulders as they did the great burden of the fighting on land; the rulers, Churchill above all, were primarily concerned to establish the closest links with the United States. Mr Taylor is the victim of neither of the illusions that these enthusiasms in turn involved. He shows how little ground there was for believing that Soviet Russia was in the war for Britain's reasons or with a similar outcome in mind; his 'Left' did not 'speak to' that 'Left' in the same language. And he is well aware that the American view of the post-war world could never be that of Churchill.

Where Mr Taylor breaks more controversial ground is in his estimate of the relationship between these facts and British grand strategy. While he accepts the view that the Russians were out to ensure their own security as they defined it, and that the Americans were prepared to use Britain's plight to enforce their own views on economic policy and improve their own competitive position, Mr Taylor does not think that the actual conduct of the campaigns was influenced by considerations relating to the power-politics of the post-Hitler world. In particular he denied (in one of those provocative notes with which so many chapters end) that Churchill's hankering after Mediterranean or Balkan campaigns had anything to do with forestalling the Russians. It would, he argues, not have helped to contain Russia even if the Americans had agreed to the use of resources for this purpose. Indeed, he goes so far as to say that a Stalin really bent on reaching the Rhine rather than the Elbe would have encouraged the British and the Americans to tie themselves down in the Balkans.

In Mr Taylor's view, it was not prescience about the future but nostalgia for chances missed in the past that affected Churchill's military judgement; to do 'Gallipoli' again and get it right. And this view – right or wrong – is again consistent with Mr Taylor's general view about statesmen and statesmanship – that the images of the past are, whether for good or evil, more influential than calculations about an unknown and always unexpected future.

In the event, as Mr Taylor shows, it was America, the one of the three major victors that suffered least, that gained the most – becoming the dominant world-power as a by-product of British and Russian resistance to the Nazis. As an anti-imperialist, Mr Taylor cannot repine at one result of this turn of events, the acceleration in British disengagement from overseas responsibilities; as one who refuses to feel superior about the spread of affluence, Mr Taylor cannot reasonably object to the material aspects of the increasing Americanization of Britain. And yet in some respects his point of view remains very little affected by the Americanization of Britain's intellectual life that has been the consequence of Britain's new dependence upon the United States. Alone among leading British scholars, Mr Taylor has continued to profess attitudes more European than American, to hold out against joining the Harvard-Princeton-Berkeley circuit that keeps most of them almost as busy in the United States as they are at home. His History remains concrete where that of the Americans looks to abstractions; his language is simple and direct without any genuflections in the direction of new concepts derived from other disciplines. In his impatience of humbug, especially of lofty humbug, in his radical warmth and unashamed humanity, Mr Taylor as an historian comes closest to those things which are in the end the most admirable things about the English. I can think of no book that more successfully challenges current academic snobberies, none less likely to be approved of by those prim young men who treat history as Haydn treated music, something only to be written in a frock-coat. But with all its faults this book is a masterpiece; and that is not something that one comes across every day of the week.

II GOVERNING BRITAIN

5 *The Machinery of Government*[1]

The phrase 'machinery of government' may have a somewhat dry and academic sound, but any consideration of the subject takes one into the heart of many of the problems that our society faces, and raises in an acute form many issues about which deep and genuine divisions of opinion undoubtedly exist. Indeed the reason why the plethora of inquiries into aspects of our constitutional and administrative arrangements have not included one into this central topic may be attributed to the fact that its political implications are so considerable. It is after all more than half a century since the famous Haldane Report on the subject.[2] An enormous volume of additional responsibilities has been taken over by the central government in that period but the machinery for coping with them has been a matter of mere improvization governed by no generally agreed doctrine and subject to very frequent changes.

If one looks at these changes it is often apparent that the reasons for them are as much political and personal as the product of serious inquiry about the most economical and expeditious manner of handling particular blocs of government business. Sometimes indeed we can stand back from some particular process of change and see how it conforms to the logic of some new development or developments in the subject matter handled by a department or group of departments; and here one can derive assistance by looking at broadly similar trends in other countries. If we take, for instance, the sphere of defence organization – and national defence is the oldest and most important function of central govern-

[1] Lecture at the University of Newcastle-upon-Tyne, 22 January 1970.

[2] *Ministry of Reconstruction. Report of the Machinery of Government Committee* (Cd 9230, 1918).

ment – we can see that the successive reorganizations of the last half-century follow upon changes in weapon technology and consequently in the raising and disposition of the country's armed forces. The introduction of the aircraft in the first world war resulted in the creation of an Air Ministry alongside the historic departments, the War Office and Admiralty; but for some years it was not clear whether this arrangement would persist, or whether Britain would follow the American pattern of allocating part of the air arm to the ground forces and part to the navy. In this case the British model was eventually followed by the Americans who created their own Secretary for Air in 1947. But almost at once in both countries a movement towards a greater concentration of authority in defence matters began to gather momentum, as the interdependence of the services, brought out by wartime experience, was made even more manifest by the new technology of nuclear weapons and ballistic missiles.

The development in Britain of the Ministry of Defence, not as a mere piece of coordinating machinery but as the department responsible for all three services, with its internal organization an increasingly functional rather than a service one, while still incomplete, has now gone too far to be unscrambled, and the unusually long tenure of its present incumbent has made possible a fairly systematic approach to the whole problem, just as in the case of Mr Robert McNamara's long reign at the Pentagon.

It is likely of course that the long-term view which suggests a rational and continuous appraisal of the demands of the new era in defence matters as a basis for the reorganization of the machinery of control conceals short-term conflicts of interest and opinion between the services in recent years no less intense and hard-fought than those which are now so well-documented for the inter-war years, or even the pre-1914 period. No one who has studied the works of Professor A. J. Marder or Captain Stephen Roskill could subscribe to the view that the Royal Navy is the 'silent service'. It is also true that the present arrangements may not necessarily be the most efficient possible. One does not have to take everything Professor Parkinson writes absolutely literally not to wonder at the enormous size of the Ministry of Defence as compared with the size of the armed forces of which it now disposes. But this is perhaps more a matter of management than of machinery. All in all, the record in respect of defence does seem to make a good deal of sense.

We can trace a similar long-term response to the changing external circumstances of Britain in respect of foreign relations and Commonwealth and Colonial Affairs. At the time when the Haldane Report was written, the British Empire had almost reached its maximum territorial extent. Three great departments, the Foreign Office, the Colonial Office and the India Office presided over its destinies, not always without some friction between them.[3] But already the growing autonomy of the Dominions made the Colonial Office an increasingly inappropriate instrument for dealing with them. The Dominions office with its own Secretary of State was thus added to the Whitehall spectrum. From 1947 onwards the scene changed again; the Empire was dismantled; the vast territories for which the Secretary of State for India had been responsible became Dominions or achieved independence outside the Commonwealth; the same happened to all but the most minuscule of colonial dependencies; the Empire translated into a Commonwealth lost all cohesion as an international system, and relations between the United Kingdom and its other members took on more and more the character of foreign relations. In response to these changes both the machinery for the formulation of policy and the external services were brought together, so that by now the Foreign and Commonwealth Office and its instrument the Diplomatic Service covers almost the whole external field.

Almost but not entirely, for quite apart from the fact that all the economic departments and more especially the Treasury and Board of Trade have their own important external interests,[4] one wholly new Department, the Ministry of Overseas Development, has come into being within the last few years. And the fact is a suggestive and revealing one. For one would think that the giving of financial or technical aid to foreign countries is as much a part of foreign policy as any other form of contact. The reasons for the constitution of a separate department must therefore be sought in the political field: the desire to convince the Government's supporters at home that particular importance was attached to this

[3] See Max Beloff, *Imperial Sunset,* Vol. 1: *Britain's Liberal Empire, 1897-1921* (London, 1969).

[4] For an earlier treatment of this aspect of the matter, see Max Beloff, *New Dimensions in Foreign Policy* (London, 1961).

subject, and originally, the creation of a post for a member of the governing party of a kind likely to please the faction within the party to which she then appeared to belong.

What is exceptional in the external sphere would appear to be almost the rule, domestically. If one studies the way in which the economic duties of government have been handled over the last decade and in particular under the present administration, the rise and fall of the Board of Trade, the establishment and demise of the Department of Economic Affairs, the advancing empire of the Ministry of Technology, the attempted transformation of the Ministry of Labour into a more positive instrument of economic action and control, one is bound to feel that the degree to which the adventitious accident of personal claims has influenced decisions in this field must be reckoned an abnormally high one. And even if one admits that a more positive philosophy of economic interventionism was almost certain to require a rather different disposition of departmental and ministerial responsibilities, changes made in such quick succession are bound to be so time-consuming and exhausting for the individuals involved at both the ministerial and the official level that something more than the convenience of a Prime Minister in the dispensation of patronage is surely necessary to justify them.

There is therefore, to go back to my original contention, a real cause for concern that no systematic attempt should have been made, or even attempted, to try to work out the principles of a system which should stand the test of time, and enable governments of whatever party to concentrate for a period upon the making of policy and upon efficiency and economy in administration rather than upon this process of continuous tinkering with the machinery itself. The omission to do so is the more striking because of the efforts that have been made in other aspects of government to preface change with substantial and systematic inquiry: two committees – Plowden and Duncan – on the overseas services; the vast Fulton inquiry into the home Civil Service; Maud on local government and even the attempt to look at some underlying elements of the Constitution by the Crowther Commission, which by investigating the current interest in regional devolution may supplement the work done on local government in the strict sense.[5]

[5] *Report of the Committee on Representational Services Overseas* under the Chairmanship of Lord Plowden. (Cmnd 2276, Feb. 1964); *Report of the Review Committee on*

I take the Commission's own view of its task rather than that of the cynics who regard the whole operation as an attempt to draw the political fangs of Scottish and Welsh nationalism in time for the next general election. Even the procedures and practices of Parliament have been the subject of repeated scrutiny by parliamentarians themselves.[6]

But all this does not add up to a comprehensive view of why we are governed in the way we are; and indeed, leaving out the structure and functions of the central departments gives a certain incoherence to all the other proposals for reform. In the words of a recent PEP broadsheet: 'the more *ad hoc* commissions are set up, the more they draw attention to the need for a national agenda with the centre of our political system near to the top of the list for critical examination.'[7] And that centre is surely the Executive itself – the Cabinet and the departments.

It may be that so far I have been labouring the obvious. But one does sometimes hear the argument or find it implicit in what is said or done that what particular minister or ministry deals with a particular body of work is a secondary matter, that the policies pursued are largely independent of the structure from which they emerge. In answer to this one can only say that this has never been the view of those engaged in the governmental process itself, and in particular in that bargaining for one's share of public expenditure which is often the heart of that process on the policy-making side. It was – to take an example already referred to – certainly not the opinion of those who struggled to get the air arm fully recognized that it would not make any difference to British preparedness for its use if its development was left to the War Office or the Admiralty. And they were almost certainly right . . . one thinks of what happened about the tank.

To take an example nearer to us in both content and time. It is now becoming evident what a difference it was going to make to

Overseas Representation, Chairman: Sir Val Duncan (Cmnd 4107, July, 1969); *The Civil Service*, vol. 1: *Report of the Committee*, Chairman: Lord Fulton (Cmnd 3638, 1968); *Royal Commission on Local Government in England*, vol. 1, Report, Chairman: The Rt Hon Lord Redcliffe Maud (Cmnd 4040, June 1969).

[6] See e.g. First Report from the Select Committee on Procedure: Session 1968-69: *Scrutiny of Public Expenditure and Administration* (HC 410, July 1969).

[7] *Renewal of British Government*, PEP broadsheet, no. 513 (July 1969), p. 656. Some of the themes of the broadsheet were previously discussed at greater length and more controversially in Max Nicholson, *The System: The Misgovernment of Modern Britain* (London, 1967).

the universities that the Government should have rejected the recommendation of the Robbins Report that the governmental responsibility for the universities should in future rest with a new Minister for Arts and Sciences.[8] Instead, the UGC was handed over in April 1964 to the new Department of Education and Science, which was in effect the old Ministry of Education. Its own major task of dealing with the non-autonomous sectors of education made it peculiarly unfitted in tradition and ethos to deal with institutions whose autonomy is their very life-blood. The consequences have been the loss by the UGC of the independence of action it enjoyed vis-à-vis the Treasury when that was its sponsoring department, the elevation of the Committee of Vice-Chancellors and Principals into a powerful additional instrument of centralization, the further burden of administration involved in the acceptance of the Comptroller and Auditor-General's access to university accounts, and the intrusion of ministers and parliamentarians into activities which they neither comprehend nor appreciate. In this way what seemed to some people at the time a mere matter of machinery has come to have fateful consequences for the country's intellectual and cultural life to an extent which it is not yet possible to measure.[9]

The question of the appropriateness of the allocation of a particular aspect of governmental responsibility to a particular department, while very important, is by no means the whole of the problem. In order to go further into it we must distinguish between what might be called the classical areas of debate, and those that have arisen more recently, without, however, diminishing the importance of the older issues. The newer ones arise directly from the increased responsibilities undertaken by modern governments and from that different and far more positive attitude to the role of government which has been the conventional wisdom of the past

[8] *Higher Education,* Report of a Committee under the Chairmanship of Lord Robbins (Cmnd 2154, Oct. 1963), pp. 250-52.

[9] See Max Beloff 'British Universities and the Public Purse', *Minerva,* vol. V., no. 4, Summer 1967. Since that article was published, things have got much worse. See for instance the preposterous proposal for a Higher Education Commission made by a sub-committee of the Select Committee on Education and Science in its Report, *Student Relations,* HC. 499-i, July 1969.

I do not think Professor Hugh Seton-Watson goes too far when he talks of a 'pogrom against culture' in this country, led by the 'demagogues in Westminster, the masochists in Whitehall and the underminers in Fleet Street' (letter in *The Times,* 10 December 1969).

quarter century in most western countries. The issues raised by the Haldane Report remain relevant today; but the report itself can now be regarded as having closed an era rather than opened one. It is true that by 1918 British central government had acquired responsibilities far wider than those contemplated by the administrative reformers of the mid-Victorian age. Even allowing for the future dismantling of the wartime instruments of economic intervention in a more radical fashion than was perhaps contemplated by those engaged in planning 'reconstruction', the welfare services were by now accepted as permanent features of the British scene. These raised in turn one of the oldest questions in administrative theory, whether what might broadly be called the welfare departments should be organized according to the services provided or the persons served. And despite the Haldane Committee's strong stand in favour of the former, it cannot yet be said that the principle has been fully applied either centrally or locally. Indeed, the fact that an individual and still more a family might require the assistance of more than one service if life were not to be intolerably fragmented, has brought about the need for a network of voluntary organizations trying to look at the individual's or family's needs as a whole, without respect to the demarcation lines of Whitehall or town hall.

The subject is too familiar to linger over. But it does serve to remind us of one fact which is too often overlooked by those responsible for addition to or innovations in the machinery of government. What appears when viewed from the heights as a perfectly rational way of dealing with things may seem extremely complex to the ordinary citizen upon whom the decisions made on the heights eventually impinge. I suspect that when the Crowther Commission investigate, as they seemingly intend to do, the fashionable modern cry for greater citizen 'participation' they will find that, barring that public-spirited element in the population which probably finds an outlet already, the real demand is not for participation at all but for an administrative system, simple enough to be understood and more responsive to what the citizen thinks are his personal rights. We have two quite different scales by which the rationality of governmental action is judged. One which is inherent in the whole idea of 'planning' is that the forces of growth and change should be canalized in the common interest, so that accurate estimates can be formed of future requirements, physical or financial. The other which is that of the citizen is based upon what he sees as legitimate expectations for himself and his family. He is concerned not with national planning but with his own property, his own savings, and the existence of some kind of relationship between his own effort or sacrifice and his ultimate reward. It may be argued that in a rapidly changing society such hopes are utopian, that the acceptance of wider responsibilities by the State and the decline in the role of the family and of the *paterfamilias* means that the ability to plan for oneself rather than to be 'planned' will become increasingly unimportant. What a society from which self-reliance had been totally eliminated would be like is a matter for the social theorist rather than for the student of public administration.

But we have some hard experience to go on which is not alien to our theme. The years that followed the publication of the Haldane Report were years of considerable economic and social upheaval, a legacy of the war and of some aspects of the peace settlement. The most important permanent result of the economic disarray of the early 1920s was the prevalence of monetary inflation. The social effects of this inflation particularly in Germany have frequently been chronicled. The psychological effects of any inflation are, however, equally disturbing since they appear to render vain all prudential calculations about the future. Yet despite these warnings from the past, Britain like the United States and other countries has obviously entered upon a new inflationary cycle, and as we shall see there is reason to believe that some degree of inflation may actually be encouraged by our governmental arrangements.

Before entering upon that argument we may note the other classical problem posed but not fully resolved by Haldane, that of the reconciliation of the multiplication of governmental functions with the needs of responsible cabinet government, since there are obvious limits to the size of the Cabinet if it is to be the effective focus of decision-making. The increase in the number of functions that government performs leads inexorably to an increase in the number of departments. Not all can be represented in Cabinet. For the coordination of policy three main types of solution have been attempted, singly or jointly, beginning with the Lloyd George experiments during the first world war. They are the cabinet committee, the 'overlord' minister giving general supervision to several ministries, and the enlarged or 'federal' ministry in which junior ministers may have special responsibilities for particular divisions of the work.

The cabinet committee has clearly come to stay, and the responsibility of the cabinet secretariat which the creation of the cabinet committee system made imperative, quite apart from the needs of the Cabinet proper, has become an essential part of the working of the system. Examples of the other two devices are to be found in the present phase of the incumbent administration; both have their well-known weaknesses and it is improbable that finality has been reached or ever can.

The large size of some departments which is a relatively new factor has led to more radical suggestions for reversing the addition of new departmental responsibilities and/or actually decreasing them. It is argued that in many departments much of the work is managerial in nature, concerned with the implementation of agreed policies, and not with the provision of advice to ministers which used to be the primary function of the classical departments. It is argued under the rather inaccurate term 'hiving off' that these sides of the work could be devolved upon independent boards or commissions or other organizations outside the department, leaving the central core around the minister a great deal smaller. Instances of this having been done here in the past or in the present – e.g. the assessment and collection of taxes is not a ministerial function – and in other countries are not hard to find.

Nevertheless there are clear limitations on the extent to which this process can be carried out. Two illusions would seem to be present. In the first place, it is too readily assumed that the distinction between mere management and policy is an absolute one; and that no political element can ever enter into the former. Indeed, the making of policy itself should normally arise directly from the experience of managing existing activities, and the institutional separation between the two may have very serious consequences. For, in the second place, it is also important to remember that the citizen himself is not purely a passive object of administration but will seek redress if his interests seem to be adversely affected or his sense of justice affronted. If the parliamentary channel is blocked because the matter is a non-departmental and hence non-ministerial one, he will seek other means of making his views known. If he fails to get redress, political resentments will build up irrespective of the precise constitutional or administrative position. One can take the neutrality of technocracy too far.

The extent to which these considerations are relevant can be seen when we pass to the first, though not, I think, the most important of the post-Haldane problems: that created by the fact that the modern State quite apart from its controlling functions in the economic sphere is also itself a major entrepreneur. When the Haldane Report was written, there had already been some state experience in the direct handling of parts of the economy, and it was assumed that nationalization would be a feature of the post-war period. The reaction against wartime methods made this assumption premature. There was some public enterprise in the inter-war period but the really dramatic advances in the public sector did not occur until after 1945.

By this time the straightforward notion that the State could run industrial or commercial enterprises as it ran the army or navy or the post office was as obsolete as the idea that state-owned enterprises could be run by the 'workers' themselves. The public corporation on the model of the London Passenger Transport Board – the Morrisonian model – was almost universally accepted as correct, and is so still; even the Post Office has now fallen into line.

Nevertheless, there are good reasons for saying that the model is a very imperfect one and the arguments advanced in its favour fallacious, like most other attempts to square the circle. For what is asked of enterprises in the public sector is impossible of fulfilment; and must always be so. In the moderate language of a Parliamentary Select Committee the 'two obligations' on the nationalized industries – to be responsive to the public interest and to operate as efficient commercial bodies – are not easily reconcilable.'[10] But it is hard to believe that the Committee's principal recommendation, which has not so far been accepted – the creation of a single ministry responsible for all the nationalized industries, would substantially improve matters. And this is because the conflict of obligations is thought of as marginal to their operations not as central to them.

It is not just the case that railways cannot make profits if they are compelled to cater at standard rates for the empty valleys of central Wales or airlines if they are to service the Scottish isles

[10] *First Report from the Select Committee on Nationalized Industries,* Session 1967-8: *Ministerial Control of the Nationalized Industries,* vol. 1 (HC 371-1, July 1968), p. 16.

adequately. These special cases, where what are called 'social considerations' enter in, could be budgeted for without too much difficulty. It is really much more than that. After all we do not normally regard it as impossible for an enterprise to serve the public and also make a profit. Messrs Marks and Spencer have done more for the welfare and gaiety of the daily lives of ordinary people than most political or social reformers, and they have made a great deal of money in the process. But private firms of this kind enjoy a double advantage over nationalized industries. They have the spur of competition and the possibility of total catastrophe if they fail. All this powerfully concentrates the mind of management. But in addition they have to a considerable extent freedom of decision-making in what is the essence of a commercial operation, pricing policy. It is true that governments are ever eager under a variety of pretexts to eliminate this freedom; fortunately for us all they have not yet succeeded.

But the nationalized industries have no such freedom. Although their statutory responsibilities may require them to keep their accounts in balance, 'taking one year with another', neither the public nor the politicians are prepared to give them the freedom they would require in order to succeed. It is somehow held that such things as railway or bus fares are not suitable for settlement in the manner in which Messrs Marks and Spencer price a particular line of goods. Somehow these are thought of as governed by different principles in which the interests of the consumer should be uppermost. And indeed if the public is not to benefit by nationalization why nationalize?

On the one hand therefore the minister will be under pressure to keep prices down; on the other hand, he is under great pressure to let costs rise. For, since the nationalized industries are an essential part of the general economic structure and indeed vital in some instances to the orderly functioning of daily life, industrial pressure within them is much harder to resist than in the private sector. The history of the attempt by the present government to frame an incomes policy and then keep to it is punctuated by surrenders on the part of the nationalized industries to wage-claims – surrenders which, given the situation in which they are placed, are perhaps not deserving of the kind of censure that the high-minded writers on *The Economist* normally mete out to the ministers responsible.

In other words, the device of the public corporation has by no means insulated the public sector from political pressures; it has merely limited the extent to which the private individual or small community can make its voice heard. A natural monopoly with the power of the State behind it but not responsible for day-to-day operations to parliamentary control through a minister is a hideous perversion of democratic as well as of commercial principles. It is responsible neither to the discipline of the market nor to the discipline of public scrutiny and debate – however hard an amateur and inadequately equipped body like the Select Committee may work. The device of the public corporation cannot fulfil the purposes for which it is set up. It may be that no public authority could do better and that nationalization itself was a crude response to certain historic inadequacies of private enterprise which is no longer necessary or desirable, and that a substantial reduction in the size of the public sector would be one obvious way of lessening the overload on the machinery of government.

But the case of the nationalized industries is only the most obvious example of what appear to be the characteristic problems of British government as it has developed since Haldane – the assumption by the Government of duties and responsibilities which are in themselves irreconcilable. The turning point came with the adoption of Keynesian economic theory and of the belief that the Government itself could organize and direct the entire economic life of the community according to recognizable criteria, and even plan its future growth and development.

Before this assumption was generally made and accepted these things, like the weather, were thought to be the product of factors outside the range of governmental control. Like the weather the economic outlook might be plotted and with about the same margin of error. It was for the entrepreneur to take warning from the danger signals that went up. The role of the State was to intervene to protect particular branches of activity, or to afford succour to individual victims of structural change or temporary depression. That was what was meant by state intervention only a generation ago. And although such intervention was imperfect, and often inadequate, it had the merit of being well understood and of possessing a high degree of internal coherence. All that has gone; and in its place we have a set of high-sounding objectives – full employment; steady growth; absence of industrial strife; a favourable balance of payments and the rest. The theoretical arguments as to the practicability of any or all of these objectives and of their reconciliation need not detain us. Nor would I be competent to pronounce on them. All we are concerned with is their reflection in our administrative and political arrangements.

Let us take a series of pairs of ministries: the Treasury responsible for the balance of payments and the curbing of inflation demands a rigid control over incomes; the Ministry of Labour, which has not been de-natured by its new, imposing but hollow designation as the Department of Employment and Productivity, is mainly concerned to avoid industrial conflict; its inevitable bias is towards appeasement, if necessary at the consumer's and so at the currency's expense, and so to inflation. Or let us take the Board of Trade and the Foreign Office. The former is instructed by the Treasury to save on imports and imposes quotas or surcharges or what-not; the Foreign Office is made miserable by the breach of international agreements or understandings, the alienation of friendly governments and so on. Take the two federal ministries; the Ministry of Technology and the Department of Education and Science. The former applies the stick and the carrot to try to improve the performance and increase the scope of the so-called science-based industries (Has anyone thought what a non-science-based industry might be?) This calls for the utilization of more scientists and technologists; and their education and training is a matter for the Department of Education and Science. But the latter which is as much an instrument of the NUT as the Ministry of Agriculture is of that other successfully conscienceless pressure group the NFU, is inhibited by this fact from taking the kind of measures that might get more children into the scientific streams, and provide more adequate teaching for them when they got there.

When then we look around the cabinet table – an exercise better indulged in in imagination than in reality – what we see is not a group of men moved by the same set of objectives and attempting to coordinate the national effort in order to bring them about, but rather a number of individuals who bring to the process of decision-making not merely the normal quota of personal rivalries and jealousies but quite genuine commitments to quite different priorities. And these are not personal quirks but the outcome of their immersion in the atmosphere and requirements of their own particular departments. The Churchill who slashed the defence estimates when Chancellor of the Exchequer in the 1920s was the same man who as First Lord of the Admiralty had fought the navy's battle for funds in the pre-war years.

This example may suggest that this is another of the classical dilemmas not something essentially new. And it is of course true that there must always be some tension between the spending departments and the Treasury. The new element lies in the much greater pretensions of government, and the consequent multiplication of differing priorities, and of different ways of looking at the world in general and the economy in particular. For the more far-reaching the involvement of Government in manipulating the social order, the more likely it is it will come up against not merely individual citizens whose objections it may be able to ignore, but also organized forces too powerful to challenge. The recent history of the proposed trade union legislation and of the incomes policy is very revealing in this respect. It may suggest that the more a government attempts, the less it can rely on being able to do. And what one wonders about concerning the machinery of government is, what gaps in information and understanding concealed the facts from ministers for long enough to oblige them to accept an ignominious retreat.

An additional factor of great importance is the difference of time-scale to which governments now work. In part this is due to the nature of some of the technological changes with which it has to come to terms so that some departments must necessarily project what they are doing some decades ahead. But some again comes from the ambitious nature of the policies themselves, as when people are asked to make contributions to pension schemes from which they themselves will not fully benefit. All saving involves the sacrifice of present enjoyment for future benefits; it is characteristic of modern government that the extent to which this is done is thought a matter for collective rather than individual decision-making, despite or because of the fact that continued inflation makes all saving less attractive.

These commitments to planning or to 'social engineering', as it is sometimes called, give added importance to something Haldane was very keen on: proper intelligence and planning machinery within the departments. And a recurring problem of organization within departments is how to ensure that planning sections are neither overwhelmed with current issues nor kept too far from where decisions are actually made; the distinction between long-term and short-term considerations is one of the hardest to draw effectively.

The new feature is that much forward thinking is not or should not be departmental in character; and the place of some central thinking apparatus presents a question that has by no means been resolved. Apart from the Central Statistical Office which seems to have found a permanent billet under the aegis of the Cabinet Office, there is very little that seems settled. The post-1945 adventures of the economic planners are familiar; from the Cabinet Office to the Treasury; to the new Department of Economic Affairs and back now to the Treasury.

It would seem not impossible that such activities should fall in what we have not so far had in this country, a Prime Minister's department. Non-departmental advice on current questions is something that all Prime Ministers must crave to some extent though they go about getting it in different ways. Sir Winston Churchill and Mr Harold Wilson have come closest to creating formal machinery for this purpose. But the only important survival is the Chief Scientific Adviser; and one has the feeling that he and his staff might come to be the embryo of a new ministry, perhaps of a proper Department of Science, free from its peculiar association with people who deal with the problems of the eleven plus, school meals and so on.

If that were so one might, as an alternative to a Prime Minister's department dealing among other things with long-range estimates of future trends and needs, divide this function between three departments dealing respectively with the scientific-technical, the economic and the social at large. And provided those engaged in such studies had sufficient historical training to be aware of the inevitable relativity of their own standpoints, some good might come out of it, and no great harm would be done.

At the one end of the scale, then, our study of the machinery of government would lead us into the whole new science or pseudo-science of futurology, which so preoccupies Americans; at the other end there is the question of the effect of rapid change, planned or unplanned, upon human communities and of whether these ought to have greater safeguards in the form of more powerful local or regional organs of government; 'devolution' to use the word that was fashionable in the Haldane era. Upon all this, as we have seen, work has been done and is being done. But unless one is aware that devolution implies a self-denying ordinance on the part of central government, and unless one is clear what functions or powers should be or could be devolved, and how this would affect the central machinery, such exercises are largely artificial. Nor must one forget the possibility that while there may be demands that central powers should be devolved downwards, a new move towards European integration on any serious scale, or even the entry of Britain into the existing European Economic Community, would mean the transfer of some powers not downwards but upwards to supranational instead of national institutions. And in this sense too we face complexities that Haldane did not.

I am conscious of having in this lecture raised far more questions than I have answered, and of having risked being accused of superficiality. But I would remind you in conclusion that all I have been attempting is to make a case for renewing after more than half a century the attempt to organize the departments of central government in a way sufficiently in accordance with our needs to assure some period of stability in Whitehall. If in so doing we find that there are aspects of our affairs with which central government need not concern itself so directly, if, in other words, we can find some way to reduce the size of the machine as well as to improve its efficiency, only the doctrinaires of collectivism will grieve. I will not ask you to grieve with them.

6 *The Planners' Place in Foreign Policy*

Whatever the view of individuals may be on the great issues of defence and foreign affairs that now confront us, there seems to be a widespread measure of agreement that we are in for a major stocktaking. Some people have raised the further question of whether we have adequate instruments for such an operation. Lord Boothby and others have suggested that it is the weakness in the structure of Parliament that most requires correction – that the customary objections to a system of specialized committees as being incompatible with cabinet government no longer apply in the external field. But in spite of obvious difficulties, it would seem logical to begin by considering not the legislature but the executive.

The future historian of Britain's foreign and defence policy over the past five years or so looks like being obliged to record a number of miscalculations as to the probable behaviour of other countries. It is generally admitted that we have made mistakes at every stage of our relations with the Common Market – first, in believing that the project itself was foredoomed; second, in believing that the problem could be met by the device of a wider free trade area; and third in underestimating the likely obstacles to our own accession to the Common Market itself, once this course had been decided upon. It is less generally accepted, but no less probable, that there has been some failure to foresee the likely course of the Kennedy Administration in the United States, both in relation to the structure and policies of the western alliance and in regard to some aspects of the role of the United Nations.

An outsider cannot, of course, be certain as to the point in the governmental hierarchy at which the break between the reality and the image actually occurs. From what we now know of the 1930s, we can see that there were occasions when the permanent officials (including ambassadors) were at fault, and others when their perfectly valid assessments were overridden or ignored by ministers. And the more that personal diplomacy of the summit or near-summit variety flourishes, the more likely it is that there will be new examples of the latter.

Nevertheless, it would not be surprising to discover that the briefing of ministers has at times been inadequate. For the problems that the Government has had to confront have been of precisely the kind that the traditional British system of handling foreign affairs is least likely to cope with adequately. The failures to adjust have helped to throw light on certain traditional and not unfamiliar weaknesses of the Foreign Office and its outlook which have remained largely unaffected by the considerable progress in other respects – for instance, in building up the new techniques of multilateral negotiation.

To put it crudely, what has been asked of the Foreign Office is an interpretation of what are in essence policies of a strongly ideological flavour, that is to say, policies based upon definite and systematic interpretations of the contemporary world scene and of a particular country's role within it. And it can even be argued that the only viable answer to such policies is to be found in making something of the same kind of analysis for oneself.

The tendency to disbelieve in the importance of ideology, to reduce foreign policy to a matter of dealing with successive contingencies of a concrete kind, has been particularly strong where our own allies have been concerned. We have underestimated first the strength of the trend for European unity, then the international implications of Gaullism, and finally, most important of all, the depth of intellectual conviction and systematic thought that lies behind the foreign and defence policies of President Kennedy.

These ideologies, which must largely reflect underlying trends in the societies that produce them, will not normally provide matter for the contents of the day-to-day telegrams from our missions abroad. Ambassadors and their staffs are there primarily to transact current business, and their tenure of a particular post (governed as this normally is by establishment considerations) may well be too short to acquire a knowledge in depth of the country concerned. By contrast, Stratford Canning's four spells at Constantinople added up to a quarter of a century. Therefore only a research and planning staff with real authority for securing the material it requires – if necessary through a voice in postings – and with direct access to the highest level of policy making can do the job.

Now it is fair to admit that all ministries of foreign affairs find it hard to settle the proper sphere of action for 'planners' and to refrain from raiding planning staffs when they require additional personnel for 'line' posts. The history of the 'policy planners' in the State Department over the past fifteen years has been a chequered one. But in the British case there are two additional difficulties in the way of expanding our hitherto modest efforts in that direction.

The intense professionalism of the Foreign Service has meant that little or no use is made of what energy or talent may be found elsewhere in the community. For while the two-way barrier between the official and the non-official worlds is a general feature of British government, it seems more rigid in the Foreign Office than in some other departments.

A review of Professor Walt Rostow's book *The United States in the World Arena* a couple of years ago suggested that the work deserved careful study in Britain because its author was clearly going to be a considerable figure on the 'New Frontier'. It would be interesting to know if the hint was taken in official quarters. But the notion that the policies of an incoming American Administration could be gleaned from a ponderous tome by a professor of economic history is one as alien to our ways of thinking as the idea that the author's previous career was an appropriate preparation for the kind of positions that he has since held in the Executive Office and the State Department.

If we reflect on the parts that Professor Rostow, Mr McGeorge Bundy and other recruits from the world outside are now playing in shaping the policies of the Kennedy Administration – and almost equally important, in expounding these policies at home and abroad – we are brought up against the second fundamental point. The policies of Kennedy and de Gaulle are the policies of the effective rulers of their countries. Their ministers, whether of foreign policy or of defence, are only executants of their purposes. The British Prime Minister has had inevitably to take similar responsibilities upon himself; for foreign policy, defence policy and financial policy are in their broad outlines inseparable.

Should the Prime Minster, then, rely on departmental briefs as made available to him by departmental ministers, or is there not a case for creating something much closer to a Prime Minister's Department? In other words, should a British Planning Staff in the field of external affairs find its proper place in the Cabinet Office rather than in the Foreign Office, or is there, on the current American model, room for one in both?

An American President will of course draw his own planners partly from men who have served him during the campaign; and whether they have so served or not their appointments are political. British tradition has for a long time been against such political appointments. There was, of course, talk in Labour circles in the early days of the need for a Labour Government to bring outsiders into the departments so as to avoid 'sabotage' of 'socialist policies'. It did not happen; and Ernest Bevin's high reputation in the Foreign Office seems to be based in large part upon his readiness to follow his officials' advice.

But, party politics aside, is there really nothing to be said for bringing in fresh minds at any but the ministerial level? Certainly one could not follow the American example at all closely; neither the Conservative nor the Labour party organizations are designed to produce individuals appropriate for appointments of this kind. But that does not necessarily mean that the present arrangements should be regarded as sacrosanct. If the House of Commons does set up a Standing Committee on Foreign Affairs it is much to be hoped that questions of organization as well as of policy will come within its purview.

7 The Plowden Report on the Foreign and Commonwealth Services[1]

The Plowden Committee was appointed for two principal reasons, one specific and one more general. The specific problem was the increasing difficulty now felt of distinguishing between the nature of the work of the Foreign Office on the one hand and the Commonwealth Relations Office on the other. The recommendations that the overseas representation of the two offices should be merged in a single Diplomatic Service has been accepted by the Government and the necessary steps towards this end are going forward. It may be argued that this process should be carried further, and indeed the Committee itself admitted that the 'logic of events' pointed towards the amalgamation of the two offices themselves : 'the unified control and execution of our external policy as a whole which would result would be a rational and helpful development'. But it was the view of the Committee, which the Government has accepted, that 'to take such a step now could be misinterpreted as implying a loss of interest in the Commonwealth partnership'.

Since the Committee reported there have been a number of issues in foreign policy where the respective roles of the Foreign and Commonwealth Relations Secretaries have overlapped, and this may well, in due course, lead to a reconsideration of this point. Such a reconsideration will no doubt be to some extent affected by the decision taken at the Commonwealth Prime Ministers Conference in July to instruct officials 'to consider the best basis for establishing a Commonwealth Secretariat'. If such a Secretariat were to come into being, and if it assumed the set of functions

[1] *The Report of the Plowden Committee on Representational Service Overseas* (Cmnd 2276, February 1964).

suggested for it in the Prime Ministers' communiqué, the argument for a separate Commonwealth Relations Office would lose much of its force, particularly since, as the Plowden Committee notes, all Commonwealth countries other than the United Kingdom already have a single Ministry of External Affairs.

The more general reason for the review of the overseas services was simply the need to see whether their organization was appropriate for the new situation brought about by the changes in the tasks before them since they came into being in their present form in 1943; for as the Committee correctly if somewhat sententiously remark: 'the world in which the overseas Services now have to operate is no longer the world of 1943 or even a world which could be foreseen in 1943.' Some of the changes have been of a magnitude affecting every aspect of the nation's life, such as those which relate to the general decline in Britain's relative economic and military strength; others, such as the multiplication in the number of sovereign states and the increasing role of international organizations, have a more obvious and immediate impact upon the size and structure of the overseas services. But if we omit a number of minor but very sensible reforms relating to conditions of service the Report is, with one exception, remarkable more for its conservatism than for any suggestion that either the Foreign Office or the two overseas services now to be amalgamated in the new Diplomatic Service need much in the way of reform.

To some extent this is a healthy sign. Much popular criticism of the conduct of foreign policy is based upon ignorance of its problems, and of the range of choices open in regard for instance to such matters as recruitment to the foreign service. But there are ways in which it might have been possible to do a little more probing of accepted orthodoxies.

One of these relates to the problem which the Report discusses under the head 'Specialization'. The Committee notes that the trend over the last forty years in Britain as well as in other countries has been 'consistently away from separate specialized and localized services such as the former regional consular services'. They would not like to see a return to such services which would among other things make recruitment more difficult and believe that it should be possible to create the necessary experts within the single wider service of today. They believe that the Foreign Office has now worked out in its planning of training and careers the means necessary for producing 'general-purpose' officers with one or more specialist qualifications of either a regional, linguistic or functional kind.

They further point out that in a world where developments in one area so easily impinge upon those of another, it may well be desirable to appoint to a post in one capital someone with relevant experience obtained in some quite different part of the globe. But they admit that the pendulum did swing too far against specialization in the immediate post-war period, and would like to assist the present movement which they detect as being in the opposite direction, by requiring officers to spend normally half their careers in the area or subject of their specialization and by lengthening the consecutive period of service in any one post to between four and five years.

Given the tendency of all establishment branches to give great weight to factors other than the actual work to be done, one may be rather doubtful as to whether even these modest recommendations will be carried out in full. But ought they to go further? Here one comes up against the very different nature of the work of the service at different levels and in different areas. From the point of view of high policy what matters most in the head of a mission must be first, the ability correctly to report upon the changing political scene in the country in which he is serving, and second, to represent to as wide a cross-section of that country's population as possible the principles and purposes of British policy. The first demands an intimate knowledge of the country concerned that only long experience and considerable opportunities for study can give; the second demands almost everywhere a knowledge of the language, not so much for purposes of negotiation (which under modern conditions are secondary anyhow) as for purposes of representation in the broadest sense. The major errors in British foreign policy, in so far as the outside layman can trace them, seem often to have had at their core a failure of British representation in one or other respect at some vital point. Perhaps this impression is a false one brought about by the 'fifty year rule' and other such impediments to a proper study of the history of foreign policy; but one may regret that a Committee with Lord Plowden's opportunities did not include any member whose interest might have led him to demand some 'case-studies' that might have shown the possible connection between the staffing of particular posts and the failure to foresee or to avert particular developments unfavourable to British interests. It is also a pity that no attempt was made to compare the attitude and practices of the foreign service in regard to 'specialization' with what is done by major business houses with large overseas interests, where the measurement of success and failure is perhaps somewhat easier.

As far as languages go one might perhaps wonder what light is thrown upon the assessment of priorities in Britain's post-war foreign policy by the fact that the number of officers who have passed the appropriate examination, and kept their knowledge up to an approved standard, is the same in Arabic as in Russian, Japanese, Chinese, Persian, Turkish and Polish *taken together*. Unfortunately the Committee does not seem to have seen fit to inquire into mastery of the so-called 'easy' languages.

The Committee were aware that to make it less difficult to put the right man in the right job, the resources of the service must not be too tightly stretched; the same is true if there is to be the development they recommend in the secondment of members of the service to other departments or for sabbatical years at universities or other institutions, where they can get advantageous experience or opportunities of study and reflection otherwise hard to come by.

For these reasons the Committee very properly suggested that a reserve of manpower was required if the system were to function properly, and suggested something of the order of ten per cent. It is perhaps indicative of the general low level of understanding of administrative problems in the country at large that this recommendation, one of the most cogent made by the Committee, was denounced as 'wasteful' by one of the 'intellectual' Sunday newspapers.

On information work the Committee did not feel it necessary to repeat the work of the Drogheda Committee, though admitting that much had happened in the decade since its Report (Cmnd 9138) was published. The main recommendation they make is to assimilate the Commonwealth Relations Service 'information class' to the new combined Diplomatic Service. The main gap is the result of the omission of the British Council from the Committee's terms of reference, though evidence was taken from its Director-General. The distinction between a branch of the government service and an 'independent body' like the BBC or the British Council, which seems normal and sensible to those who understand how

Britain's affairs are run, remains incomprehensible to most foreigners. (Even Britain's private institutions are hard to grasp; one finds foreigners who have had a good deal of contact with this country still talking about *The Times* as a semi-official newspaper.) If the British Council is not thought of as part of Britain's official representation abroad, its utility in many countries will be curtailed. Furthermore in so far as its work is a branch of information work – and why else should the taxpayer support it? – it should be, in the Committee's own words on information work, 'essentially an activity designed to further policies'. For both these reasons the subject ought to have been looked at, and serious consideration given at least to giving the Council's principal officer in each capital the rank of cultural attaché as is done in some places, and making the Head of Mission broadly responsible for the Council's activities in respect of the country concerned.

The Committee have been bolder in dealing with the problems at the centre than in tackling those in the field. They are worried at the burdening of senior members of the Foreign Office with day-to-day responsibilities at the expense of time that should be devoted to long-term issues. They approve of the latest version of the planning staff but think it undermanned. They applaud the present practice of holding seminars in connexion with outside bodies. They propose that use should also be made of 'panels', that is to say of groups including outside experts who should not merely discuss a particular theme but also 'take a hand in drafting reports and recommendations'. To do this on any scale, which would involve outside experts having access 'to appropriate confidential information, subject to normal security requirements', would imply a very considerable shift towards something much more familiar in the American context. And it will be interesting to see whether this recommendation is taken up. Had there been a university teacher on the Committee he might have persuaded his colleagues that there is really no comparison in effectiveness between a 'seminar' and a 'panel study'; it is only when responsibility for actual drafting comes in that the outside experts are likely to give of their best.

But the Committee goes further. Not only would it advocate an amalgamation of the planning staffs of the Foreign Office and the Commonwealth Relations Office, even while the two departments remain separate, but it also wishes to see a further development of the idea of a Cabinet Office Committee with responsibility for coordinating planning in all the overseas fields including defence as foreshadowed in the 1963 White Paper on Defence Organization (Cmnd 2097). It would want to see such a Committee having a staff of its own apart from the regular Cabinet Office secretariat, and to be empowered if necessary 'to borrow personnel with experience of policy from other Departments, including the Foreign and Commonwealth Relations offices', once more using 'outside experts on a consultant basis'.

What the Plowden Committee does not appear to have gone into, is the place of such a Committee in relation to the Cabinet. In some respects the formula resembles that of the old Committee of Imperial Defence in its pre-1914 heyday but at the official level. That was in the days before there was a Ministry of Defence and when coordination between the two armed services could be carried through, if at all, only in the Cabinet. The key to the CID system was the chairmanship of the Prime Minister. Is the chairman of the new body to be the secretary to the Cabinet with that special relationship between himself and the Prime Minister which was so notable a feature of the Macmillan-Norman Brook era? How this question was answered would make a lot of difference to the weight of the planning machinery. As it is the Plowden Committee merely says 'we believe that a Cabinet Office Committee reinforced in the ways we have proposed would be in the best position to bring to Ministers [which Ministers?] interdepartmentally agreed papers and specific recommendations for action on matters of foreign policy, defence and finance'.

Once again the secrecy that envelops the decision-making process in British government renders the task of an external commentator difficult. How was it decided that successive bases in the Mediterranean, East Africa and the Middle East were indispensable? How was it decided that some of them were not? How is the indispensability of the remaining ones kept under review, and how are the fruits of such periodic review (if made) rendered effective in the policies of the departments? Only case-studies can ultimately affect one's opinions about 'planning machinery'; blueprints may tell one something at lower levels of administration but hardly at the top. But one certainly cannot assess proposals for changes in the planning machinery without some account of what its recent shape has been.

8 Defining the Limits of Official Responsibility

The Government's refusal to accept Lord Robens' offer of resignation raises, as did the offer in the first place, issues that are not only much wider than the personal ones, but are closely linked with a whole series of events and developments in British government that call into question what has been for at least two centuries its principal and most cherished feature – political (as opposed to criminal) responsibility for all acts of state.[1] Such is still the accepted doctrine in the political world itself and in the schools. But it bears increasingly less relation to practice or to what the general public thinks.

Most people as they read in the press of major instances of government extravagance and lack of foresight would come to the conclusion that in their personal and professional lives they are wholly at the mercy of a vast governmental machine against whose arbitrary dictates they are quite helpless. They would come to this conclusion perhaps after reading of the Government's dealings with the aircraft industry; of schemes affecting important areas of private and public concern adopted without any clear procedure of investigation and debate – as with the Stansted Airport project; or of obvious cases of mismanagement in budgetary procedures such as that which has allowed universities to build laboratories and recruit students only to find the equipment for them unavailable; of the overriding of local authorities in what had been thought to be a principal area of delegated power, as in the former

[1] The Report on the Aberfan tip disaster of 21 October 1966 was published on 3 August 1967 and the Government's acceptance of it was announced on the same day. On 7 August, Lord Robens, Chairman of the National Coal Board, offered his resignation which was declined by the Minister of Fuel and Power, Mr Richard Marsh on 6 September.

Secretary of State's refusal to sanction schemes of secondary education that did not fit in with his particular interpretation of that catch-all phrase of the passing moment, the 'comprehensive principle'; of the sudden growth of the National Board for Prices and Incomes into an extra arm of government with powers reaching into almost every aspect of the economy.

Confidence that Parliament can protect people is waning – and they are encouraged in their disbelief in its utility by the utterances of an increasing number of its back-bench members. The 'Ombudsman' may redress a grievance here or there, but is precluded from dealing with 'policy'. The courts are neither equipped nor willing to play an independent role unless there has been some plain violation of statute law. Only publicity tempers the despotism of the State. The Englishman's only remaining political right is to send a letter to the Editor of *The Times*.

At least two dangerous consequences flow from the breakdown of constitutional responsibility; government is less effective and less economical than it should be; and a growing disillusionment with its operation weakens popular adherence to democratic institutions, and provides fertile ground for those who seek their destruction.

It can of course be argued that there is no escape from the present position, that the extension of the responsibilities of the State into such vast areas of activity means that a constitutional system moulded in an era of relative *laisser faire* could not hope to survive in the new circumstances, that where the State is responsible for everything no one can be responsible for anything. But it is also possible to say that provided we keep the principles of constitutional government in mind, and insist that institutional innovations are governed by them, it is possible to devise appropriate remedies for present discontents, that what one needs most of all is greater awareness of the problem and a greater concern for the political and constitutional consequences of economic and administrative expedients.

What are the principles? First that every act of Government is either collective – as where a basic question of policy is concerned – or that of an individual department. In the latter case, except under special circumstances of a narrowly circumscribed kind the Minister who heads it has the responsibility for its actions and can neither push them downwards on to his permanent officials nor diffuse them among his ministerial colleagues. The second principle that flows from it is that all that central government does that is substantive in character – that is to say of a non-judicial or arbitral kind – must be organized within departments whose heads are seated in Parliament, and are there amenable to parliamentary inquiry and control. The third principle is that Parliament should be organized in such a way as to make this controlling function a meaningful one.

In recent years most attention has been concentrated on the last of these questions. '*The Economist* has', it tells us, 'long maintained that Parliament should be given its job back, through a system of investigatory committees empowered to examine how government really goes about its business.' But if one looks at the existing committees – including the not unsuccessful Committee on Nationalized Industries – one can see that within the present constitutional and parliamentary structure, the contribution of developments of this kind could only be limited. To go further, and to reorganize Parliament around strong well-staffed committees on the lines of Congress, would be to have not only a different kind of Parliament but a different kind of parliamentarian. We would have to consider the impact of such changes of what is in our system (though not in the United States) a still more important function of the legislature, the selection and training of aspirants for ministerial office. It may be that for a modern state, the American system – or some variation of the system of the separation of powers – is more desirable than our own monolith, but it is dangerous to assume that one can borrow individual features of it at will.

It is more important to get the structure of government organized so that responsibility can be made effective. Lord Robens' letter of resignation showed how far from such a situation we are. While admitting that maybe the doctrine of ministerial responsibility did not 'strictly' apply to himself he felt, after a lifetime in the public service, 'bound by its rules'. Such sentiments do Lord Robens credit, but reflect a complete misunderstanding of his position.

The 'rules' of the public service are not the same for every member of it. Ministers have rules; backbenchers have rules; civil servants have rules; the military have rules; so have members of boards of nationalized industries and their chairmen – but they are all different. The nationalized industries are not quasi-government departments; the persistence and strength of the idea that a public corporation is the best way to organize each of them is a tribute to the persuasive powers of the late Lord Morrison rather than to any inherent necessity. But this device, adopted in order to square the circle of commercial enterprise and public accountability, was not intended to exclude the nationalized industries from the operation of the principle of ministerial responsibility in major matters.

It is hard to believe that the safety of the public in mining areas is not a major matter, or one not affecting the 'national interest'. A proper code of safety for tips, for the absence of which Lord Robens blamed himself, was something upon which the Minister could have given directions under Section 3 (1) of the Act. The Minister is therefore ultimately responsible for this lacuna, and if anyone should have resigned, he was surely the individual indicated. On the other hand if the organization and staffing of the board displayed errors that in the Minister's view pointed to a lack of capacity on the part of the board or its chairman, then it was open to him to ask for their resignations, though the regulations would seem to preclude their dismissal except on grounds of 'unfitness' to continue.

In the case of the University Grants Committee, the twilight zone of responsibility created and maintained in the name of academic freedom has made the apportionment of responsibility between itself, the Secretary of State and the Chancellor of the Exchequer an almost impossible task. But the decision to make the Permanent Secretary of the Department of Education and Science the accounting officer must now have settled this question in favour of the Secretary of State being responsible for questions of university finance, and answerable to Parliament for the kind of miscalculation that has recently been called to public attention. We must assume that the position of power without responsibility enjoyed by the UGC will now come to an end; and indeed the future of the institution itself must be regarded as highly doubtful.

Where departments themselves are concerned, the usual criticism is that the Minister cannot be expected except notionally to know what his civil servants are doing and cannot therefore be held personally responsible, for instance, for cases of extravagance. *The Economist* again has its solution – making civil servants 'truly answerable for what they do – answerable with their jobs if necessary, just like senior managers elsewhere'. But this of course means an end to the anonymity of such civil servants. To some extent such a development would be in line with what is already happening. The career civil servant still retains his importance and the prescriptive right to certain posts – but clearly not to all the important ones. Lord Robens (whatever he may think) is closer to a civil servant than to a minister in the nature of his responsibility – and so too (whatever he may think) is Mr Aubrey Jones.

It may well be arguable that what one needs is a separation of the managerial functions of the civil service as a whole from its advisory and administrative ones: indeed this view has not gone unexpressed of late. Or it may mean a new form of delegation in all those aspects of central government where 'management' is involved, new criteria for appointment and new sanctions for the proper performance. Similarly revived local or new regional governments might be given full instead of nominal responsibilities for some public tasks.

There is much to inquire into and much to do; but the danger with our close-knit political-administrative network is that most inquiries are so manned that they turn out to be nothing but the system looking at itself, and finding more to admire than to blame.

9 *Examining the Working of Whitehall*[1]

Though not many people share the view that the whole fabric of our democratic institutions is in danger there is widespread agreement that we are badly and inefficiently governed and that our traditional institutions demand a fairly radical overhaul.

In this connection, the report of the Fulton Committee on the Civil Service which is to be published next week[1] will be examined both for the recommendations it makes and for the evidence it gives of the way in which the mind of Whitehall is moving. For the composition of the committee, dominated by two powerful civil servants and containing no known radical critic of our governmental structure, suggests that what we shall get is likely to be fairly orthodox by current criteria.

The terms of reference of the committee, 'to examine the structure, recruitment and management, including training, of the Home Civil Service and to make recommendations', would themselves seem to imply a somewhat restrictive notion of the area of inquiry unless the word structure had been very widely interpreted. Furthermore, the Prime Minister, when announcing the setting up of the committee, was at pains to point out that there was no intention of altering 'the basic relationship between Ministers and civil servants' and that 'civil servants, however eminent' would 'remain the confidential advisers of Ministers, who alone were answerable to Parliament for policy'.

From these assurances and from the memorandum submitted by the Treasury in May 1966 one might forecast a report dealing with a number of issues by now both familiar and not particularly

[1] The Report of the Committee on the Civil Service under the Chairmanship of Lord Fulton was published as Cmnd 3638, June 1968.

controversial. The principal change would be the recognition of the fact that changes in the educational system have made it inevitable that the service's basic recruitment is bound to be of graduates leading, if the Treasury's view be accepted, to an amalgamation of the administrative and executive classes.

We shall get some greater opportunity for scientific and other professional civil servants to reach the higher rungs of the administrative ladder. We shall have further attempts to limit the 'Oxbridge' element in the higher posts – difficult though it is to see how this can be done in a centralized university system where nothing now prevents the gravitational pull of the older universities from having its full effect on the ablest would-be students. We shall have a lengthening and elaboration of training both at the initial and at later stages, though if the French model is followed too closely, recent events might create some scepticism about proposals of this kind.

We may get some new ideas on how to facilitate an increasing interchange of personnel between the Civil Service public and private enterprise, the universities and so forth, and perhaps some suggestions as to how limits can be placed on the high degree of party and private patronage which has marked the development of this process under the present Government.

All this will be very useful; so, too, though more controversial, will be the recommendation (should this be the way the committee's mind has moved) to separate the management of the civil service from the Treasury and place management, recruitment and training in the hands of a new and more powerful public service commission or even ministry. But if this is all it will not be sufficient to allay public disquiet.

For the assumptions of this approach is that what needs correcting has something to do with civil servants as individuals and not with the system which they operate. In fact, however, this is not the primary problem. There is really nothing much wrong with the calibre of British civil servants, as most foreign observers are ready to testify. Indeed, the average member of the administrative class is abler, better trained and better equipped to shoulder responsibility than the average Member of Parliament; the average Permanent Secretary is likely to be the intellectual superior of his Minister. That is part of the problem. But the advantage which this gives to the bureaucracy over the politically responsible element in government is made even greater by the weaknesses in the machinery which they have to handle.

When the Prime Minister talked of the basic relationship between ministers and civil servants he was moving in a world of platonic ideas rather than of reality. Ideally indeed a Minister with a grasp of the principal political implications of his department's work and advised by highly qualified experts on the matter in hand best corresponds to the requirements of democracy and efficiency.

But this happy situation is rarely attained today. One reason is well outside the terms of reference of a committee on the Civil Service, and that is the appalling speed of the ministerial merry-go-round. How can the transient and embarrassed phantoms who come and go at such a rate hope to acquire enough knowledge of and feel for a major branch of government business to exercise any realistic measure of control?

The reasons why ministers are shifted so often are to a large extent political, and there is no one to teach the parties wisdom. But they are partly to be found in the irrational distribution of departmental responsibilities, the multiplication of ministerial posts and the perpetual alterations in the relationship and balance of departments. There are at least two ways in which, without unduly stretching their terms of reference, the Fulton Committee may well have something useful to contribute. They could see that what was offered the Minister on the departmental side was genuine expertise, and this means going back to the practice of having the permanent heads of departments and their deputies chosen from within the department and not imported from outside. As Lord Salter has powerfully argued in his *Slave of the Lamp,* we need to get rid of the *damnosa hereditas* of Warren Fisher and of the over-centralization which he did so much to fasten on the Civil Service, with its consequent denigration of expertise in favour of some mystical overall administrative competence.

In the second place they could recommend a large-scale delegation of routine management functions without political implications away from the central departments, so as to free ministers and their advisers from some of the clogging weight of business and get the departments down to a more manageable size. This process would demand considerable alterations in the traditional forms of financial control and new safeguards for the public. But as Mr J. H. Robertson showed in the evidence he submitted to the Fulton Committee in December, 1966, modern management techniques enable these problems to be faced.

But unless the committee has transcended its terms of reference in a most unusual way, it must be assumed that they will certainly not have taken the final and necessary step which would be to investigate the main source of trouble and difficulty – to which Mr Robertson's evidence also pointed – namely, the actual distribution of work and responsibility between departments. For what we have here is analogous to the position on the last occasion when this whole vast subject was looked into – the Haldane Committee on the Machinery of Government in 1917–18. We could without difficulty apply verbatim to our present situation paragraph 4 of the Haldane Report itself:

> there is much overlapping and consequent obscurity and confusion in the functions of the Departments of executive Government. This is largely due to the fact that many of these Departments have been gradually evolved in compliance with current needs, and that the purposes for which they were called into being have gradually so altered that the later stages of the process have not accorded in principle with those that were reached earlier. In other instances Departments appear to have been rapidly established without preliminary insistence on definition of function and precise assignment of responsibility. Even where Departments are most free from these defects, we find that there are important features in which the organization falls short of a standard which is becoming progressively recognized as the foundation of efficient action.

At the time the Haldane Committee was working the problems were those created by the beginnings of the welfare state and by the improvised machinery on the supply and industrial side brought about by the war. As the recent assignment of new supervisory functions to Mr R. H. S. Crossman suggests, we have not yet solved the problem of organizing the social services side of government in an effective or acceptable way. But far more important is the complete chaos that exists in relation both to the post-Keynsian macro-economic controls and to the more recent attempts at direct state-intervention in industry.

When one looks at the changes in the roles and relationships over the last few years of the Treasury, the Board of Trade, the Department of Economic Affairs (since its creation) and now the new inflated version of the Ministry of Labour under its new style of Ministry of Employment and Productivity, not to mention the lesser and more specialized departments, one cannot detect any serious thinking about their respective roles nor indeed anything at all but a series of hurried improvisations in response to quite immediate pressures of a personal or party political nature.

When one further recollects that much of the work of direct intervention has local implications, and that regional organization has been subject to almost as much improvisation, it is not surprising that there should be the feeling abroad that whatever the State does is bound to be done badly. A would-be investor or an existing entrepreneur faced with the need for getting governmental approval for some project, or needing some estimate of the government's likely handling of some issue, is going to find it hard indeed to locate the point at which the relevant decision is likely to be made, or the department or office to which he would need to have recourse.

No business which was not intent upon bankruptcy would conduct its affairs by a series of departmental reorganizations every few months and a simultaneous or even more rapid shifting of key personnel. So when businessmen see the affairs of the State conducted in this fashion it is not surprising to find business confidence lacking. The same confusion of procedures and the inability to deal in an open and regular fashion with individuals or groups affected by public decisions – Stansted, the British Museum, the Post Office directories affair – contribute to the same feeling of helplessness among even wider circles of the population.

Too much attention has been paid to the purely parliamentary aspect of governmental reform and too little to the shape of the executive itself. It is incoherence in the executive, added to by the proliferation of free-floating irresponsible agencies – the Prices and Incomes Board, the Land Commission and so on – that needs tackling first. If Lord Fulton has not done this job – and it would be surprising if he had done it – the way will be open for a new 'Haldane.'

10 *Making the Civil Service more Outward-Looking*

One of the explanations of the manifest failure of the Fulton Committee on the Civil Service to produce a satisfactory general appreciation of the problems with which it was confronted, and for the somewhat inchoate set of solutions which it offered to them, must be found in the exclusion from its terms of reference of the vast and increasing area of government business that involves relations with other countries. For one of the principal features of the post-war decades has been the way in which aspects of policy once decided wholly within the governmental machinery of individual countries have been recognized as legitimate subjects for discussion with international organizations or with representatives of foreign states.

As the Shell Petroleum Company pointed out to the Fulton Committee: 'The need for Civil Servants other than those of the traditionally overseas departments to become internationally-minded is already pressing and, should Britain be accepted as a member of the Common Market, will be paramount.'

Nothing that the Fulton Committee has suggested will do much to meet the need and some of the proposals might go even further to produce what the Shell Company referred to as 'Civil Servants who are somewhat over-orientated to the United Kingdom and whose appreciation of the problems of an international group is correspondingly impaired.'

The notion that it is still possible to separate the whole question of representation overseas, including its economic aspects, from the administration of Britain's domestic concerns is also apparent in the latest attempt to deal with the former problem: the committee under Sir Val Duncan, which is now looking into the staffing of British embassies and other posts abroad, including the work of specialist attachés and advisers – military and civil – and the British Council. The fact that Sir Val Duncan's Committee has been set up only a little more than four years after the report of the Plowden Committee on the same general subject is evidence, not merely of the extent to which Britain's place in the world is presumed to have changed, but also of the inadequacies of the Plowden Report itself.

The Plowden Report like the Fulton Report was a testimony to the almost inevitable tendency of government-sponsored inquiries to examine everything except their own major premises. Instead of looking first at the purposes that the institutions they investigate are supposed to serve, and proceeding from there, they normally do little more than register the current conventional wisdom of Westminster and Whitehall. In this case such wisdom has already been encapsulated for them in Sir Con O'Neill's article in *The Times* of October 22, which would seem to accept the current denigration of the importance of political and information work and to suggest the possibility of some curtailment of the Service should economy so dictate. It may be said, however, that Sir Con does not subscribe to the other fashionable and dangerous notion which one must hope the Duncan Committee resists to the uttermost: the notion that commercial work should now play an increasing part in the work of the foreign service, with the inescapable consequence that the appropriate training for a diplomat would be much the same as that for a super-salesman.

That the commercial world should feel that the Foreign Office and diplomatic service neglects its interests is natural enough; people always feel their own affairs are the most important. Sir Val Duncan and his colleagues will be getting exactly the same kind of accusations about the Foreign Office indifference as were being made a century ago and which have been explored at length in Mr D. C. M. Platt's important book, *Finance, Trade and Politics in British Foreign Policy 1815–1914*. And just as in the past, some of the complaints will arise from an unjustifiable belief that government action can make up for industry's own deficiencies. But it is true also that in an age when the relations between governments and commerce everywhere are more intimate, and when government policy may so largely dictate what will be bought and where, a much greater degree of involvement in the details of commercial and financial transactions on the part of the British Government is inevitable.

But this does not necessarily mean that commercial work should be, or can profitably be, that of the Foreign Service. The first duty of the Foreign Service must remain one of reporting as accurately as possible upon developments in the countries to which its members are accredited. The fact that Britain's relative weight in the world has declined does not make political intelligence any less important; rather the contrary, since it is harder to intervene actively to set right the results of neglect. The external critic is at a disadvantage when trying to document this claim because he has no access to reports from overseas posts of recent date. But if one looks for instance at the successive rebuffs that Britain has suffered in her European policy over almost two decades, it is hard to believe that the Government was invariably as well-informed as it should have been.

The Middle East would seem to be another area where many miscalculations of trends and of the relative strength of different local forces seem to have been made. A defence might be that diplomats can report but that ministers are not obliged to listen; but where this happens (if it does) the Service cannot wholly evade responsibility.

If it be admitted that reporting is the principal function of a properly diplomatic kind that falls upon the Service, and that negotiation (which depends upon correct information) is the second most important one, it is very important indeed that it should be abilities in these directions that the Foreign Service should endeavour to recruit and train. It is also important that the Foreign Service should regard the general prestige of the country abroad as its particular concern, and that it should therefore take responsibility for information work, and probably supply much of its personnel.

When we look at these functions it is obvious that the traditional subjects of study for diplomats are the correct ones – history and languages; some might add social anthropology and other aids to the understanding of societies from outside the common matrix of European-North Atlantic civilization. In the interests of making people equally fit to do commercial work these priorities have been somewhat neglected. Furthermore, work of this kind, if it is to be done well, requires a fairly high degree of specialization by country and region. The Foreign Office's habit of shifting people about from one country to another at very short intervals is as dangerous as a similar tendency to shift people too often between different types of work, as the Fulton Committee noted in the Home Civil Service. The evidence of the *Guardian* to the Fulton Committee is well worth the Duncan Committee's attention:

> The work of the British Foreign Service suffers, we think, because career diplomatists seldom stay in one country long enough to learn about it. The second-in-command of the French Embassy in London has been here since the war. He is in a better position to keep his Government accurately informed than is his opposite number in the British Embassy in Paris. . . .

A Foreign Service designed solely for these functions could of course be smaller than that of today, and the feeling of inflation that Sir Con O'Neill detects would diminish. What it would mean is that all commercial work would revert to the responsibility of the relevant home departments – notably the Board of Trade – and that other home departments, agriculture, transport and so on, would furnish the manpower for all special inquiries and actual negotiations, as they largely have to today in the international organizations.

The proposal is not one of reverting to a separate Commerical Diplomatic Service nor to a separate consular service, though there might be room for rethinking what the responsibilities of consuls ought to be. What would happen would be that the Foreign Service would consist of specialists in the field of political (and economic) reporting, and that most of the rest of the work would be done by members of the Home Civil Service, who would have to accept as normal quite considerable periods of being posted to embassies abroad, and of frequent shorter foreign tours of duty. The departments of government would be distinguished by their functions, not by the degree to which these functions were carried on outside the United Kingdom. In the age of the jet-plane and radio a 'Home' Civil Service as contrasted with a 'Foreign' Service is not easy to justify once one starts to think about it. One hopes that the Duncan Committee, for all the element of urgency surrounding its setting up, will be given time to think as well as to inquire.

11 Comes the Devolution?

The history of ideas is not a very British branch of learning, though it has a persuasive advocate if rather rare practitioner in the person of Sir Isaiah Berlin. One thing that is clearly lacking is a history of intellectual fashions, notably on the political scene. The strong enthusiasms that pull now towards individualism, now towards collectivism have been recorded by Marxists and others, but there are lesser styles that still baffle the observer. Some fashions appear to spread uniformly through a large number of countries more or less at the same time. Others go in counterpoint; thus American administrative reformers have tended to admire the British model while the British Civil Service is now being recast on somewhat American lines. (In another field 'comprehensive education' is the fashionable thing in this country – while the Americans show an increasing interest in selectivity.)

These reflections are prompted in part by the debate of the last two decades over the correct size of the unit of government in the modern world. Twenty, and even ten, years ago, the accent was all on *size*. We were harangued continually by people who proclaimed that the nineteenth-century nation-state, quite apart from its war-making proclivities and other disagreeable habits, was much too small for its people to enjoy the full benefits of modern technology and a modern economy. Federations, unions, amalgamations were all the rage. And even though not all of these schemes came off – even though the European idea itself has been 'demonnetized' – it was still the case that the national governments themselves, extending their functions from controlling the economy to planning it, have been extending and hence centralizing their authority all the time.

There seemed no end to the benefits that could be sought through conferring increasing powers on the governments of large units. In Switzerland, alone in Europe, considerable powers were retained by the lesser units – cantons and even communes – but this was regarded as a pleasant anomaly, like the survival of San Marino or Andorra, and not as something to be taken seriously by the student of politics.

And now quite suddenly, the landscape has changed! Since increasing centralization has not produced its expected benefits, there has been a sudden reaction against the whole idea, a sudden rediscovery of the virtues of the smaller unit. Sometimes this is strengthened or pervaded by nationalism – the Scots, the Welsh, the Bretons, the Basques, the Flemings show signs of taking up again the course of their national histories, interrupted centuries ago by conquest and apparent assimilation – just as the Poles and the Czechs and the Slovaks, and other peoples of Eastern Europe did half a century ago. But nationalism is not essential to this feeling that too much is decided too far away. Cornish nationalism, perhaps; but Mercia, Northumbria? – that is really too much to ask. 'Regionalism' – a neutral word – covers most of the debate where England is concerned; with Scotland and Wales the argument is more open.

The reversal, one hastens to add, is a reversal in intellectual fashion and to a limited degree in popular debate. It has not yet made much of an impact on national politics, still less upon the entrenched panjandrums of Whitehall. Indeed just at the moment when the virtues of local initiative are being stressed once more, successive Ministers of Education have been doing their best to force on often reluctant local authorities a new uniform pattern of secondary education (of, incidentally, a highly undesirable kind). Since education is the most important of state functions still largely in local hands, this is not without significance.

Nevertheless, one must assume that an intellectual fashion of this kind will in the end have its effect on both Westminster and Whitehall. To some extent the issue is being forced upon the Government. The failure of existing units of local government to conform to the present distribution of population and consequent requirements for services made necessary the appointment of the Redcliffe-Maud commission (in 1966) with powers to recommend major changes in both the units and the structure of the system. The unexpected delay in producing its report suggests that the difficulty of getting agreement has been more considerable than was thought. Meanwhile, the possible electoral consequences for the Labour Party of Scottish and Welsh nationalism has been responsible for the creation of a constitutional commission under Lord Crowther which is expected no doubt to take a decently long interval before saying that nothing can be done, in the hope that by then no one will mind any more.

For intellectuals, however, the moment to get on the regionalist bandwagon is clearly now, and from Mr John Mackintosh (a professor of political science turned Labour MP) we have as a harbinger no doubt of a great deal more, an admirably succinct statement of the problem and a set of answers of impeccable sobriety – unlike some of his compatriots in the Labour Party, Mr Mackintosh is not to plead deafness when the siren-voices of Scottish nationalism are heard. He listens – and is, on the whole, unmoved.[1]

Mr Mackintosh, broadly speaking, comes out for a series of regional governments – nine for England and one each for Wales and Scotland – with roughly identical powers, and ruled as to the powers devolved upon them by elected councils (in the case of England) and elected assemblies (in the case of Wales and Scotland). Financial autonomy for the regions would be assured primarily by a system of allocations from the central pool of taxation, the spending of which would (within certain limits as to the provision of equal services) be at the discretion of the regional councils or assemblies and the ministries responsible to them. Mr Mackintosh thus repudiates both the extended conurbation as the best unit for tackling England's problems and, except in token fashion, special treatment for Scotland and Wales along the lines claimed by the nationalists.

On the whole, I find Mr Mackintosh's arguments quite convincing. If something like this were put into effect, I would not go to the barricades to stop it; nor, however, would I go to the barricades in support of his plans. Would anyone?

The difficulty is to know how far this new fashion for devolution is a serious reflection of what people want as distinct from what they occasionally say they want, or more often say they do not want. When they ask for devolution, does it mean they want the benefits of centralization and think they can get devolution and

[1] J. P. Mackintosh, *The Devolution of Power: Local Democracy, Regionalism and Nationalism* (London, 1969).

still retain all of them, which seems on the face of it improbable? Or does it mean that they want devolution so much that they would make some sacrifices in order to get it?

The argument in economic terms is an old one. There have always been those who disagree with the contention that poorer regions benefit by being joined to richer ones. We are confronted with the classic case of the Italian *mezzogiorno*. But even if it can be shown that the economy of southern Italy suffered when Italy was united, the opportunity for southern Italians to migrate northwards in search of their fortunes has been important for them, just as the high road to England was in Dr Johnson's view the fairest prospect that could face a Scotsman. Now capital comes back from northern into southern Italy, from England into Scotland – who can say where advantage lies? The matter is further complicated today by the coming of the welfare state. Is anyone prepared to accept a lower scale of state benefits, educational provision, and so on because he happens to live in a poorer part of the country? One doubts it. Mr Mackintosh's treatment of Northern Ireland is germane to this issue. In the case of Scotland, it is possible to argue (as Mr Gavin McCrone has shown in a recent article in *New Outlook*) that Scotland – if it contracted out of national defence and overseas aid – could make up for the subsidy of her services that now comes from the British Exchequer. On the other hand, her dependence on the British market would so severely restrict her capacity to have separate economic, fiscal, or monetary policies as to make the talk of independence in any genuine sense almost empty of content. That is to say, of course, provided the Scots are not prepared to accept major economic upheavals and (at any rate, for a time) a lower standard of living.

But suppose, the Scots, the Welsh, the Cornish, the Mercians, *et al.*, are prepared to accept Mr Mackintosh's *via media,* do any difficulties still remain to be faced? Mr Mackintosh is right to see that at this level it is men not money that is at the root of the matter. But he is optimistic that the giving of greater powers to the regions than are available to the existing organs of local government would get over this. He may be right; but the problem may be harder to solve than he thinks.

At one stage in his argument, Mr Mackintosh produces figures to show how deficient existing local authorities are in qualified manpower – whether we are thinking of architects, town-planners, child-care officers or of other skilled persons. Mr Mackintosh's view is that this deficiency arises largely from the fact that most units are too small to provide an adequate career structure and that regional authorities would have more to offer, as indeed they would. But one reason may be that the pay offered is too poor to attract the necessary people – and that this may be due to parsimony rather than to poverty. Would larger units do better?

Or the cause may be even more deep-seated; the people may just not exist, and may never exist. One of the other intellectual fashions of the last few years which is about to take a severe jolt is what might be called the 'Robbins fallacy'. Provided you give people higher education – it does not much matter in what – the 'hidden hand' will see that the necessary skills will emerge in the right proportions from the uncontrolled operation of consumer choice. It seems odd that the economics of Adam Smith, rejected in so many respects, should reappear in this one. But suppose the hidden hand does not operate – suppose we go on getting more sociologists and too few child-care officers, more philosophers and too few welfare workers, more experimental psychologists and too few mental health workers? It just could be.

But even if we believe that the problems of professional staffing will resolve themselves, who are to man the new regional councils and the new ministries? Mr Jo Grimond, Prime Minister of Scotland, perhaps; but Mr Enoch Powell, Prime Minister of Mercia? And what about the South West? – Mr Jeremy Thorpe, say the Devonians – Dr A. L. Rowse, presumably, say the Cornishmen . . . it seems highly improbable. The principal trouble with local government is that the calibre of council members is so weak. (Some critics would say that this goes for Parliament too.) Mr Mackintosh appears to think that high-powered people could be recruited for regional councils for sessions totalling up to two-and-a-half months in the year. What part-time (or near part-time) occupations does he know of from which these people could be drawn?

One must assume that Mr Mackintosh would not want his councils to be composed exclusively of the superannuated, or of middle-aged, middle-class housewives. Does he not, like so many other reformers who go on adding to the number of public functions that have to be performed, also go on cherishing the view that the country is still full of large numbers of people with no need to earn their livings, available at need to do what the public good commands?

Given the likelihood that the personnel of the regional councils will be weak and those of the necessary second-tier authorities even weaker, are we quite sure that we want regional governments or any other kind of devolution? It may be that those closer to the scene are better able to assess local need; but it may be that the only needs they can assess are those of the moment. If we look at the record of destruction of English amentities, rural and urban – natural beauty and historic buildings, not to speak of sewage in the sea – are we convinced that what we would get from devolution would be better government than what the mandarins of Whitehall give us now?

The real fashion of the moment may not be devolution at all – it may be 'participation'. If it is, if people want to participate, and not watch television, pub-crawl, fill in the football pools, and all the other things they apparently do want, then we shall move away from centripetal politics and get devolution all right. In every Briton there may be a Swiss screaming to get out. But I would want more than a couple of 'in-words' among intellectuals to convince me on that point. It would be great fun if one were wrong.

III FOREIGN AFFAIRS

12 *Lucien Wolf and the Anglo-Russian Entente 1907-1914*[1]

Lucien Wolf's most positive memorial, apart from his own historical writings, is to be found in the record of his work for the Jewish Community of this and of other countries. Yet he was more than a communal leader of unusual distinction. As a political journalist he exercised responsibilities which went beyond the bounds of his own community. This aspect of Lucien Wolf's career must in retrospect bear the air of frustration. Not only were the policies he assiduously advocated rejected, but the historical record suggests that his analysis of the European scene was indeed largely at fault. Events themselves and later disclosures suggest he underestimated the strength of German militarism, and the irresponsibility of Austria's rulers. But the period during which Lucien Wolf was one of the leading journalistic critics of Sir Edward Grey's foreign policy is of such major interest, and the issues involved of such permanent significance, that I make no apology for my choice of subject.[2]

From Lucien Wolf's own point of view his activity as a British journalist was in no sort of contradiction with his abiding concern for the welfare of Jewry. No notion of divided loyalties entered into his thought; and he was prepared to justify every position that he took up according to the interests and traditions of Great Britain.[3] Yet in one important aspect of his attitude to foreign affairs – his

[1] Lucien Wolf Memorial Lecture, 7 March 1951.

[2] For a brief survey of Lucien Wolf's career, see the 'Memoir' prefixed to *Essays in Jewish History* by Lucien Wolf, edited by Dr Cecil Roth (London 1934).

[3] See 'Lucien Wolf' by D. Mowshowitch, in Supplement to *The Jewish Chronicle*, 26 August 1932.

opposition to closer association with Tsarist Russia – the connection with his Jewish affiliations and sentiments could not be overlooked. His enemies in the more bellicose and anti-German section of the press inevitably ascribed his antipathy to Russia to his devotion to the cause of Russia's persecuted Jews, when they did not descend to alleging that he was in the pay of pro-German Jewish financiers. It is indeed a curious and tragic irony that the same accusations of bias were made against Lucien Wolf for calling attention to the iniquities of the Russian government, as were to be made a quarter of a century later against those who argued that Germany's treatment of her Jews was a portent and a threat. It is true that the accusations brought against Wolf were brought against him by a minority of more or less anti-semitic tendencies. They did not prevent him from occupying, throughout the period with which we are concerned, an honoured place in his profession. But in a country such as Great Britain the problem of conscience involved must always be a real one, because of the homogeneity of the mass of the population and the consequent strength of national sentiment.

In the United States over the same period, a very different state of affairs existed. There was a country in which large groups of recent immigrants and their descendants were prone to exert their influence on foreign policy in all matters that seemed to them of special importance. The political system of the country and the lack of strong party ties gave minority groups considerable power. In Great Britain, the minority group was an anomaly and the Jewish vote, if it existed, a matter of indifference. The fifteen or so Jewish Members of Parliament showed little disposition to differ from their party colleagues on policy towards Russia, and obediently followed the party whip.[4]

The most clear-cut illustration of this fact concerns a relatively minor aspect of the whole Russian issue. This was the attempt by

[4] In 1880 there had been 10 Jewish candidates for Parliament of whom 5 were elected; in 1886, 7 were successful out of 8; in 1892, 7 out of 11; in 1895, 8 out of 17; in 1900, 12 out of 21. In the election of 1906, 16 Jews were elected to Parliament; 12 as Liberals and 4 as Conservatives, out of 32 candidates. (*Jewish Chronicle,* 12 January, 2 February 1906). In the election of January 1910, there were again 32 Jewish candidates of whom 14 were elected; in December 1910, 15 were elected out of 31. (*American Jewish Yearbook.*) The fidelity of Jewish MPs to the Government's policy of friendship with Russia gave Hilaire Belloc a chance to make some sarcastic remarks at their expense in the debate on the King's visit to the Tsar in 1908. *House of Commons Debates,* 4 July 1908.

the Russian Government to refuse visas to foreign nationals who were Jews. Activity on the part of American Jewish organizations stimulated the protests of the American Government against this claim on the part of a foreign state to discriminate between different classes of American citizens. Failing to get satisfaction, the American Government gave notice, in December 1911, to terminate the treaty of 1832 which regulated relations between the two countries, and with support from both political parties refused to negotiate a new one unless its wishes on this point were met.[5] In Great Britain a similar campaign had no effect upon the Foreign Secretary. Grey continued to take the view expressed by his predecessor Lord Granville in 1881, that so long as British Jews were not subjected to disabilities greater than those affecting Russian Jews, no violation of the Anglo-Russian Treaty of 1859 had taken place and no intervention could be undertaken without an unwarrantable interference in Russia's internal affairs.[6]

The passport question has an intrinsic importance greater than might at first sight appear, since it did provide an opening for calling attention to general anti-Jewish discrimination in Russia which might otherwise have been lacking. And of course, the defence of the Government's refusal to intervene, that the issue was an internal one for the Russian Government, was made with even greater firmness when the major question did manage to protrude into Parliamentary discussions on foreign policy.

The attacks made upon the British Government for associating so closely with a government which practised towards its Jewish subjects, and to an only lesser degree towards the other minority peoples of its Empire, a policy of legal discrimination and brutal repression, raised one of the permanent problems that confront democratic governments in their foreign policy. In the words of Professor A. V. Dicey, writing in 1912, 'The wrong wrought upon the Jews under the law and still more under the administrative system of Russia, is all but incredible to Englishmen . . . Russian law and Russian officials apply to the treatment of Jews methods of barbarism or of medievalism . . . Russian statesmanship forces Russian Jews to make a choice between exile on the one hand and

[5] See for American diplomacy in this matter, Cyrus Adler and Aaron M. Margalith, *With Firmness in the Right* (New York, 1946), chapter ix.

[6] Lucien Wolf: *Notes on the Diplomatic History of the Jewish Question,* (London, 1919), pp. 80–83. Cf. *Darkest Russia,* 15 July 1914.

on the other hand either ruin or (what is still more terrible) the permanent degradation of every Jew who inhabits the Russian Empire.'[7] It is true that some apologists for Russia – Sir Bernard Pares and Maurice Baring for instance – tried to deny or explain away the most serious allegations. Others argued that it would discourage the forces making for reform in Russia if the country found itself ostracized by the great Liberal Powers. But fundamentally, the answer put forward by the Government and its supporters on this issue, was a straightforward one of *raison d'état*. National safety demanded the Entente with Russia, and the unpleasant aspects of one's partner were not to be inquired into too closely. For Grey, for diplomats like Sir Charles Hardinge and Sir Arthur Nicolson, and for much of the press, those who called attention to the evil deeds of Tsarism were acting against the best interests of the country, for reasons that were at best merely matters of sentiment. Nor could they be persuaded that an unreformed Russia was a weak reed on which to lean, that its social and political order could not survive the ordeal of a major war – prophecies that were to receive an abundant and terrible justification in 1917. This perennial conflict between the demands of humanitarianism and considerations of sheer power had rarely been so sharply defined, although Lucien Wolf, with his real feeling for the continuity of English history, found many parallels in the past. On the publication of the text of the Entente, Wolf wrote:

> On the ethical side I persist in believing that it is deplorable. It is a reversal of all the ideas which have inspired British foreign policy since the time of Pitt. For the Liberal Party, which boasts its descent from the Puritans, to conclude such an Agreement, especially under present conditions, seems like a freak of topsy-

[7] *The Legal Sufferings of the Jews in Russia*, edited by Lucien Wolf, with a preface by A. V. Dicey (London, 1912), p. 1. At a meeting held on 28 October 1913, at Memorial Hall, Farringdon Street, under the auspices of the English Zionist Federation, with Sir F. A. Montefiore in the Chair, Dicey moved a resolution of protest against 'the re-crudescence of the utterly baseless and wicked Blood Ritual charge against the Jewish people', in connection with the Beilis case. Four other non-Jewish speakers took part and many messages of sympathy were received from ecclesiastical dignitaries and other prominent persons. *The Times*, 29 October 1913. Dicey expressed disappointment, however, that of all the known sympathizers, only the novelist Anthony Hope (Hawkins) actually attended: 'I did wish that some one of the numerous Bishops and other orthodox divines who expressed their sympathy could have made an effort to be present.' *Memorials of A. V. Dicey*, ed. R. S. Rait (London, 1925), pp. 220–1.

turvydom. I am not altogether a believer in the quixotry of Puritanism, but Liberalism has travelled very far from its ideas in foreign policy as they were once expounded by Harrington: 'The duty of a free Commonwealth is to relieve oppressed peoples and to spread liberty in other lands to the intent that the whole world may be governed with righteousness.' I venture to commend this dictum to the Nonconformist conscience.[8]

And as might be expected there are references in his writings to Canning and Lord John Russell as robust defenders of the tradition of sympathy with liberal causes.

There is another reason why the study of Lucien Wolf's share in the campaign against the Entente is of particular interest; and that is because it was only in the press that the critical attitude that many people had towards the new departures in British foreign policy, could be fully expressed. When Mr James Joll writes in the introduction to his collection of documents relating to British foreign policy in the period between 1793 and 1940 that 'throughout the period it is particularly noticeable with regard to foreign policy how adequately every viewpoint has been represented in Parliament even if only by a limited group of members',[9] he by no means reflects the view taken by Grey's critics during the period with which we are specially concerned. On the contrary, it was a commonplace among them that Parliamentary procedure did not permit a full discussion of the momentous changes that were taking place.

The main reason for this was the fact that the two front benches were at one on the main issues of policy and that Parliamentary procedure under the control of the whips gave the minority of dissenters very little opportunity to express their views. One of the Liberal opponents of Grey's policy put it succinctly in an article written in 1912:

The difficulty has, in fact, been due to a real and fundamental change in our politics affecting the whole character of the House of Commons. It has been due to the 'Party machine' absorbing more than ever the life of Parliament, and to an increasing

[8] *The Graphic*, 28 September 1907.
[9] *Britain and Europe*, ed. James Joll (London, 1950), p. 25.

belief that questions of defence and international relations ought, as far as possible, to be kept outside the party system.[10]

As another of them put it:

The Foreign Office Vote is the one opportunity for a special debate on our foreign relations. But even this is dependent on the request of the Opposition. In recent years a Session has been known to pass without the Foreign Office vote being taken at all. The small minority – and it is very small – of Members, on both sides of the House, who are specially interested, and who may from their knowledge foresee difficulties and changes ahead, is practically powerless if it desires to have a debate.[11]

Asquith, when asked by the Select Committee on Procedure of 1913–14, agreed that foreign affairs were less often discussed than formerly and attributed it to lack of a sharp difference between parties. He pointed out that private members did not put down motions on foreign affairs for the ballot. The Speaker, Lowther, also thought foreign affairs were less discussed than formerly; not so Balfour.[12]

Opposition on foreign policy in the parliaments of the pre-1914 decade was confined to the Labour Party and the Radical pacifist group of Liberals. With the exception of their hostility to Tsarist Russia and their desire to see greater efforts made to come to terms with Germany, the adherents of these groups had little in common with Lucien Wolf who, though calling himself a Liberal, was for nineteen years on the staff of a Conservative newspaper, and who had no sympathy with pacifism, unilateral disarmament, compulsory arbitration, or other radical panaceas.[13]

In particular, he did not share the pacifist opposition to naval armaments, since he fully recognized that naval supremacy for

[10] Philip Morrell, MP: 'The Control of Foreign Affairs: The Need for a Parliamentary Committee', *Contemporary Review*, November 1912. Morrell was elected chairman in 1913 of the 'Foreign Affairs Group' of the Liberal Party, which had been organized in the previous year by Noel Buxton and Arthur Ponsonby with the support of about 70 of the Party's left wing. Ponsonby succeeded Morrell as chairman in 1914. T. P. Conwell-Evans, *Foreign Policy from a Back Bench* (London, 1932), pp. 81–2.

[11] Arthur Ponsonby, *Government and Diplomacy* (London, 1913), pp. 50–51.

[12] *Select Committee on House of Commons (Procedure) Report*. Parliamentary Papers: House of Commons 378 (1914), pp. 160–1; 212–13; 89.

[13] He expressed himself once as being worried to find himself (on the Persian question) on the same side as *The Nation*. *The Graphic*, 22 June 1912.

this country was a matter of life and death.[14] Indeed by early in 1913 he was referring to an Anglo-German naval agreement as 'that fascinating will-o'-the-wisp of the Pacifists'.[15] Before the important debate on foreign affairs after the Agadir crisis in November, 1911, Wolf had lamented that it was 'a pity it should have been left to the rag-tag and bobtail of Radical Pacifism to place before the country the very serious issues involved in this debate'.[16] But on the issue of greater Parliamentary control, Wolf was at one with the members of this group, quoting Canning effectively on the danger to a minister 'who should undertake to conduct the affairs of this country upon the principle of settling the course of its foreign policy with a Grand Alliance, and should rely upon carrying *their* decisions into effect by throwing a little dust in the eyes of the House of Commons'.[17]

Earlier in the same year we find Wolf dealing with the whole subject of Parliamentary control of foreign policy in an article entitled: 'The Sacrosanct Foreign Office'. He advances the highly un-Radical view that one of the main causes was the fact that the Foreign Secretary now sat in the House of Commons, as well as the Under-Secretary (Mr T. McKinnon Wood). And he refers disapprovingly to Sir Edward Grey's 'extraordinary arrangement that he should only come down to the House of Commons to answer questions twice a week and that then he should be spared the heckling which is the common lot – and a very stimulating lot too – of all Ministers'. In the House of Lords, he writes, sit the only two living ex-Foreign Secretaries and all the five ex-Under-Secretaries as wall as prominent retired members of the diplomatic service, so that debates there on foreign affairs are uniformly of superior quality. There is also, he adds, 'the movement, led by *The Times,* for discountenancing public criticism of the Foreign Office on patriotic grounds. In this way the vicious circle has been completed, for the vigilance of Parliament, chloroformed by the Cabinet has thus been denied the saving stimulus of an alert public opinion'.[18] In a later article, Wolf returned to this theme when in deploring the passing of the Parliament Act, he asserted that if 'the emasculation of the House of Lords' had any effect on the manage-

[14] *The Graphic,* 20 March 1909.
[15] Ibid., 15 February 1913.
[16] *The Graphic,* 25 November 1911. The Debate took place on 27 November.
[17] *The Graphic,* 25 November 1911.
[18] Ibid., 22 April 1911.

ment of foreign policy 'it would rather be in the direction of giving a looser reign to the Foreign Office than of democratizing it'.[19]

Such criticism was without effect and down to the outbreak of the first world war, the position of the Foreign Secretary remained not very different from the way in which it was described in 1913 by a French student of British affairs:

> The British cult of discretion shelters the Foreign Office from idle chatter. The national religion of the monarchy protects it against attacks. In the shadow of the King, and assisted by the permanent Under-Secretary of State, the Minister of Foreign Affairs retains at the beginning of the twentieth century, the power of an autocrat. Established on the poop, well above the agitation of the democratic mob, the pilot, his hand on the wheel, his eyes on the compass, steers the ship, lending only an unattentive ear to the distant murmurs of the parliamentary passengers between decks.[20]

In such circumstances, the role of the journalist was more than usually important; and the study of his contribution to the formation of the public mind is an important element in the task of an historian of this period. It is not one that has hitherto been performed.[21] Because he stood aside from the main group of the Government's journalistic supporters as well as from that of its journalistic critics, Lucien Wolf's role appears to deserve particular attention. By the time of the negotiation of the Anglo-Russian Entente, Wolf was already known for his contributions (often under the pseudonym 'Diplomaticus') to a number of important

[19] Ibid., 19 August 1911.

[20] Le culte britannique de la discrétion met le foreign office à l'abri des bavardges. La réligion nationale de la monarchie le protège contre les attaques. Couvert par le Roi, aidé par le sous-secrétaire d'Etat permanent, le ministre des Affaires étrangères conserve au début du XX siecle, la puissance d'un autocrate. Campé sur sa dunette, bien au dessus des agitations de la foule démocratique, le pilote, la main sur la barre, les yeux sur la boussole, guide le navire en ne prétant qu'une oreille distraite aux lointaines clameurs des passages parlementaires, qui, là-bas pérorent dans l'entre-pont.' Jacques Bardoux, *L'Angleterre Radicale,* (Paris, 1913), pp. 394–5. Cf. F. Gosses, *The Management of British Foreign Policy Before The First World War* (Leiden, 1948).

[21] Cf. E. L. Woodward, *Great Britain and the German Navy* (Oxford, 1935), p. 9. A beginning has been made in the book by Oron J. Hale: *Publicity and Diplomacy with Special Reference to England and Germany, 1890-1914* (New York, 1940). On certain examples of anti-British bias in the handling of this author's material, see *The History of The Times,* vol. III (London, 1947), pp. 806–809.

periodicals. He had been since 1890 the foreign editor of *The Daily Graphic,* a post which he resigned in 1909, though continuing to contribute signed articles, and the weekly illustrated journal, *The Graphic,* contained in almost every issue a signed article by him on foreign affairs, usually, though not always, under the heading 'The Foreign Office Bag'.[22]

At this point it is necessary to digress in order to deal with a most extraordinary series of statements made about Wolf in a book published a few years ago by a Mr Comyns Beaumont who was, for a short period, the editor of *The Graphic*. According to Mr Beaumont he objected to the anti-Russian tendencies of Lucien Wolf's articles and secured his replacement as the author of 'Foreign Office Bag'. Mr Beaumont's narrative is vague on the subject of dates; but he would appear to place this incident somewhere around 1909. It is true that Lucien Wolf's file of these articles has a considerable gap for the latter half of that year which may spring from his differences with Mr Beaumont. But they were certainly resumed before the end of the year and continued without a break until after the beginning of the war. Mr Beaumont's 'victory' seems to have existed only in his imagination.[23]

Mr Beaumont follows this up with the statement that on a subsequent visit to Berlin with Lord Northcliffe (whose employment he entered on leaving *The Graphic*), he discovered that Lucien Wolf was acting as an *agent-provocateur* for the German Government, who found it useful to have someone in London who could insert bellicose articles in the English press and thus justify to the German public the Kaiser's plans for naval expansion.[24] Why the Germans should have needed this kind of thing when they had far more bellicose journalists such as J. L. Garvin and Leo Maxse busily beating the big drum is unexplained. In fact Lucien Wolf, as we now know, did not bask permanently in the sunshine of the Wilhelmstrasse's approval, despite the attention paid in Berlin to his writings. Indeed in September 1906, it was reported to the

[22] The 'Memoir' confuses the two papers. *Essays in Jewish History,* p. 5.

[23] Comyns Beaumont, *A Rebel in Fleet Street* (London, n.d.), pp. 45–8.

[24] Ibid., pp. 50–51. In the same category comes the allegation made by J. W. Robertson Scott that Wolf was a director of the German Wolff Telegraph Bureau. See his article, 'Who Secured the Suez Canal Shares?', *Quarterly Review,* July 1939. This seems to have been withdrawn in a footnote on p. 212 of the same writer's *The Story of the Pall Mall Gazette* (London, 1950). I owe these references to the article by Alexander Behr, 'Lucien Wolf: a recollection', *Jewish Monthly* (August, 1950).

German Foreign Office that Lucien Wolf had been excluded from contact with the German Embassy because of the anti-German tone of his articles in *The Times*.[25] In reply to a query of the Kaiser's on 4 October 1911, about the veracity of an article by Wolf on Agadir, it was described by Kiderlen-Wächter as a mixture of truth and falsehood.[26] And on 11 June 1912, von Kühlmann wrote to Bethmann-Hollweg that, although Wolf had formerly been favourable to the Triple Alliance, he had now turned himself into a zealous advocate of the 'Alliance' with France.[27] On 17 June 1914, Lichnowsky called attention to an article by Wolf applauding the Anglo-German agreement on the Baghdad Railway.[28]

As a final touch, Mr Beaumont adds the information, picked up in Berlin, that Wolf received 'a very considerable retainer from a most influential and wealthy international finance house in the City, whose tentacles stretched to all the world money centres', and which was 'opposed to any close military alliance between Russia and Britain for reasons no doubt well known to its partners' so that it became 'Mr Wolf's object to throw a spanner into the works'.[29] Mr Beaumont does not pause to reflect that the line Wolf's German paymasters are alleged to have wanted him to follow would be exactly the contrary of that which would be followed by outright opponents of Russian Alliance whose writings would hardly justify increased German armaments. Nor of course does he suggest any reason why it should have been necessary to bribe Wolf to attack a policy which was being as vigorously opposed (if on dissimilar lines) by such obviously high-minded and unbribable journalist as C. P. Scott and Herbert Sidebotham in *The Manchester Guardian,* H. W. Massingham and H. N. Brailsford in *The Nation,* A. G. Gardiner in *The Daily News,* and F. W. Hirst in *The Economist*.[30] The whole of Mr Beaumont's confused

[25] *Die Grosse Politik der Europaischen Kabinette* (Berlin, 1921–27), vol. xxi, p. 416 fn. There are further references to Wolf's writings of a less unfavourable character vol. xxiii, pp. 475, 509–10.

[26] Ibid., vol. xxx, pp. 71–3.

[27] Ibid., vol. xxxi, pp. 506–510.

[28] Ibid., vol. xxxvii, p. 449.

[29] Beaumont, op. cit., p. 51.

[30] For the opposition to Grey's policy in the bulk of the Liberal press, see J. L. Hammond, *C. P. Scott* (London, 1934), pp. 149–176. On Grey's failure to keep in touch with the press, see R. C. K. Ensor, *England, 1870-1914* (Oxford, 1936), pp. 572–574.

and insubstantial allegations seem no more than a reflection of the campaign against Wolf, carried on at the time of these events by Leo Maxse in the *National Review*.

It would be an error, as we have seen, to regard Lucien Wolf as in any sense consistently pro-German during his journalistic career; and it would be equally untrue to suggest that he was consistently anti-Russian. His first published work (at the age of twenty) was a small pamphlet composed in 1877 and calling attention to the danger of Russian designs in Central Asia.[31] But later his attitude seems to have modified. Indeed by 1896, when Britain appeared to be faced with the growing hostility of France, Lucien Wolf was advocating an entente with Russia as the best safeguard: 'Of all the Powers Russia is the one which competes with us the least and with whom we have most in common.'[32] In an article entitled 'The Russian Ascendancy in Europe', he argued that the great schemes of internal development and reform – including religious toleration – would occupy Russia's best energies for years to come and were the best guarantee of her continuing to pursue a peaceful policy.[33]

The revelation by Bismarck in October of the same year of the existence of his secret reinsurance treaty with Russia of 1887 and the circumstances of its non-renewal in 1890 led him to point out to the French that 'La Belle Russie' had turned out to be 'no better than she should be, a lady with a past, a sort of second Mrs Tanqueray on a very large scale' and that they would do better to try to improve relations with Great Britain. Nevertheless, Russia is not directly attacked and Germany appears as the more dangerous power.[34]

Thus fifteen years after the new acute phase of Russian antisemitism had begun with the anti-Jewish outbreaks of 1881 and the May Laws of 1882, Lucien Wolf by no means regarded implacable hostility to Russia as one of the guiding principles of his

[31] Lucien Wolf, *The Russian Conspiracy, or Russian Monopoly in Opposition to British Interests in the East* (London, 1877).

[32] 'The Two Eastern Questions' (article signed W.), *The Fortnightly Review*, February 1896.

[33] 'The Russian Ascendancy in Europe' (article signed 'Diplomaticus'), ibid., October 1896.

[34] 'Prince Bismarck's Secret Treaty' (unsigned article), *Fortnightly Review*, December 1896.

writings on foreign affairs.[35] Indeed, writing in 1907, he placed the abandonment of his previously Russophile attitude at a quite recent date:

> Eight years ago I was afforded an opportunity of seeing things rather more closely than other people, and I came reluctantly to the conclusion that the theory of the incurable bad faith of the Russian Government was no mere superstition. Since then my doubts have been confirmed, and I question very much whether we should be safe in any undertaking we might negotiate with the Tsar's Foreign Office.

And he went on to refer to Russian hostility to Britain as manifested at the time of Fashoda crisis, during the Boer war and in her Afghan and Manchurian policies.[36] There is no doubt that this correctly represents the development of his views.[37] At the same time Wolf was increasingly in contact through political exiles with the beginnings of Jewish political activity in Russia – the year 1897 saw both the first Zionist Congress and the formation of the Bund, the Jewish Revolutionary Labour Organization – and

[35] See on Russian anti-semitism at this period, Wolf's own article 'Anti-Semitism' from the eleventh edition of the *Encyclopaedia Britannica* reprinted in *Essays in Jewish History*, pp. 411ff. Cf. James Parkes, *The Emergence of the Jewish Problem* (London. 1946), pp. 212 ff.

[36] *The Graphic*, 12 January 1907.

[37] The first evidence of a new anti-Russian tone is to be found in an article signed 'Diplomaticus' and entitled 'Count Muravieff's Indiscretion' in the *Fortnightly Review*, December 1899. This deals with Muraviev's sounding of the Powers with regard to joint intervention in the South African War. Wolf wrote that France and Spain had been sounded and that the former had refused. He declared his informant to be 'a diplomatist of the highest standing whose business it is to know such things'. It has been suggested to me by Mr A. W. Palmer of Oriel College, Oxford, that the information may have come from the Foreign Office since it tallies with that given in Sir Charles Hardinge's telegram from St Petersburg of 27 October 1899 (FO 65/1582 no. 104), and in his despatch of 20 October (FO 65/1580). Muraviev saw the article and summoned the Ambassador, Sir Charles Scott to deny that he was 'the arch-conspirator of the somewhat silly type that "Diplomaticus" had made him appear' (FO 65/1580, no. 369. 14 December, 1899). An example of Wolf's belief that anti-semitic movements on the Continent were inimical to British interests is found in his reference to 'the Christian Socialist movements with its (*sic*) adjunct of anti-semitism, both invented by the clericals as a means of attaching anti-capitalism to the Church, and both everywhere a powerful vehicle for the anti-English propaganda among the lower classes', loc. cit., p. 1037. An article in the *Fortnightly Review* for October 1900, dealing with Far Eastern affairs and entitled 'Count Lamsdorff's First Failure' is also very anti-Russian in tone.

his concern for the fate of his Russian co-religionists came in to fortify his political suspicions.[38] Nor was it only anti-semitism in Russia itself that Wolf reacted against; he was also concerned at Russian's influence in favour of anti-semitism in other countries, arguing for instance that it was Russia's influence that prevented the enforcement of the obligations concerning her minorities imposed upon Roumania by the Berlin Treaty of 1878.[39]

In 1903, after the Kishenev pogrom, Wolf visited Russia and paid a much criticized call upon the arch-reactionary minister of the interior Plehve.[40] This visit confirmed him in his views as to the implacable nature of Russia's official anti-semitism.[41]

When the Russo-Japanese war and the revolutionary outbreaks it aroused made the status of the regime a matter of international concern, Wolf had a platform for his views in *The Times* which at that time was on bad terms with the Russian Government, following the expulsion in 1903 of its correspondent D. D. Braham – himself a Jew.[42] In March 1905, *The Times* printed and supported editorially a series of articles expressing doubts of Russia's national credit and of the wisdom of extending to her further loans from abroad. *The Times* did not accept the Finance Minister Kokovtsev's challenge to its Editor to come and inspect its gold reserve in person.[43]

But the coming of peace and the apparent inauguration of an era of constitutionalism tended to silence foreign misgivings, and in March 1906, a new loan was floated. In July, the first Duma was dissolved. The connection between these two events and the fact that the Tsarist regime was propped up by foreign capital later became an important argument in the arsenal of the radical

[38] I am indebted to Dr Mowshowitch for information on Wolf's contacts with Russian Jewry.

[39] For the question of Roumanian Jewry, see *Notes on the Diplomatic History of the Jewish Question*, pp. 23–52. Cf. Parkes, op. cit., pp. 91 ff.

[40] His own account of the visit is reprinted in *Essays in Jewish History*, pp. 64–69.

[41] A full study of Lucien Wolf's activities on this matter would clearly demand a study of his work in the Jewish press, and for Jewish organizations in this country such as the Anglo-Jewish Association and the Board of Deputies, which were concerned with attempts to mitigate the sufferings it brought about. But such a study would more properly be a prelude to a discussion of Wolf's work during the War itself and at the Paris Peace Conference which falls outside the scope of this paper.

[42] *History of The Times*, vol. III, pp. 382–8.

[43] Ibid., pp. 408–9 and cf. Behr, loc. cit.

opponents of the British Entente with Russia.[44] Meanwhile the British Liberal Government was becoming convinced that the only way of averting war with Russia over one of the many points of friction in Asia was to complete the Entente with France, which they had inherited from their Conservative predecessors, by a similar arrangement with Russia. To the achievement of this object British indignation over Russia's internal conditions was a standing obstacle.[45] And the pogroms at Odessa and Bielostock in October 1905, and other outbreaks as well as the obvious inability of the Duma to remove any of the legal disabilities of the Jews provided ample material for such indignation to feed upon. On 7 October 1905, Wolf had begun the publication of a weekly news sheet called *The Russian Correspondence* described as 'issued by sympathizers with the Russian struggle for freedom'.[46] Its first issue had contained an interview with Wolf in which he denied that the Peace had restored Russian solvency. By this time the question of supplying arms to the Jews with which they could resist the murderous attacks of the armed bands organized against them by, or with the connivance of, the administration had become a crucial issue. On 17 November the paper published an appeal from the Bund for aid, and a letter from Lucien Wolf regretting that more had not been done by Jews outside Russia to assist the campaign for self-defence.

Lucien Wolf himself took no part in the collection and dispatch of arms to Russia which was carried on by exiled members of the revolutionary parties.[47] But he came to be regarded by the Russian Government as the very embodiment of the movement. Following the Björkö Agreement of July 1905, Lamsdorff, the Russian Foreign Minister, proposed that the new arrangement should be given an anti-semitic bias by the conclusion of an agree-

[44] See e.g. H. N. Brailsford, *The Fruits of Our Russian Alliance* (London, 1912), pp. 8, 27–28. Cf. Brailsford, *The War of Steel and Gold* (London, 1914), pp. 45, 221–2; 225–30.

[45] Campbell-Bannerman strongly deprecated English Liberals demonstrating against the Tsar's Government despite his well-known 'indiscretion' (in 1906): La Douma est morte: Vive la Douma. See G. N. Trevelyan, *Grey of Fallodon* (London, 1937), pp. 184, 190.

[46] The 'Memoire' is thus in error in giving 1906 as the date when this publication began. The last copy in Wolf's file is dated 17 February 1906. Similar publications were edited in Berlin and Paris. Wolf's file of *Correspondence Russe* is almost complete for the year 1907.

[47] So I am informed by Dr Mowshowitch.

ment providing for joint surveillance of the Jews on the model of a previous secret agreement concerning the surveillance of anarchist activities. In the course of this memorandum (which met with the Tsar's full approval) Lamsdorff wrote that in England in June, 1905, 'an Anglo-Jewish Committee for collecting donations for the equipment of fighting groups among Russian Jews was openly organized with the most active co-operation of the well-known Russophobe publicist Lucien Wolf.'[48]

In October 1905, Sir Charles Hardinge was recalled from St Petersburg to become Permanent Under-Secretary at the Foreign Office and from then until May 1906 the Embassy was in the charge of Cecil Spring-Rice. In November, on Lord Lansdowne's instructions, he not only joined in the general remonstrances of the diplomatic corps against the pogroms, but also saw Witte to explain to him that such events would infallibly provoke the hostility of the foreign press.[49] On 20 January, however, he wrote to a Jewish friend, Oswald Simon, to point out the objection to using consuls to distribute the relief funds that had been collected because it would be said that foreign governments were distributing funds that would be used for the purchase of bombs and arms. He also argued that foreign intervention on the Jews' behalf would only worsen their situation; if demands for emancipation were granted, the Government itself would be unable to check the popular movement against them.[50] It was an argument that was to be reiterated constantly during the course of the subsequent years. In a letter to Grey on the prospects of Revolution, on 29 March, Spring-Rice wrote: 'The attitude of the Jews is quite uncompromising and they can, of course, provide the necessary means.'[51]

This view was accepted by Grey who wrote on 3 October 1906, to Sir Arthur Nicolson who had now become Ambassador to Russia; and who had been stiffly received by Izvolsky when he went to make representations after the Bielostock pogrom,[52]

[48] *Notes on the Diplomatic History of the Jewish Question*, pp. 54–62.

[49] Letter from Cecil Spring-Rice to Mrs Theodore Roosevelt; *The Letters and Friendships of Sir Cecil Spring-Rice*, ed. Stephen Gwynn (London, 1929), vol. 2, pp. 12–13.

[50] Ibid., pp. 27–29.

[51] Ibid., p. 31. Stolypin made use of this argument in a conversation with Sir Arthur Nicolson. Viscount Grey of Fallodon, *Twenty-five Years* (London, 1925), vol. 1, p. 157.

[52] Harold Nicolson, *Sir Arthur Nicolson, Bart, First Lord Carnock* (London, 1930), p. 221.

> I . . . realize that you can do nothing by representations about pogroms, and I shall not ask you to make any, though we may send you from time to time the apprehensions that are expressed here. In some parts of Russia there is apparently civil war, carried out by bombs on one side and pogroms on the other.[53]

Nicolson himself realized how hard the antipathy to Russia's proceedings was going to make his mission.[54] But according to his son and biographer, his own attitude towards the Russian problem in general altered after his arrival; he came to have more respect for the Russian Government, and in particular, 'his sympathy for the Jews was also somewhat damped by a closer study of the problem.'[55] Mackenzie Wallace, we are told on the same authority, 'with his expert knowledge of past conditions and events, subsequently convinced him that Russia could only be reformed gradually and from above.'[56] On the other hand, he was not too easily won over to optimism about the progress of events in Russia, and in a dispatch on 1 January 1907, expressed justifiable doubts as to whether Stolypin intended to deal with the Jewish question in a really liberal spirit.[57]

In view of the arguments that were being offered against foreign intervention, it was important to convince the public abroad that the real instigators of the pogroms were to be found in the Russian Government itself, and that the popular masses were merely their dupes, being encouraged to treat the Jews as the scapegoats for their wrongs. This was done in a book by a Russian, Semenov, published in Paris in 1906 and in London in January 1907. Anatole France wrote a preface to the French edition, while Lucien Wolf provided an introduction for the English translation.[58] In this introduction Wolf stated that he had been prevented

[53] Grey, op. cit., vol. I, p. 156.

[54] Nicolson, op. cit., p. 207.

[55] Ibid., p. 225.

[56] Nicolson, op. cit., p. 224. Sir Donald Mackenzie Wallace, the most noted British authority on Russia and a former *Times* correspondent in St Petersburg, arrived in St Petersburg soon after Nicolson, as the latter's unofficial adviser, having been pressed to do so by King Edward. The increasingly favourable tone taken by *The Times* towards Russia leading to an official 'reconciliation' at the end of 1096, after which the paper again had a fully recognized correspondent in Russia was due to Wallace's influence. *History of The Times*. vol. III, pp. 479 ff.

[57] Nicolson, op. cit., p. 231.

[58] E. Séménoff, *Une page de la Contre-Révolution Russe* (*Les Pogromes*) (Paris, 1906). *The Russian Government and the Massacres, A Page of the Russian Counter-Revolution* (London, 1907).

from attacking the loan of March 1906, on the lines of his articles in *The Times* in the previous year, because agents of the Russian government had intimated to his friends that the result would be 'reprisals in the shape of "pogroms" '. But what Wolf was chiefly concerned with was the indifference shown by the governments of Western Europe, and their consequent responsibility for what had taken place.

> The whole moral consciousness of the free nations of the West – and not least of England herself – is being degraded by this officially nurtured apathy ... To me as an old Liberal, born with the echoes of 1848 ringing in his ears, and piously reared on the traditions of England's unswerving and unfaltering championship of oppressed peoples, the policy pursued by Sir Edward Grey in this respect has been profoundly disheartening. I say this without any consciousness, and I believe without any trace, of specifically Jewish feeling, for it is not my co-religionists alone who are being outraged and massacred.... Properly speaking there are no Jews in this great struggle.... Hence my feeling on this subject is exclusively that of an Englishman and a Liberal.[59]

A Russian Entente was not a Gladstonian legacy, it was forced upon the Government by clamour about the Germany bogey, and the wish to forestall the coming together of Russia and Germany.

> But even were Germany the danger she is represented it would surely be better to risk an intrigue on her part with Russia than to incur the ignominy of complicity in the blackest crime of modern times ... We could well afford – indeed it would be a good business even from the point of view of Mr Maxse – to make Germany a present of the alliance of the bankrupt and eternally discredited Government of the Tsar, so long as we secured the permanent affections of the Russian people whose eventual triumph is as certain as the day succeeds the night.[60]

But the battle was a losing one. In January 1907, Wolf noted that the more favourable attitude of *The Times* was thought on

[59] *The Russian Government and The Massacres*, pp. xxvii-xxviii.
[60] *The Russian Government and The Massacres*, pp. xxxii-xxxiv.

the Continent to herald an Entente with Russia.[61] Two months later he expressed scepticism as to the possible success of the reported arrangement, and attacked the rumoured suggestions for solving the dispute over Persia.[62] Here he showed prescience; for it was on Russia's conduct in Persia that most later criticism of the Entente was to concentrate. But for the time being, public opinion was much moved against Russia's internal policies and a powerful list of signatures was appended to a letter to *The Times* giving a reasoned argument against the proposed agreement.[63] The Dissolution of the Second Duma gave point to their fears; but by mid-July Wolf was writing that signature was imminent.[64] The Convention was in fact concluded on 31 August 1907. Wolf commented that the public would probably be told gradually what it contained. Meanwhile he reiterated his refusal to accept the official thesis that Russia's internal affairs were no concern of Britain's. Grey 'might just as well say we have no concern with the private lives of the people we place on our visiting lists'. The agreement was 'the most notable example of the unextenuated adoption of the *raison d'état* in Liberal foreign policy of which we have any record. . . . From the time of Earl Grey it has never hitherto entered the mind of a Liberal statesman to negotiate an *entente* with a despotic Government, especially one at war with its own subjects.'[65] The publication of the text of the agreement did not improve Wolf's opinion of it. Not only was it ethically reprehensible, but Russia was in fact so weak that she could not afford to quarrel with Britain anyhow; 'It follows that we had no necessity to negotiate a special Agreement, and bribe Russia with expensive *pourboires* in order to secure the peace of Asia.'[66] A week afterwards Wolf pointed out how cool the reception of the agreement had been except in *The Times* and *The Westminster Gazette*

[61] *The Graphic,* 12 January 1907. Wolf's last article in *The Times* was one entitled 'The Northern Question' on 3 December 1907. Although this dealt with Russian and German policies in Northern Europe, it was not markedly anti-Russian. From 1908 to 1913, letters from Wolf appeared from time to time in the correspondence columns of *The Times.* Some dealt with general topics; others were concerned with the persecution of Jews in Russia. For the latter see *The Times,* 25 December 1908; 27 June 1911; 10 May 1912; 28 May 1913; 3rd July 1913.

[62] *The Graphic,* 10 March 1907.

[63] *The Times,* 11 June 1907. It is reprinted in Joll, op. cit., pp. 207–9.

[64] *The Graphic,* 3 July 1907.

[65] Ibid., 7 September 1907.

[66] *The Graphic,* 28 September 1907.

which, under the editorship of J. A. Spender, was the only Liberal paper of the period consistently to support Grey. On the other hand, the glee of the Russian press and some of its interpretations of the agreement showed what a bad bargain Britain had made.[67]

But public opinion gradually settled down to an acceptance of the new policy – foreign affairs rarely held the centre of the stage for long – and even Wolf's weekly article during the next few months only rarely contained any reference to Russia's conduct. On the extreme Left however, the Russian Entente was bitterly condemned. Keir Hardie described it at the Huddersfield Conference of the ILP in 1908 as 'giving an informal sanction to the course of infamous tyranny' which had 'suppressed every semblance of representation' and had condemned great numbers of their Russian comrades 'to imprisonment, torture, and death'.[68] F. W. Jowett in *The Clarion* described Russia as 'a cruel despotism which is abhorrent to all thinking men'.[69]

This opposition was chiefly aroused by the visit of the King to the Tsar at Reval in June 1908. Ramsay MacDonald publicly described the Tsar as 'a common murderer' and objected to the King 'hobnobbing with a bloodstained creature'.[70] In an article entitled 'Consorting with Murderers', Keir Hardie maintained the familiar thesis that the First Duma had been called merely to get a loan. Now Russia was bankrupt.

> Financial reasons therefore, probably explain why King Edward has been advised by his responsible advisers to pay this official visit to a monarch reeking with the blood of his slaughtered subjects. The Stock Exchange hook needed to be baited. Two years

[67] Ibid., 5 October 1907. Lord Curzon made this point in his speech on the Agreement in the House of Lords on 7 February 1908. See his letter to *The Times*, 14 February 1908, where he refers to Wolf as 'that well-informed writer on Foreign Affairs' and Wolf's comments in his letters of 15 February, 29 February and 21 March. Curzon had been concerned only with the Persian question, expressly disclaiming the view that any regard should be had to internal Russian questions. Wolf had answered some of the Agreement's supporters in a further article in *The Graphic* on 12 October 1907.

[68] W. Stewart, *J. Keir Hardie* (London, 1921), pp. 261–2.

[69] Fenner Brockway, *Socialism over Sixty Years: The Life of Jowett of Bradford* (London, 1946), pp. 120–1. Jowett was later forced to abandon his campaign against the Government's foreign policy, as Robert Blatchford made *The Clarion* an organ of extreme jingoism. Ibid., pp. 122 ff. The latter's propaganda during the election of January 1910, was criticized by Lucien Wolf in *The Graphic* on 8 January 1910.

[70] Sir Sidney Lee, *Life of King Edward VII* (London, 1925–7), vol. II, pp. 586–7.

ago the bait was a popularly elected Duma; this time it is a Royal Crown. Truly Kings have their uses.[71]

The matter was raised in Parliament by James O'Grady and Victor Grayson in the debate on the Whitsun adjournment on 4 June.[72] Although the visit was also attacked in *The Daily News* and *The Nation,* Wolf thought that it should not be criticized so long as no exaggerated significance were attached to it. There was no common ground for a cordial understanding between the two peoples such as existed between the British and the French: 'The Russian people are as much in fetters as ever they were – and between the British democracy and the Russian autocracy there can be no sense of affinity and consequently no bond of enthusiastic cordiality'.[73] Talk of a new triple alliance was foolish and Germany had no justification for fearing the creation of such a coalition against her.[74]

On 3 June the King had received a letter from the Rothschilds asking him to use the occasion to intercede for the Jews and the subject was mentioned by the King to Stolypin on 13 June. He received assurances that ameliorative legislation was proposed and informed the Rothschilds of this through Hardinge.[75] This was no doubt the basis for the unwontedly optimistic tone of Wolf's article in *The Graphic* on 27 June in which he declared that Stolypin had made a highly favourable impression at Reval. 'I understand', he wrote, 'that M. Stolypin contemplates the abolition at an early date of the Field Courts-Martial and a measure of partial relief for the Jews.'

This optimism was not generally shared and the matter was again debated on the Foreign Office Vote on 4 July 1908. O'Grady moved the reduction of the vote and specifically referred to the continued persecution of the Jews. The Irish Nationalist member for Tyrone E., a Mr Kettle, speaking in his support, declared that 'the visit was not a fraternization with the people of Russia but a fraternization with the hangman of liberty in Russia'. Keir Hardie

[71] Stewart, op. cit., pp. 264–5.

[72] As a result of the criticisms of the Reval visit, Keir Hardie, along with Victor Grayson and Arthur Ponsonby, was omitted from the list of invitations to the Royal Garden Party, The Parliamentary Labour Party passed a resolution in support of Keir Hardie over this affront. Ibid., pp. 226–7.

[73] *The Graphic,* 6 June 1908.

[74] *The Graphic,* 13 June; 20 June 1908.

[75] Lee, op. cit., vol. II, p. 594.

declared that 'For the King of Great Britain to pay an official visit to the Tsar of Russia was to condone the atrocities for which the Russian Government and the Tsar personally must be held responsible'. The word 'atrocities' was ruled out of order. The argument on both sides followed conventional lines.

Grey declared that any policy involving intervention in the internal affairs of another country must lead to war and quoted the Cadet Leader Milioukov as favouring the Royal visit. One should balance the pogroms against the revolutionary violence on the other side. Progress had been made and there was no question in Russia of a nationality oppressed by foreign rulers as had been the case in some past instances of British intervention abroad. Balfour gave the Opposition's full support to Grey declaring that accepting the critics' case involved the rupture of diplomatic relations: 'It is not the business of this House to register protests with regard to the internal administration of countries with which we have no treaties justifying our interference.' On a division, 59 members voted against the Government whose supporters mustered 225.

After the Bosnian Crisis in 1909 and the apparent success of Germany, Nicolson suggested to Grey that the only hope of preventing the Central Powers dominating Europe was to convert the Entente with Russia into an Alliance. Grey in reply showed that the discussions of the previous year had not been without their effect. Opinion was hostile to far-reaching British commitments: 'Russia too must make her internal government less reactionary – till she does, liberal sentiment here will remain very cool, and even those who are not sentimental will not believe that Russia can purge her administration sufficiently to become a strong and reliable Power.'[76]

The feeling on this subject was again aroused by the visit of the Tsar to Cowes in August. Asquith was so alarmed that he wanted Izvolsky and Stolypin left behind at Cherbourg so as to make it look a purely family affair. But he was overruled.[77] Grey had no sympathy with the critics. 'The question for us', wrote Grey in a private letter, 'is whether we will help Stolypin and the Duma by being civil to the Czar as long as he stands by them; or whether we will play straight into the hands of the reactionaries by insulting the Czar as Keir Hardie and Co. want us to do.' But such argu-

[76] Nicolson, op. cit., pp. 304–7.
[77] Ibid., pp. 314–15.

ments could not be stated publicly. The Czar after all was 'a kind, moral family man, who as an English squire would be much respected in his parish.'[78]

The matter was aired not only by the left wing in Parliament, but also by Lucien Wolf in a signed article in *The Morning Leader*. Burke, he reminded his readers, had railed at the altruism of Fox in foreign politics as childish futility: Sir Edward Grey 'edging more and more towards the shadowland of the Old Whigs' seemed to voice 'this unhappy humour of Burke' in chiding the critics of the Tsar's visit for their 'futile folly'. In a long examination of the historical tradition of Whig and Liberal foreign policy, he found it to have been best formulated by Gladstone in his West Calder speech of 27 November 1879, when he had stated that there should be 'a sympathy for freedom, a desire to give it scope, founded not upon visionary ideas, but upon the long experience of many generations within the shores of this happy isle'.[79] He pointed out that the arguments that Grey had been advancing were inconsistent, for having declared that Russia's internal affairs were no affair of Britain's, he proceeded to use the argument that the Tsar would be remembered for the introduction in his reign of the beginnings of constitutional government.[80]

> The Empire is virtually in the hands of a bloody assize of drumhead courts martial. The prisons everywhere are crowded to suffocation, and their condition and administration may not unfairly be compared with the dungeons of King 'Bomba'. Executions are carried out on a scale which would have made Jeffreys and Cumberland shudder. The Tsar himself has patronized and subventioned a Reactionary League which has openly organized massacres, and has connived at the assassination of prominent Liberals. In the Western provinces five millions of Jews are submitted to an oppression which has no parallel in history and which M. Stolypin acknowledged two years ago to be intolerable. Finally, Finland is still condemned to see her people dragooned, and her ancient rights trampled upon.

Even if moral grounds were insufficient to condemn the

[78] G. M. Trevelyan, *Grey of Fallodon,* pp. 190–3.

[79] Extracts from the speech are in Joll, op. cit., pp. 185–7.

[80] The criticisms were directed against Grey's speech in the House of Commons in the debate on The Foreign Office Vote, 22 July 1909. The Opposition on that occasion mustered 79 votes.

Entente, events had shown that it was equally condemned by *raison d'état,* 'It has brought us a bitter quarrel with Austria, it has created suspicion against us in Turkey, and it has completely compromized us in Persia.' As for the apologists who declared that its critics were 'playing the game of the Wilhelmstrasse and the Black Hundreds', that, said Wolf, 'is not an argument – it is blackmail.'[81]

In reply, Sir Bernard Pares wrote an article in which he held up the Third Duma to admiration, as the representative of the happy middle ground between the extremes of reaction and revolution; and declared that the boycott of the Tsar's visit desired by the adherents of the latter school would play into the hands of the former. Wolf declared that he himself did not advocate a boycott of the visit now that it had unfortunately been arranged, and pointed out that Sir Bernard Pares's argument that the number of executions had not been high compared with that of the terrorist outrages, overlooked the victims of the pogroms, organized with official connivance. Pares in a second reply said that he adhered to the views he had expressed in an article on the progroms in 1906, that they were the work of the police, and that he utterly condemned their perpetrators;[82] but they were a reprisal for the murders of policemen; and a disproportionate number of the murderers had in fact been Jews.[83] Anyhow, it was wrong to identify the present Russian government which had put down the pogroms with the reactionaries whom he had always opposed. As for the Tsar's visit, prominent Russian Liberals had deplored the boycott; what good could it do?

In the following years the debate continued on the same lines. Wolf's weekly articles in *The Graphic* continued from time to time to raise the issue of Russia's conduct at home in Finland or abroad in Mongolia, Tibet, Persia, and other areas where Liberal opinion was critical of her actions, without himself making any new frontal attack upon the Entente. Public opinion, as he recognized with regret, was indifferent to foreign affairs and absorbed by the greater excitement of party politics at home. But the affair of the Stepney anarchists to which Mr Winston Churchill gave such

[81] *Morning Leader,* 30 July 1909.
[82] Ibid., 3 4, 6, August 1909.
[83] The reference is to his article 'The Russian Massacres', *Quarterly Review,* October 1906.

prominence, gave him a chance by calling attention to the fact that Russian tyranny at home had as one result, the flooding of Europe with desperate exiles.[84]

He was always on the watch for signs that the Russians were mending the wire to Berlin. And when the Potsdam agreement on the Baghdad Railway was announced, he pointed out that others besides himself were now beginning to have their doubts about the results of British policies, whereas, previously, it had been suggested that Wolf must be in German or Austrian pay as the only possible explanation for 'the mental obliquity of a writer who declined to subscribe to the infallibility of Sir Edward Grey and the chivalrous good faith of the Russian foreign office'.[85] In a series of articles in December 1911, he claimed that Britain had been duped over Persia, that Russia, now reviving the Dardanelles question, was not to be cured of covetousness by an Entente Cordiale. Britain now faced a prospect of heavier and heavier armaments. There were only two courses open – either back to 'splendid isolation' which was what Wolf himself would have preferred, or forward to real alliances which would at least bring some element of stability into a situation by making it clear where the Powers stood.[86] To this argument he subsequently returned with increasing frequency.[87]

At the same time, nothing inside Russia seemed to be justifying the arguments of those who, like Pares, had expected her to imbibe liberal influences from her Western associates. As Ramsay MacDonald put it in the House of Commons:

> The whole justification for our friendship with Russia lies in the liberalizing of Russian institutions – and that has not happened. As a matter of fact one of the results of our pro-Russian policy has been to encourage the Russian bureaucracy to stamp out Parliamentary institutions as much as they possibly can in Russia.[88]

This two-sided attack on the Entente infuriated some of its advocates. The *National Review* in January 1912, spoke of a cabal of

[84] *The Graphic,* 14 January 1911.

[85] Ibid., 21 January, cf. ibid., 4 February. On 26 August, Wolf wrote of his anxieties about the newly announced Russo-German agreement on Persia.

[86] *The Graphic,* 2, 9, 16, 30 December 1911.

[87] Ibid., 24 February 1912; 1 June 1912; 8 June 1912.

[88] Debate on Foreign Affairs, 27 November 1911.

'cosmopolitan Jews' and talked of 'the heavy Hebrew control' of several 'British' newspapers:

> Unsuspecting Radical journalists who are engaged in this disgraceful business, which discredits them and dishonours their profession, should ask themselves what they hope to gain by joining hands with the International Jew who is permanently on the war-path in the interests of the most reactionary and dangerous Power in Europe.[89]

This abuse did not deter Wolf from his course. Indeed in the same month he began the publication of a new four-page journal entitled *Darkest Russia – A Weekly Record of the Struggle for Freedom,* in which the polemic against the Entente with Russia could be carried on by means of giving full publicity to the repressive policies of the existing regime. The first disputed issue was that of the visit of a British Parliamentary delegation to Russia, under the guidance of Sir Bernard Pares, described by Wolf as the 'Cook's tour man' of the party.[90] A pamphlet on the persecution of the Jews in Russia was prepared for the illumination of the British delegation and in his preface, Lucien Wolf pointed out that legalized persecution was not confined to Jews but extended to Lutheran Finns, Roman Catholic Poles, and even Russian Dissenters.[91] Later in the year, a more elaborate account of the position was prepared and published.[92]

[89] During the same month, W. T. Stead in the *Review of Reviews* made similar charges about Jewish Russophobe war-mongering. But by March he had himself veered into an anti-Russian position. See *Darkest Russia,* 27 March 1912. On 17 January the British Ambassador wrote from St Petersburg that the Tsar had complained of the doubts cast as to Russia's sincerity over Persia, in *The Daily Graphic* and other British papers. The Ambassador assured him that the extreme left-wing papers in Britain (a description that hardly fitted *The Daily Graphic*) were not more hostile to the Entente than the extreme right-wing papers in Russia. Buchanan to Grey, 17 January 1912, *British Documents on the Origins of the War,* vol. x, pt. 1, pp. 897–8.

[90] Ibid., 17 January 1912. In his own later account of this visit, and of a preceding visit of members of the Duma to England, Pares ignored their controversial aspects. Sir Bernard Pares, *My Russian Memoirs* (London, 1931), chaps ix and xi.

[91] *Under the Duma and the Entente: The Persecution of the Jews in Russia,* with a preface by Lucien Wolf (London, 1912).

[92] *The Legal Sufferings of the Jews in Russia,* edited by Lucien Wolf with an introduction by A. V. Dicey (London, 1912). I am informed by Dr Mowshowitch that the groundwork for this publication was prepared by Leo Motzkin. *Darkest Russia* also gave publicity to the activities of an Anglo-Russian Committee located at the same address which was likewise engaged in calling attention to internal

In 1912 the financial aspect of relations with Russia again assumed importance as the question of a new loan was in the air. *Darkest Russia* accused two leading London newspapers of ignoring Reuter reports of famine conditions in Russia.[93] And when an appeal for aid to the starving Russian peasantry was made in the early summer, Wolf pointed out in *The Graphic* that this was at a time when Russia was herself exporting large quantities of grain. This was being done in order to pay interest on past loans and the whole thing was symptomatic of Russia's financial rottenness.[94]

Darkest Russia could afford to be more combative than Wolf could manage to be in *The Graphic;* and particular scorn was poured on Russian supplements of *The Times* and the *Illustrated London News* for their over-optimistic pictures of Russian conditions. *The Times* also came in for criticism over an article in its issue for 10 August, entitled 'The Duma and England', as being misleading about Russian public opinion, and a number of anti-British pronouncements from the Russian press were quoted to prove the point.[95] More specific was the attention paid to the accusation of ritual murder against a Jew in the notorious Beilis trial which aroused much public comment in England and led to some highly polemical contributions to the British press from the Russian Consul-General in London.[96] Another case that did not

conditions in Russia and in which an important figure was C. Hagberg Wright, the celebrated librarian of the London Library. It was this organization that printed the pamphlet by Brailsford already referred to. This propaganda assisted no doubt but was not identical with the 'very active propaganda by financiers, pacifists, faddists, and others in favour of close relations with Germany' described by Sir Arthur Nicolson to M. Paul Cambon on 15 April 1912. Nicolson, op. cit., p. 369.

[93] *Darkest Russia,* 6 March 1912; cf. ibid., 8 May 1912; cf. supra, p. 11, fn. 4, for Brailsford's part in the anti-loan campaign.

[94] *The Graphic,* 11 May 1912. Some Jews despairing of any other form of protest were apt to put too much faith in the possibilities of financial pressure. In an article by Cyril Picciotto, it was pointed out that international law provided no basis for a state's interfering on behalf of oppressed persons who were not its own nationals: 'therefore we can only hope that wherever possible Jewish financiers may bring financial pressure to bear upon the Governments concerned, and that this pressure may effect that for which the law supplies no remedy.' He referred to Norman Angell's then recently published *The Great Illusion* as proving the modern state's dependence on finance. *Jewish Review,* March 1912. Wolf himself despite the importance he attached to the question of the Russian loans was highly sceptical of the general pacific implications of Angell's theories. See *The Graphic,* 20 December 1913.

[95] *Darkest Russia,* 29 May; 21 August 1912.

[96] See e.g. *The Times,* 8 May 1913.

improve the Russian Government's standing was a heavy prison sentence passed upon a young woman (Miss Malecka) who happened to be a British subject, for having seditious dealings with Polish socialists. And the fact that she was pardoned by the Tsar did not satisfy *Darkest Russia* that Sir Edward Grey's attitude to the affair was energetic enough. On the whole *Darkest Russia* lived up to the motto on its editorial page that Wolf had chosen from Cromwell's fiery intervention on behalf of the Vaudois – one of his favourite historical incidents – 'To be indifferent to such things is a great sin, and a deeper sin still it is to be blind to them from policy or ambition.'

It was at this time that the so-called passport controversy, to which reference has already been made, reached its climax. And on the occasion of Sazonov's visit to Balmoral in September, an attempt was made by the Conjoint Foreign Committee of the Anglo-Jewish Association and the Board of Deputies to get Grey to receive a private deputation. Grey refused to do so, on the grounds that the purpose of such a deputation would be to suggest intervention in the internal affairs of another country.[97]

Wolf in *The Graphic* was concerned with the wider political implications of the visit, making the point that Russia retained the upper hand in the Entente through her policy of never breaking with Germany, whereas Britain had no such alternative policy open.[98] At least it was satisfactory to record that the Anglo-Russian convention had not been amplified.[99] For at this time of increasing Balkan tension Wolf held that the danger spot of Europe was not Thrace but St Petersburg.[100] Nevertheless on 30 November 1912, he expressed the conviction that Russia would not go to war for Serbia as she was much too weak. Disaffection internally had never been more pronounced.[101] As for Russian

[97] *American Jewish Yearbook,* 1912–13. See also *Darkest Russia,* 23 October, 27 November, 4 December 1912. Earlier in the summer Nahum Sokolow came to England on behalf of the Zionist Executive, his mission being partly for the purpose of seeking British support in getting the Russian Government to modify its anti-Zionist attitude. See the article by S. Rawidowicz: 'Nahum Sokolow in Great Britain', *The New Judea,* May 1941. I owe this reference to Dr O. Rabinowicz.

[98] *The Graphic,* 21 September 1912. See also *The Graphic,* 21 December 1912; *Darkest Russia,* 25 September 1912.

[99] *The Graphic,* 5 October 1912.

[100] *The Graphic,* 26 October 1912; 11 January 1913.

[101] *The Graphic,* 30 November 1912.

sympathy for oppressed peoples as a motive for her foreign policy, the sponsor of *Darkest Russia* could not be expected to take this very seriously. Russian talk of Turkish oppression was answered there in a leader entitled 'The Kettle and the Pot' and when Kokovstev talked in the Duma on the Balkans, Wolf carried the attack into *The Graphic:* 'With M. Kokovstoff a solicitude for subject races is essentially an article for exportation; for Macedonia was never worse off than the southern Caucasus, or that inferno of oppression and misery, the Pale of Settlement.'[102]

Wolf's position on the general question of British foreign policy in the last eighteen months before the war can be seen in the commendation that he gave to a remarkable article on the subject in *The Edinburgh Review* for January 1913, which points out that no stability in the map of Europe was to be looked for and British intervention in European affairs should be limited to where Britain's own interests were challenged, or a treaty violated to which Britain itself was a party.[103] Even this writer, however, does not go as far as Wolf 'in thinking that "Splendid Isolation" is still possible, and would even be advantageous'.[104] On the other hand, Wolf has to admit that 'today public opinion in England, is not only for the French Alliance, but is for it on terms which muzzle the tactless inquisitiveness and criticisms of Parliament and give the minister of the day all the freedom he needs to hold his tongue and pile up obligations in regard to our popular ally'.[105] Yet the Entente was still too indefinite to contribute to the balance of power; and with Germany obviously arming fast and in a sabre-rattling mood, it would be folly to be weak.[106]

[102] *The Graphic,* 28 December 1912. In its issue of 24 December, *Darkest Russia* included an article by the well-known Russian Menshevik, Theodore Dan, arguing that Russia was far too weak to go to war.

[103] 'European Reconstruction and British Foreign Policy', *Edinburgh Review,* January 1913.

[104] *The Graphic,* 18 January 1914. Commenting on a lecture by Lord Esher at the Sorbonne he wrote: 'there was to be no half-way house between Splendid Isolation and a definite Alliance.' Wolf described himself as preferring the former, but not to the point of fanaticism. The latter would be better than the prevailing uncertainty. *The Graphic,* 4 April, 18 April, 25 April 1914. Esher had pleaded for the conversion of the Triple Entente into a Triple Alliance in order to provide a framework within which a disarmament agreement could be negotiated. Wolf found Esher's appearance as the 'Apostle of Norman-Angellism . . . in many respects a piquant performance'. Cf. *The Journals and Letters of Reginald Viscount Esher,* vol. III (London, 1928), pp. 93, 111–12, 157, 159–160.

[105] *The Graphic,* 28 June 1914.

[106] *The Graphic,* 17 January 1914—an article on the 'Zabern incident'.

Meanwhile, as the columns of *Darkest Russia* showed, there was little comfort to be found in the East. Goremykin's appointment to the premiership in February 1914 appeared as a notable victory for Russian reaction.[107] In such circumstances there could be no welcome for Sazonov's proposal for a standing committee of the Triple Entente,[108] or for other suggestions for strengthening the links with Russia. *Darkest Russia* pointed out that two years previously the Archbishop of Canterbury had declared that the Union of the Churches was rendered impossible by Russian atrocities.[109] The same view should be taken of an Alliance. Russia's 'romantics' would embroil her in Europe for the sake of the Slavs; her 'realists' would bring about a clash in the Far East.[110]

Indeed the position could be summed up in more general terms:

> There is always in Europe an atmosphere favourable to anti-Russian agitation, and that atmosphere, in spite of the combinations of the Chancelleries, tends every year to become more sensitive. It is due in part to the aggressive tendencies of Russian foreign policy, and in even greater measure to a detestable internal policy which not only outrages great masses of European public opinion, but is for ever exporting legions of victims, who in the freer communities of the West become permanent and effective agents of Russophobia.[111]

Darkest Russia could be even more outspoken than *The Graphic,* and those who praised Russia, from H. G. Wells on the one hand to *The Morning Post* on the other, received uncompromissing rebuttals.[112] (*The Morning Post* had deduced the greatness of Russian civilization from the magnificence of the Russian

[107] *The Graphic,* 21 February 1914.

[108] *The Graphic,* 7 March 1914.

[109] Wolf slightly exaggerated the Archbishop's view. A motion had comes before Convocation on 3 July 1912, welcoming the formation of a Russian society for promoting closer relations between the Anglican and Orthodox Churches. The Archbishop supporting the resolution spoke of the dangers of going too fast in the matter of Church Union and added that 'it was impossible to read the records of Russian life in its social aspect without feeling that they should shrink from identifying themselves with the civil life which took some steps with regard to the government of the people which they would naturally reprobate'. *The Times,* 4 July 1912.

[110] *Darkest Russia,* 22 April 1914.

[111] *The Graphic,* 21 March 1914.

[112] *Darkest Russia,* 11 March 1914; 22 April 1914.

ballet.) But the worst offender was *The Times* with its panegyrical Russian 'supplements' and a growing tendency to overt anti-semitism – a tendency that Wolf blamed on the increasing influence in its affairs of Wickham Steed.[113] The latter tendency came to a head in a leading article 'Russia and her Jews' on 5 June 1914 – an article that Wolf himself rightly remarked used arguments that professed anti-Semites themselves had long abandoned.[114]

The Sarajevo murder did not then for Wolf at any rate come out of a clear sky; and he was the more affected since he had placed some hopes for the future in the rumoured liberalism of the murdered archduke. Throughout the crisis, he took a strongly pro-Austrian line. His article on 1 August 1914, contained the rhetorical question: 'Is there a single instructed person who honestly believes that Austria wanted this or any other war?', and blamed the Serbs for playing into the hands of 'Muscovite hate'[115] In hoping to the end that Britain could keep out of war, Wolf was not alone among leading journalists. *The Manchester Guardian, The Daily News,* and *The Liverpool Post,* were frankly opposed

[113] *Darkest Russia,* 21 January 1914; 29 April 1914; 10 June 1914; 17 June 1914. Henry Wickham Steed, after posts in Berlin and Rome, became *The Times* Vienna correspondent in November 1902, and in November 1913, had come to work in the London office, being made foreign editor in January 1914. Steed's own writings suggest that Wolf's diagnosis had some bearing. For during his stay in Vienna he had become convinced that the crucial question of freeing the non-Germanic and non-Magyar races of Central Europe from the permanent yoke of a pan-German empire was being kept out of the international arena largely through Jewish influence. 'Jewish idealism', he later wrote, 'strengthened for a time the pro-German and pan-German tendencies of Jewish finance by bringing Jewish hatred of Imperial Russia into line with Jewish attachment to Germanism.' Henry Wickham Steed, *Through Thirty Years* (London, 1924), vol. II, pp. 390–3.

[114] *The Times* leader after voicing its disapproval of a recent measure excluding Jews from sitting on the boards of joint-stock companies that owned land went on to discuss the demand that 'all restrictions on the rights of Jews in Russia should be removed', and the claim that they were entitled to equal privileges with their Christian fellow-subjects'. This demand betrayed 'utter ignorance of the real situation'. Were the Jews emancipated they would through usury and the sale of drink 'eat up' the tillers of the soil. 'It is a question for Russia and for Russia only, to decide what degree of emancipation from the fetters which now bind them may safely be given to the different classes of the Jewish community within the Empire.' *The Times* 5 June 1914. Lucien Wolf's hand may perhaps be detected in a dignified letter in reply signed by the Presidents of the Board of Deputies and the British Association, which pointed out *inter alia* that the sale of liquor was now a state monopoly with well-known consequences in the form of widespread drunkenness and that the peasants were worse off since the 'May Laws' had removed Jewish competition from the markets. *The Times,* 9 June 1914.

[115] *The Graphic,* 1 August 1914.

to intervention; the *Daily Chronicle* and *Yorkshire Post* and even *The Westminster Gazette* were 'pacific in their utterances'.[116] The list of prominent advocates of neutrality goes well outside the confines of radical pacifism. 'Everyone,' wrote G. M. Trevelyan, later Grey's admiring biographer, in the *Daily News* on 3 August, should 'do something to demonstrate his or her antagonism to the participation of England in the European crime'.[117] Most of the Conservative press however moved rapidly over to the extreme interventionist position held originally only by *The Times* and the other Northcliffe papers. Only *The Daily Graphic* and *The Standard* remained unconvinced.[118]

Like many others, Wolf was won over to full support of the war by the attack on Belgium. The cause was just.[119] And support of the war meant abandoning attacks on our Russian ally. *Darkest Russia* was one of the war's first victims. This did not save Wolf from the bitter attacks of the now justified jingo press. In an article in September 1914, the neutralists of the recent crisis were flayed by Leo Maxse: 'the symposium of cranks was headed by G. M. Trevelyan who has written too much of the Europe of yesterday to have any knowledge of the Europe of today.' *The Nation,* the *Manchester Guardian,* 'the most frantic of all our Potsdam papers', and even *The Westminster Gazette* were not spared. The Jews – with some exceptions – did not escape. The public was warned against a recrudescence of Russophobia stimulated by them for purely Jewish ends. And in the section on the press, pride of place is given to *The Daily Graphic* – 'the Kaiser's favourite newspaper' – which 'under the amiable auspices of Mr Lucien Wolf . . . appears to endeavour to express precisely the opinions which Germany would desire to see prevalent on this side of the channel'. These opinions were substantiated by quotations from *The Daily Graphic* during the crisis, including a signed article by Wolf on 27 July, 'The Coming War and England's Duty' and further articles on 31 July, 1 August, and 4 August.[120]

[116] Hale, *Publicity and Diplomacy,* pp. 462-3.

[117] Ibid., p. 464. Cf. Trevelyan's footnote on his own position; op. cit., p. 254 fn.

[118] Hale, op. cit., p. 466. Prof. Hale is in error in stating that Wolf was still foreign editor of *The Daily Graphic.*

[119] 'The Casus Belli', *The Graphic,* 15 August 1914. On 8 August, Wolf had written a long article arguing that no one power was responsible for the war which had been brought about by misplaced notions concerning the balance of power.

[120] 'The Fight Against Pan-Germanism', the *National Review,* September 1914. Wolf wrote to Maxse on the 14th to point out that *The Daily Graphic* was not 'under

But however unfair and prejudiced such attacks might be, Wolf's authority as a commentator on foreign affairs was bound to be shaken. And the outbreak of war marks for this as well as for a more personal reason, the virtual end of this aspect of his many-sided career. Yet to retrace it should not be without its value; and the last word may be left to an author who cannot be accused of pro-Jewish leanings:

> The gulf that severed Western Europe from Russia during the latter half of the nineteenth century [wrote Wickham Steed], was dug and kept open chiefly by Jewish resentment of Russian persecution of the Jews.[121]

There were in fact many other reasons for that gulf; but its existence is undeniable and from its existence have followed many of the forces that shape our own world. Any investigation of it should be a contribution as much to general as to Jewish history.[122]

his auspices, and that he had not for the past six years been a salaried member of its staff.' He also offered to send to Maxse his articles on the Agadir crisis in order to dispose of some of Maxse's charges. Maxse failed to print this rectification and Wolf wrote to him again on 16 October, and again without result. (Copies of letters in the possession of the late Dr D. Mowshowitch.)

[121] Steed, op. cit., vol. II, pp. 390–1.

[122] I wish to record my indebtedness to the late Dr David Mowshowitch for lending me his valuable collection of press cuttings etc., dealing with Lucien Wolf's career.

13 The Sixth of February

In an article dealing with the question of why so little historical work has been done on the problems of the last two decades of the Third Republic, M. René Rémond points out that no satisfactory book has been written so far on the 'Sixth of February' and its consequences.[1] The present essay is in no sense an attempt to fill this gap. It originated indeed in quite another way, in a seminar devoted to the study of Revolutions and *coups d'état* that have been successful: but what about the ones that did not come off? There might be something to be gleaned from them too. But of course, if a Revolution or *coup d'état* fails how can one know that it was ever seriously intended? In this respect the Paris riots of 6 February 1934 provide a good example. Even today, over twenty years later, the only thing people seem agreed upon is that they were important. But this importance is looked for in two directions. In the first place, there is the light thrown by the riots on the immediately pre-war phase in the long French tradition of anti-Parliamentary movements of the Right. As a French historian has written recently: 'It is in the perspective of the street riots of the

[1] René Rémond, 'La Fin de la Troisième République', *Revue Française de Science Politique*, vol. VII, no. 2 (1957). The main source for the riots themselves is to be found in the reports of the Commission of Enquiry (including the minutes of the evidence). This Commission was set up by the Chamber of Deputies. Its Documents were published as *Chambre de Députés Rapport de la Commission d'Enquête sur les Événements du 6 février 1934* (15th Legislature, Session of 1934. Documents 3383–3393). The Commission's findings were summarized by its President, M. Laurent Bonnevay, in his book *Les Journées sanglantes de février 1934* (1935). The most useful account is still that written immediately afterwards by Mr Alexander Werth in his *France in Ferment* (London, 1934). (All books mentioned are published in Paris unless otherwise stated.) See also an unpublished thesis 'Les Groupes Anti-Parlementaires Republicains de Droite en France de 1933 à 1939' by H. Maizy, presented at the Institut des Sciences Politiques in 1952. For assistance with documentation, I wish to thank two French friends, M. Jacques Kayser and M. Mattei Dogan.

Boulangist movement, of "Panama" and of the Dreyfus case, much more than in that of the march on Rome or the Munich putsch that one should undoubtedly place the "day" of 6 February 1934.'[2] In the second place, the riots are seen as one in a series of events which led to the formation of the Popular Front, in reaction against the danger, real or alleged, of the setting up of a right-wing dictatorship.

But were the riots themselves part of a deep-laid plot against the Republic? Did the rioters have any clear intentions, or did a political demonstration of a familiar kind get out of hand because of clumsiness or worse on the part of the authorities? It is obvious that answers to this question were coloured at the time, and later, by other events to which those of the Sixth of February were regarded as having some relation. It is pointed out that riots on a lesser scale on 22 July 1926 helped to bring about the failure of M. Herriot to get a majority for a government just formed by him.[3] But more important is the fact that in the end France did finish up at Vichy with a non-Parliamentary regime which seemed, both ideologically and in its composition, to fulfil some of the desires attributed to the rioters of the Sixth of February.

The most authoritative and vigorous exposition of this view is to be found in the discussion of the penetration of authoritarian ideas into France in the inter-war years in Léon Blum's evidence before the post-war Parliamentary Commission that inquired into the events leading up to the fall of France in 1940, and into the conduct of the Vichy regime.

> These same elements, conservative elements for whom the evils of dictatorship counted for little by the side of the benefits of national discipline, and elements of the Left, tempted by the idea of a dictatorial authority applied to the revolution – one finds them joined together and confused on the Sixth of February as at Vichy and under the Vichy régime. . . .
>
> There was a tendency immediately after the 'day' of the Sixth of February to make it at one and the same time ridiculous and odious. It was presented as a sort of scuffle, minor but nevertheless bloody, started off by hot-headed youths who had shed

[2] Raoul Girardet, 'Notes sur l'Espirit d'un Fascisme Français 1834–9', *Revue Française de Science Politique* (1955).

[3] E. Beau de Loménie, *Le Mort de la Troisieme République* (1951), pp. 26–7. On the crisis of 1926, see E. Herriot, *Jadis*, vol. II (1952), pp. 248–51.

French blood without any danger menacing French institutions. . . .

This is by no means my opinion. The Sixth of February was a redoubtable attempt against the Republic, and even today I continue to wonder how it was that it did not succeed, for logically it should have succeeded. . . .

If the handful of 'gardes mobiles' which was defending the pont de la Concorde had not held, and if above all, in circumstances which to my knowledge have never been cleared up, the column advancing on the left bank under the orders of Colonel de La Rocque had not stopped in front of the slender barricade of the rue de Bourgogne, there can be no doubt but that the Assembly would have been invaded by the insurrection. . . .

I think that there can be no doubt for anyone who was a witness to these events but that the insurrection had within the Chamber itself both representatives and leaders. Their tactics were, I believe, to bring about the fall of the cabinet as rapidly as possible and to ask the Chamber to disperse. It would have been in the empty Chamber that a provisional government would have been proclaimed, as had happened in the same place in 1848 and on the Fourth of September. . . . Who would have composed it? I am ignorant, it goes without saying, as to what the relations of Marshal Pétain may have been with the organizers of the riot. But I believe that his name would have been found on the list of the government, and that the names of Pierre Laval and in any case of André Tardieu would have been found there likewise.[4]

President Lebrun in his turn, though not going into the same detail, referred in his evidence to 'this assault against Republican institutions.'[5]

This evidence was accepted by the Commission. Its rapporteur-general wrote:

The sixth of February was a revolt against the Parliament, an attempt against the régime. The intention was, by means of a popular uprising, to disperse the deputies, to take possession of

[4] Rapport au nom de la Commission chargée d'enquêter sur les événements survenus en France de 1933 à 1945. Assemblée Nationale première legislature, session de 1947, Doc. 2344, vol. 1, p. 122.

[5] Ibid., vol. 4, p. 95.

the Chamber and to proclaim an authoritarian government at the Hôtel de Ville of Paris. The march of the columns of demonstrators towards the Palais-Bourbon, the leaflets distributed in the different quarters of the city and the instructions given to the nationalist leagues leave no room for doubt. It was not a question of a spontaneous demonstration but of a genuine insurrection, minutely prepared.[6]

Recent historians of these years remain, it would seem, unmoved by these arguments. As might be expected, those of the Right deny the charge outright. 'To talk of a fascist plot on the occasion of the "days" of January or February is an absurdity', writes one of them.[7] More authoritative is M. Rémond's own considered judgement:

> Undoubtedly the intentions of the organizers are even today not fully known: what character was assigned to the demonstration in their plans – a simple movement in the streets designed to bring pressure on the Chamber in order to force it to bring about the resignation of the Daladier government, or a forcible *coup* against the actual institutions? An overthrow of a majority, or of the régime? – one does not know, but in the present state of the question, there is nothing that invites us to accept the second interpretation in preference to the more limited hypothesis of an agitation that turned out badly. Nothing proves that a precise objective had been determined upon in concert; on the contrary, each organization had its assigned rendezvous and put jealous care into avoiding getting mixed up with the others. . . . The Sixth of February was not a putsch, not even a riot, only a street demonstration which history would already have forgotten if it had not turned to tragedy and if the course of events had not given to it retrospectively an importance out of proportion to its true significance.[8]

Another historian has put it more sharply still: 'To tell the truth, neither the inspirers of the operation, nor the parliamentarians who had lost their bearings had worked out what they wished to bring about.'[9]

[6] *Événements en France*, Doc. 2344 (Report of M. Charles Serre), vol. 1, p. 13.
[7] Jacques Debu-Bridel, *L'Agonie de la Troisième République* (1948), p. 235.
[8] René Rémond, *La Droite en France* (1954), pp. 207–8.
[9] Beau de Loménie, op. cit., p. 27.

There is also force in the argument of M. François Goguel that anyone wanting to bring about a real *coup d'état* would hardly have concentrated so exclusively upon the Parliament buildings: 'The ministries, the telephone and telegraph exchanges, the radio stations, gas and electricity works, stations and airports, petrol dumps, the headquarters of trade unions, newspaper printing-offices – these are the principal objectives in our times of anyone attempting a *coup d'état*.'[10]

But even so, there remain one or two curious aspects of the events of the Sixth of February that are hard to reconcile with the view that nothing precise was envisaged. Nor indeed can one say that if people were thinking of more than a simple change of ministry they were doing so without good reason. The political situation at the beginning of 1934 was in itself something about which a more patient people might well have been indignant. France was facing the worst of the economic depression which, although delayed in its action in France as compared with the United States, Germany and Britain, had now produced the familiar pattern of unemployment and industrial and economic decline. French prices remained above world prices, although often unremunerative to the producers themselves, and the devaluation of the franc which some regarded as the only means of righting the situation was ruled out by the bitterness engendered by the devaluations of the 1920s and their effect upon savings.

In international affairs, France's position was far from brilliant. Despite (or perhaps because of) the evacuation of the Rhineland before the treaty date, German nationalism had triumphed: Germany under Hitler had left the Disarmament Conference and it was only a matter of time before her recovery as a military Power would be manifest. There had been in the summer of 1933 a failure to come to terms with Italy from which France had long been estranged; and although the prospects of a renewal of the German threat had brought about some willingness to put out feelers towards Soviet Russia, fear of Communism was still a powerful deterrent against seeking her alliance.[11] But it must be added that although the example of Mussolini was already an important in-

[10] François Goguel, *Politique des Parties sous la Troisième République*, vol. 2. (1946), pp. 243–4.

[11] See M. Beloff, *The Foreign Policy of Soviet Russia*, vol. I (London, 1947), part II.

F

fluence upon the anti-parliamentarian 'leagues' in France, and although the triumph of Hitler may have given them encouragement, foreign affairs did not occupy the centre of the French political stage. Indeed, in retrospect, the general neglect of the external menace would seem remarkable, were it not for the fact that the other major democracies, Great Britain and the United States, were equally unalive to it.

It is unnecessary to rehearse here the reasons why the French political system in the last decades of the life of the Third Republic failed so abysmally to provide the country with the stable and resolute leadership that both the internal problems that were acutely felt, and the external problems that were not, equally

THE CHAMBER OF DEPUTIES 1932

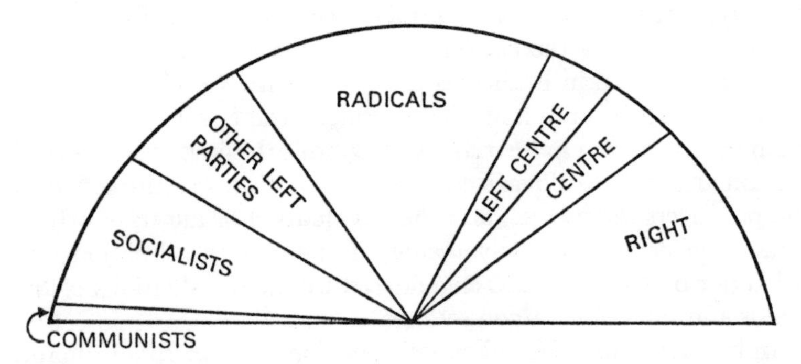

called for. The Chamber elected in 1932, in elections that seemed on the face of it a victory for the Left, was incapable in fact of producing a coherent majority because of the fictitious nature of the Left itself. If the range of opinion in the Chamber be represented by the familiar hemicycle, it is clear that a stable government must command a segment whose opposite sides are divided by an angle of well over 45 degrees. If an effort is made to carve such a segment out of the hemicycle in the diagram the difficulties are at once apparent. The Socialists were an essential part of any majority of the Left and the Radicals were committed not to rule 'against them'. On the other hand, the Socialists refused out of principle, and because of competition from the Communists on their Left, to participate in the governmental combination, and their support could at any time be withdrawn. Without the Socialists, the Radicals, with whom no possible coalition could arithmetically-speaking dispense, could only govern in harness with elements drawn from the Right. This is a decisive element in a political explanation of the crisis of February 1934.

In the Chamber elected in 1932, the seats were divided as follows: Communists 10; Socialists 131; other Left Parties 83; Radicals 157; Left Centre 46; Centre 48; Right 131. Originally this provided a majority for a government based on the Radicals, but with Socialist support, of about 390 to 150. But two governments, those of Herriot and Paul-Boncour, lasted six months and a month respectively. At the end of January 1933, Edouard Daladier, a leading Radical, formed a government which early next month got a majority of 376 to 181 on a vote of confidence. But in the course of the summer, the Socialists began to weary of supporting governments committed to a deflationary policy. This produced a schism in the party, the neo-Socialists, headed by Déat and Marquet – whose later course was to bring them over to the far Right – and numbering about 25, continuing to support the Government while the Socialists moved over to opposition. In October 1933, the Government was deserted by all its Left supporters except the 'néos' on a motion to reduce the salaries of government servants, and was defeated by 329 votes to 241. Daladier's successor, Albert Sarraut, survived for barely a month and was succeeded on 27 November by yet a third Radical, Camille Chautemps. Thus when the 'Stavisky' scandal broke at the beginning of 1934 France was enjoying its fifth ministry in eighteen months. It is not altogether surprising that anti-parliamentary movements which had been endemic almost throughout the Republic's life, and which had had one spurt of activity during the financial and governmental confusion of the mid–1920s, should have enjoyed a new lease of life.

The report of the Commission of Inquiry divided the groups that took part in the events of 6 February into four categories: various political groups of the Right; the Communist Party on the extreme Left; ex-servicemen's organizations; and finally, a certain number of members of the Municipal Council of the City of Paris.[12] For our present purposes only the first and third of these require notice.

Of the Right wing political groups, the Action Française had

[12] *Commission d'Enquête,* Doc. 3383, pp. 1230–95.

had a stormy career ever since its foundation in 1905. Despite the appeal of its intransigent nationalism and anti-parliamentarianism to certain intellectual circles, it was not at the time a large body, numbering some 60,000 adherents in the country at large of whom some 8300 were residents in Paris. Nevertheless this was the most active of the bodies concerned in the intermittent rioting and demonstrations in Paris between 9 January and the eve of the Sixth of February itself.[13] The Solidarité Française had been founded in 1933 by M. François Coty, the eminent perfumer whose important role in the financing of Right wing anti-parliamentary activity in the inter-war years has not yet been fully explored. This body claimed 180,000 adherents, 80,000 of them in the capital.[14] A smaller but senior body of the same kind was the Jeunesses Patriotes founded in 1924 by the Paris deputy Pierre Taittinger and now about 90,000 strong with 6400 members in Paris. There were also two minor anti-semitic groups, both styling themselves Francistes, which were later to become more or less overtly Hitlerite. The larger of these, that led by a certain Bucard, founded in 1928, numbered only 1500 with only 300 members in Paris. All these organizations had in common an antipathy to the regime and preference for some more or less dictatorial substitute of distinctly nationalist leanings. More amorphous and without any clear-cut organization was the Fédération des Contribuables, a sort of ancestor of the post-war Poujadists, which owed its foundation in 1928 to the industrialist Lemaigre-Dubreuil, another figure whose activities in the period and later await their historian. This body claimed about 700,000 members. It did not play any individual role on 6 February but it was indicated to its Parisian members on that day that they should join any group of demonstrators with which they found themselves in sympathy.

More interesting and more important in the width of their appeal, as well as in their role on 6 February, were the ex-servicemen's organizations. In France, as in other countries, after the war, a number of ex-servicemen's organizations were developed ostensibly for the protection of the interests of ex-servicemen, but often coming to take an active part in politics, and nearly always on the

[13] Ibid., p. 1252. For the Action Française, see also Robert Harvard de la Montagne *Histoire de l'Action Française* (1950).

[14] When this body with the other 'Leagues' was dissolved by the Blum government, a part went into Jacques Doriot's PPF, the nearest thing to a genuinely 'fascist' party in pre-1940 France.

political Right. As M. Rémond has written, 'while "apolitical" in its formal claims, the "ex-service" spirit is in fact one of the modern components of the psychology of the man of the Right.'[15] This generalization is hardly vitiated by its one exception in France, the existence of one strongly left-wing and communist-dominated organization, the ARAC (Association Républicaine des Anciens Combattants) founded in 1917, and numbering in 1934 about 20,000, a quarter of them in the Paris region.

The largest of the French ex-servicemen's organizations, the Union Fédérale, with about 900,000 members, played no direct part in the events of February 1934, but the Union Nationale des Combattants, which was only slightly smaller and more 'bien-pensant', was directly involved as were the Association Nationale des Officers Combattants, and the Association des Membres de la Légion d'Honneur décorés au péril de leur vie.

An ambiguous role somewhere between another ex-servicemen's organization and a political 'league' was that of the Croix de Feu. This had been founded in 1927 or 1928 as a non-political organization but had begun to show the usual right-wing political tendencies before its leadership was captured by Colonel de La Rocque in 1931. This event was followed by a rapid rise in the organization's membership from 16,000 in 1930 to 60,000 in 1933. The Commission reckoned its membership at the time of the February riots as about 50,000, of whom 18,000 were in Paris. To these should be added the total membership of two associated bodies, the Volontaires Nationaux and the Fils des Croix de Feu which was about the same size.

It seems clear that to some extent the movement was exploited by right-wing elements outside the ex-service movement proper. François Coty seems to have been involved in its foundation, and de La Rocque himself had been employed by the electricity magnate Ernest Mercier who was one of the inspirers of the technocratic movement of the 1920s.[16] On the other hand, de La Rocque himself does not easily fit into the part of a hero of the militant Right. According to a disgruntled ex-follower, he had contacts with Eugène Frot before the events of 6 February, and with

[15] R. Rémond, 'Les Anciens Combattants et la Politique', *Revue Française de Science Politique* (1955), p. 290. See also P. Frédérix, *Etat des Forces en France* (1935). On the individual organizations see *Commission d'Enquête,* Doc. 3387.

[16] Rémond, loc. cit., p. 287. On Mercier and the 'Redressement Français' see Beau de Loménie, op. cit., p. 23.

Doumergue afterwards. He himself is said to have passed the crucial evening in the flat of an elderly 'rentière' of the Faubourg St Germain, where he had had a telephone installed some time before.[17] But the most significant point of all is the fact that after the dissolution of the Croix de Feu in 1936, de La Rocque meekly accepted the transformation of the organization into a more or less conventional right-wing political party, the PSF.

The story of the Stavisky case itself does not directly concern us here. The belief that the career of the swindler had been aided by complicity in the highest circles of the Radical party was not the cause of the anti-parliamentary movement nor even of its growth at this time. But it did provide the movement with an excellent platform for its claim that France's rulers were not merely incompetent but corrupt, and the fact that previous scandals had not left right-wing reputations unscathed was overlooked. In a very real sense the Radical Party was the Republic and the Republic was the Radical Party. This was ground that both sides had in common.

The preliminaries can be stated briefly. Stavisky was first mentioned in the press on 30 December 1933 when a warrant was issued for his arrest in connection with frauds concerning the bonds of the municipality of Bayonne. On 3 January the *Action Française* published letters connecting Stavisky with Albert Dalimier, the Minister for Colonies in the Chautemps government. On 7 January the *Action Française* called upon Parisians to demonstrate under the slogan of 'A bas les voleurs!' On 8 January the death of Stavisky was announced and Dalimier resigned. On 9 January the official news agency Havas issued what purported to be an account of Stavisky's suicide. The official version was alleged by the right-wing press to be fraudulent; the police having, it was insinuated, murdered Stavisky in order to prevent further complicities in high circles from coming to light. On the same day street rioting began in Paris with the Action Français being first in the field. Riots continued on the 11th, the Action Française being now joined by the Jeunesses Patriotes and the Solidarité Française. (One may mention in passing, the allegation by the historian of the Action Française, Robert Havard de la Montagne, that so far from there being any collusion between these

[17] Paul Chopine, *Six Ans chez les Croix de Feu* (1935). On the approaches by Frot to de La Rocque, see *Commission d'Enquête*, Doc. 3387, pp. 125–8.

movements the others only joined in in order to prevent the Action Française getting all the glory.)

On 12 January Chautemps refused a demand for an inquiry. On 18 January Philippe Henriot, one of the most vigorous of the right-wing politicians and later to become notorious as a collaborationist, accused Raynaldy, the Garde des Sceaux, of being involved in the Stavisky affair; and these attacks were renewed by him in the Chamber on 18 January.[18] Meanwhile intermittent rioting mainly by the Action Française was continuing; according to his friends it was only the firmness of Jean Chiappe, the Corsican Préfet de Police, which prevented this taking an uglier turn. On 27 January Raynaldy resigned, and without waiting for a vote in the Chamber, Chautemps handed in the resignation of his government.

The ex-President Doumergue declined the premiership as did Jeanneney the President of the Senate and Fernand Bouisson the President of the Chamber. Daladier, Chautemps' Minister for War and a Radical (though not a Freemason), then accepted the task of forming a government. Untouched by the Stavisky scandal, and generally (though wrongly) believed to be a man of much determination, Daladier seemed to have a reasonable chance of getting things to quieten down.[19]

Like other Frenchmen called to office at difficult moments, Daladier tried to create as broad-based an administration as possible – one of 'all the talents'. But his attempt to get a 'national government' failed. The Socialists would not join; nor would the 'néos' on any terms Daladier was willing to accept – their leader, Marquet, was already showing himself alarmingly authoritarian. The extreme right-wing Ybarnégaray also refused to participate.[20] In the end, apart from the eminent but somewhat aloof figure of Henri de Jouvenel as Minister for Overseas France, as

[18] For Henriot's version see: Philippe Henriot, *Le 6 février* (1934).

[19] The most vivid account of the political aspects of the crisis is to be found in Georges Suarez, *Les Heures Héroiques du Cartel* (1934) but it is not that of an impartial witness.

[20] Xavier-Vallat, *Le Nez de Cléopatre: Souvenirs d'un homme de droite, 1918-1945* (1957), p. 113. According to Suarez's account, Marquet insisted upon the Ministry of the Interior, and Daladier was influenced against him by J. L. Malvy, the powerful chairman of the Chamber finance commission. Frossard (the principal socialist Daladier had in mind), Marquet and Ybarnégaray had wanted Chiappe suspended pending an inquiry into the Stavisky case. Suarez, op. cit., p. 167; Xavier-Vallat, op. cit., p. 113.

the Colonial Ministry had suddenly been renamed, there were only three newcomers from the Centre: the new Finance Minister, Piétri, the new Minister for War, Colonel Jean Fabry, and one under-secretary, Doussain.[21] Piétri had been Minister for Defence when Tardieu created this post in 1932. (It was abolished by Herriot when he became Prime Minister in June of that year.) Fabry, a one-legged ex-officer, had played an important role in the Commission d'Etudes of the Conseil Supérieur de la Defense and in the Army Commission of the Chamber.[22] Otherwise it was almost the same Radical mixture as before only with some of the old faces replaced by new and relatively unknown men – notably the 'young' Radicals, Mistler, Guy la Chambre, Martinaud-Deplat, Pierre Cot. Daladier combined the Foreign Ministry with the Premiership (a fact unpleasing to the Right) and the key position, politically speaking, of Minister of the Interior went to an ex-Socialist, Eugène Frot.

Any hope that Daladier may have had as to the possibility of securing a broadening of his majority in the Chamber by bringing in Fabry and Piétri was dashed when the former was expelled from his group, the Centre Républicain, at the insistence of Tardieu and Paul Reynaud.

It was now clear that if Daladier was to get a majority when Parliament met on 6 February it could only be with Socialist support unless he did something to win over the confidence of the Right. If we are to believe Fabry, the position of Chiappe was the test. There were allegations that Chiappe himself was involved in the Stavisky affair. Chiappe was admired and trusted by the Right because (according to Alexander Werth) he had treated their demonstrations with relative gentleness while cracking down ruthlessly on all disturbances from the Left.[23] It was also obvious that part of the difficulty of clearing up the Stavisky affair lay in the old rivalry between the Paris Préfet de Police and the Sûreté Générale. The Chautemps government had taken the 'service des renseignements' away from the former and attached it to the latter.[24] The Conseil Municipal had (with the

[21] A full list of ministers is given in Suarez, op. cit., pp. 185–7.

[22] See Jean Fabry, *De la Place de la Concorde au Cours de l'Intendance* (1942). The book is dedicated to the memory of Tardieu and Chiappe.

[23] Cf. L. Bonnevay, *Les Journèes sanglants de février 1934* (1935), p. 55.

[24] If we are to believe Stavisky, Chiappe was not his friend: 'Du côte de la Sûreté, ça va,' he is reported as saying, 'mais à la Préfecture, j'ai des ennemis.

exception of its Communist members) voted its confidence in Chiappe. Now Gaston Bergery, then in a left-wing phase of his distinctly erratic political career, was threatening the Daladier government with a hostile interpellation on the subject of Chiappe. If Daladier got rid of him, the hostility of the Right would become absolute; but if he refused, would the Left sustain him?

Meanwhile there were a few days of peace while the public assumed that Daladier was engaged in clearing up the Stavisky affair. The streets were quiet. Indeed Daladier was later to claim that the riots of 6 February were entirely new and had nothing to do with the disorders that took place before his assumption of office.[25]

Daladier had his first meeting with Chiappe on Wednesday, 30 January, the day after the Cabinet was formed. On Thursday, 1 February, he received a report on the Stavisky case, and on the following evening a report on the situation within the Paris police, the 'Mosse' report. Meanwhile Chiappe succeeded in getting the Union Nationale des Combattants to put off a demonstration that they had scheduled for the Monday, allegedly by saying that otherwise he would resign.[26]

It was presumably on the night of Friday-Saturday that Daladier decided that he would have to get rid of Chiappe, perhaps under pressure from Frot who is said to have threatened his own resignation.[27] Saturday, 3 February, was a day of confusion. At 9 a.m. Daladier saw Frot and informed him of his proposals for an administrative reshuffle involving the heads of both the rival police forces. At 9.15 he telephoned Chiappe to ask him to come to see him; but the latter, claiming that he was immobilized in bed with sciatica, said that he could not do so. And it was therefore over the telephone that Chiappe was told by Daladier that he wished him to leave the Prefecture of Police and accept instead the highest post in his gift, that of Resident-General in Morocco.[28] Chiappe, however, showed no desire to be kicked upstairs and protested with

Chiappe n'est pas chic avec moi.' Joseph Kessel, *Stavisky: L'Homme que j'ai connu* (1934), p. 67.

[25] *Commission d'Enquête*, Doc. 3383, vol. I, p. 272.

[26] Suarez, op. cit., p. 207.

[27] Ibid., p. 212; cf. Henriot, op. cit., pp. 91–2 where this is linked with the charge that Frot wanted Chiappe out of the way because he was an obstacle to his own plans for a dictatorship of the Left.

[28] Apparently Chautemps had already made the same offer. Suarez, op. cit., p. 216.

some violence. It was now that there took place the altercation that was to be the subject of so much controversy at once and later. Did Chiappe say, as Daladier claimed: 'Je refuse; vous me trouverez dans la rue,' or as he himself said later to the press: 'Je serai ce soir en veston dans la rue,' or did he say, as he told the Commission of Inquiry on 7 March: 'Je suis entré riche a la Préfecture de Police, j'en sortirai pauvre: je serai à la rue'? In other words, did he threaten the Prime Minister with leading a riot himself, or did he merely allude to his past services and complain of being thrown out jobless on to the streets?

And how does a man with sciatica lead a street riot?

The dismissal of Chiappe was fought by Fabry and Piétri in the Cabinet, the one a Paris deputy, the other a Corsican; and there is some suggestion that Daladier might have reconsidered his decision had he not been pressed by Frot to stand firm. But by the evening, Chiappe was definitely out and the dissenting Ministers, together with Doussain, had resigned. The new Prefect of Police was Bonnefoy-Sibour whose lack of experience and unwillingness to command was one admitted reason for the turn of events on 6 February.[29] Daladier had decided to move the Procureur Général, Chautemps' brother-in-law, Pressard, to a post on the Cour de Cassation and to remove Thomé, the director of the Sûreté Générale. The element of farce in the situation was increased when he was appointed Director of the Comédie Française, the incumbent, Emile Fabre, being in disgrace for having produced a highly anti-democratic play – Shakespeare's *Coriolanus*. Ponsot, the incumbent Resident-General Morocco, was apparently to go to Brussels as Ambassador in the place of Paul Claudel.[30]

On Sunday, 4 February, the resignations were made public as was a letter from Chiappe announcing his refusal to accept the post in Morocco. Renard, Prefect of the Seine, who had apparently been offered an overseas post, resigned and was replaced by Villey, the Prefect of the Rhone.[31] Piétri and Fabry were replaced by Marchandeau and Paul-Boncour. On the other hand, it was announced that the Chamber would be asked to constitute an inquiry into the Stavisky case – the issue over which the Chautemps government had fallen.

[29] Bonnevay, op. cit., pp. 178–9; cf. Le Clère, *Histoire de la Police* (1947), p. 120.
[30] Suarez, op. cit., p. 226.
[31] Ibid., p. 216.

But the Stavisky case was no longer the centre of political interest. On the evening of 5 February the Croix de Feu, coming into the picture for the first time, demonstrated in the areas of the Champs Élysées and the Madeleine with the Ministry of the Interior and the Parliament as the focus of attention. There was some scuffling with the Garde Mobile and the unfortunate M. Bonnefoy-Sibour was surrounded by a crowd outside the Ministry of the Interior, shouting 'Vive Chiappe!'

On the same day M. Daladier repented of one of his gestures: the nomination of Thomé to the Comédie Française was cancelled and M. Fabre left undisturbed. But the right-wing press had now launched a campaign against the new Government, accusing it of wishing to stifle public opinion by the use of tanks, black colonial troops, and so on. Rumours of this kind gained considerable currency on 5 and 6 February.

On 6 February, the day on which Daladier was to meet the Chamber, a series of demonstrations was called for by the different right-wing and ex-service groups. The method of summons and

INSTRUCTIONS TO ORGANIZATIONS

Organization	*Method of Summons*	*Time of Meeting*	*Place*
Croix de Feu	Instructions to Chefs de Section	Evening	South of Chamber
UNC	Individual instructions	8 p.m.	Clemenceau's statue
Solidarité Française	*Ami du Peuple*	7 p.m.	Grands Boulevards between Opéra and rue Drouot
Jeunesses Patriotes	Published appeal	7 p.m.	Place de l'Hotel de Ville
Front Universitaire	Published appeal	6.30 p.m.	Boulevard St. Michel
Action Française	Action Française	end of working day	In front of the Chamber
ARAC	Published appeal, *Humanite*	8 p.m.	Champs Elysées, Rond Point

the time and place of meeting are set out in the attached table. It is clear that two alternative comments are possible on the significance of these dispositions. One can argue, as those opposed to the idea of a genuine plot tend to do, that the dispersal of the meeting places suggests no collusion. Or one can take the view, which commended itself to the Committee of Enquiry, that there was some

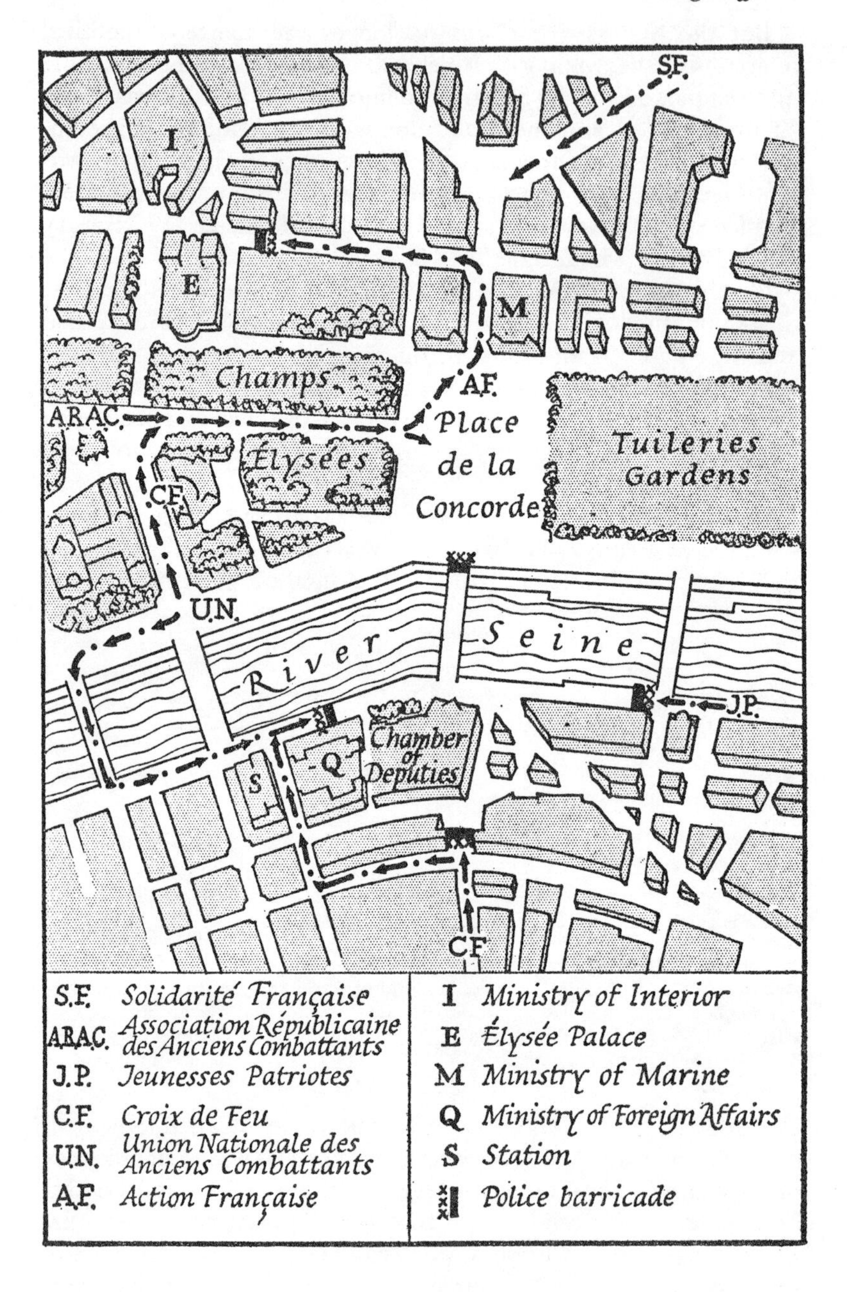
S.F.
I
E
M
A.F.
Champs
Élysées
A.R.A.C.
Place
de la
Concorde
Tuileries
Gardens
C.F.
U.N.
River Seine
J.P.
Chamber
of
Deputies
Q
S
C.F.
S.F. Solidarité Française
A.R.A.C. Association Républicaine des Anciens Combattants
J.P. Jeunesses Patriotes
C.F. Croix de Feu
U.N. Union Nationale des Anciens Combattants
A.F. Action Française
I Ministry of Interior
E Élysée Palace
M Ministry of Marine
Q Ministry of Foreign Affairs
S Station
Police barricade

significance in the fact that these meeting places formed something in the nature of a circle with the Chamber of Deputies as its centre. A police note written at the time and submitted to the Committee pointed out that although the hours of meeting did not coincide, since the UNC were only asked to meet at 8 while the Jeunesses Patriotes were asked to meet at 7, the fact that there were different distances to cover between the meeting points and the Place de la Concorde meant that between 8 and 9 in the evening there might be several thousands of demonstrators assembled in the Place.

The Chamber of Deputies met at 3 o'clock and its session was punctuated by violent demonstrations.

Activity on the Place de la Concorde began at about 4 o'clock, when it started filling up both with members of the leagues which had been summoned for later in the evening and with curious onlookers. Mixed with the crowd were a number of Camelots du Roi, the activists of the Action Française, and, so it was later alleged, a number of Communists as well. There was a marked predominance of young people. At 4.45 the Commissaire of the Arrondissement, M. Rottée, took up a position on the bridge leading across the Seine to the Chamber with 70 policemen – Gardiens de la Paix – armed with automatic pistols, 100 Gardes à Pied, and 25 Gardes Républicaines. He was later to assert, though this was denied by Marchand, the Director of the Municipal Police, and also by the new Prefect of Police, Bonnefoy-Sibour, that he had orders to hold the bridge at all costs.

By about 5 o'clock things were warming up, and young men were demonstrating at the passage of troop-carrying lorries. The mounted Gardes cleared a space on the south side of the Place and a barricade of lorries was put across the bridge, leaving a space on either side. The police and the Gardes then took up positions around this barricade.

By about 6 the crowd had become more violent and the police were being pelted with stones, garden chairs, railings, and so forth; at least two were seriously injured. The mounted Gardes again cleared a space but were received with a hail of missiles; three Gardes were hauled off their horses and had to be rescued. Henceforward charges of this kind took place every few minutes, the crowds giving way but then returning and continuing to bombard the defenders of the bridge. Razor blades on sticks were used against the horses and squibs and marbles were thrown under their hooves. Tempers got uglier and the police increasingly made use of their batons, and later of iron bars seized from the rioters. Reinforcements were sent for and began to arrive in driblets as follows:

6.30: 75 gardes mobiles
6.55: 100 gendarmes
7.00: 70 policemen (gardiens de la paix)
7.10: 100 gendarmes
7.25: 50 cavaliers de la garde républicaine
7.40: another 50
7.50: 100 gendarmes
7.55: 75 suburban policemen.

At about half-past six the rioters themselves began to build a barricade and were attacked by the police with considerable violence; 22 policemen had to be removed to have their injuries attended to. Further police charges were made eastwards towards the Tuileries Gardens and westwards down the Cours de la Reine. The rioters captured a bus and set fire to it, thus providing additional illumination. Another barricade was set up by the rioters at the eastward entrance to the Champs Élysées and policemen arriving there in ignorance of the development of the situation were attacked by Camelots du Roi. Paris had still not awakened to the seriousness of the situation and a certain amount of traffic was still going on, even in the areas directly affected by the rioting. Marchand himself now arrived at the Place with a trumpeter and formally ordered the rioters to withdraw, but was forced to retreat himself under a hail of stones.

We now approach the time for the set meetings of the different organizations, who duly assembled and started proceeding towards the Chamber. This can be followed in the next table, and in the map.

At about 7 the police made three more charges which helped the cordon to extend further out from the bridge, but the mounted police in particular sustained further serious losses. Thenceforwards a few shots were heard from the crowd and at 7.10 a mounted policeman was hit by a bullet. Rottée made two more 'sommations légales' but without effect.

At 7.15 a group of right-wing municipal councillors left the Hôtel de Ville, were joined by a number of demonstrators from the Jeunesses Patriotes, crossed the river by the Pont du Carrousel and proceeded along the Quai d'Orsay. At 7.30 they were turned back with some violence at a barricade by the Pont de Solférino.

Organization	*Place of Meeting*	*Hour of Meeting*	*Route taken*
Action Française	Boul. St. Germain, between the Chappe statue and the rue de Bellechasse	?	Boul. St. Germain
Solidarité Française	Carrefour Richelieu-Drouot	7 p.m.	Towards la Madeleine
Front Universitaire	Intersection of the Boul. St Michel and the Boul. St Germain	6 p.m.	Boul. St. Germain
Jeunesses Patriotes	Place du Châtelet	7 p.m.	Quai d'Orsay
Croix de Feu	Intersection of rue de Grenelle and rue St Guillaume	7 p.m.	Rue de Várenne, rue de Bourgogne, rue de Constantine
UNC	Cours la Reine, behind the Grand Palais	7–8.30 p.m.	Champs Élysées, Concorde
ARAC	Rond Point des Champs Élysées	8 p.m.	Champs Élysées, Concorde

At the same time the column of the Solidarité Française arrived in the Place de la Concorde from the Grands Boulevards, having got through the barricades there without much difficulty. They showed every intention of pressing on across the bridge to the Chamber; hoses which were used against them were seized and turned against the police. The rioters got right up to the barricade and a few squeezed through the gaps between the barricade and the parapets. The Gardiens de la Paix and the Gardes Mobiles lost their nerve and fired in self-defence. Marchand and other officers tried to stop the firing and finally succeeded. Marchand, Rottée and another senior officer were injured, the two latter seriously; 6 rioters were killed and 40 more injured in the firing, and a woman in the Hôtel Crillon was killed by a stray bullet. The rioters now withdrew and the Gardes Mobiles cleared a space in the Place reaching to about 100 metres from the end of the bridge. The right-wing opponents of the government were later to make a great deal of this firing, their case being that unarmed and patriotic citizens had been brutally shot down by the police acting under the orders of the sinister M. Frot. On the other hand, the evidence that the first shots came from the crowd, and that the police only fired in self-defence, was accepted by the great majority of the Commission.[32] Curiously enough, a recent right-wing account of these events declares that Frot did not give the order to fire, that he left the Chamber early and did not reach his Ministry till midnight, though this account adds that it was from Frot's personal entourage that the orders came during the night to bring military reinforcements into Paris.[33]

At about 7.30 also a column of Croix de Feu, advancing towards the Chamber from the other side down the rue de Bourgogne, was stopped by the police at its intersection with the rue St Dominique and turned away to the left towards Les Invalides. Meanwhile, the disorder inside the Chamber was continuing and some Deputies were slipping away. It was at this point that, as Mr Alexander Werth records, some humorist put outside the door of the Press Gallery the notice 'Avis à MM les Manifestants – Ici il n'y a pas de Députés'. At about 8 o'clock some of the Conseillers Municipaux had got to the Palais Bourbon and were seen by Daladier in an interview of which no agreed account is forthcoming.[34]

At about 8.15 an attempt was made by the police to clear the rioters. The terraces of the Tuileries Gardens, which, since they overlook the Place, have great nuisance value on such occasions, were cleared, but shots were fired and stones thrown and another bus was set on fire. At some time during the next half hour the two groups of Croix de Feu from the north and south of the river joined up to the west of the Chamber and were repulsed by the police outside the Ministry of Foreign Affairs. Back at the Place an attempt was made at 8.15 to set the Ministry of Marine on fire by throwing burning rags, etc., through the windows. Rioters broke in from the rue Royale but obeyed an order to leave given by a naval officer in uniform. Despite attacks on the fire-fighting apparatus and engines, the fire was rapidly extinguished. At 8.45 the procession of the UNC arrived at the Place de la Concorde from Clemenceau's statue further down the Champs Élysées. They had also been joined by the 'Association des Décorés au péril de leur vie' and by members of the ARAC. There were now five or six thousand ex-servicemen in all marching in good order and this for the moment silenced the disturbances. The main body

[32] See the book by its chairman: Bonnevay, op. cit., p. xi.

[33] Xavier-Vallat, op. cit., pp. 115–16.

[34] According to one account: Daladier fut assez brutale: Que prétendez-vous? Venger la morale? Ici il y a une dizaine de corrumpus. Combien chez vous?' J. Paul-Boncour, *Entre Deux Guerres* vol. 2, (1945), p. 304.

turned left from the Place into the rue Royale, but the 'Décorés' turned towards the bridge together with some Communist elements. They were not, however, in a violent mood, and negotiated with Marchand who was now back at his post. They decided to give up the idea of trying to cross the bridge and followed the other ex-servicemen into the rue Royale. Once the ex-servicemen's column had left the Place, disorders broke out again.

What happened next with regard to the ex-servicemen is one of the disputed points. It is not clear whether they really intended to abandon the demonstration but were prevented from leaving the area by the police at the entrance to the rue du Faubourg St Honoré, or whether – as one of the leaders later claimed – they really meant to go to the Élysée to present a petition to the President. Anyhow, there was a fairly severe scuffle with the police in the narrow rue St Honoré.

At 9 o'clock the session in the Chamber ended, Daladier having won in a series of votes by 300–217; 302–204 and finally by 360–220, and the deputies smuggled themselves out through various side entrances. Some of them were recognized and threatened by demonstrators; one of these was Herriot, who received a blow from a stick on his leg and heard cries of 'à la Seine! à la Seine!' He rather gloomily records his sense of humiliation at the thought that a Mayor of Lyons might end his life in a river other than the Rhône.[35]

At 10 the frustrated ex-servicemen returned to the Concorde in an ugly mood and proceeded towards the bridge. They got past the first obstacles and arrived up against the real barricade. The police charged with batons and used their hoses, and were showered with missiles in return. By 11.30 there had been some fifteen or thirty further mounted charges by the police, and there were occasional shots from the rioters. There was a direct attack on the bridge at about 11.30, causing a new panic among its defenders, though this was halted by the Prefect of Police. Nevertheless, some of the police fired once more; again, it seems, spontaneously. At any rate, the rioters made off and the barricade was re-established. Rioting, a new effort by the mounted police and spasmodic shooting occupied the next hour or so, but between midnight and 1 o'clock the crowds dispersed and the demonstra-

[35] Herriot, op. cit., vol. II, p. 376. For a slightly different version see Bonnevay, op. cit., p. 166.

tion came to an end. By 2.15 a.m. it was possible to remove the police from the bridge.

At a meeting of ministers at around midnight it was decided in principle, on the insistence of Frot, to proceed to a large number of preventive arrests, though it was pointed out that this would be unlawful. Since Parliament was sitting it was not possible to proclaim a state of siege.[36]

The casualties of the evening were later reckoned by the Commission of Inquiry: On the sides of the rioters, 14 killed; 236 taken to hospital and 419 other injured; on the side of the police, 1 killed, 92 taken to hospital and 688 other injured.[37]

The question of what the Prime Minister and government should do in the event of a renewal of similar scenes, or in order to prevent them, was again discussed by Daladier and others on the following morning. Frot seems to have changed his mind during the course of these discussions. Only Blum advised Daladier to resist all pressure in the name of Republican legality. In the end Daladier decided to resign at once. His reasons for doing so were given to the post-war Commission as follows:

> I resigned at the insistent demand of M. Lebrun and on the advice of M. Frot, the Minister of the Interior, of the President of the Chamber and of the President of the Senate. Only M. Léon Blum advised me to remain in office; but I was not in agreement with him; for I claimed that in order to avoid having scuffles every day whose result would be that misguided and anyhow innocent Frenchmen would fall in the streets of Paris it was necessary according to the provisions of the Constitution of 1875 itself that Parliament should be adjourned for a month.[38] I was not in agreement with M. Léon Blum who said: 'on the contrary the Chambers should sit daily; that is a necessity'.[39]

[36] Bonnevay, op. cit., pp. 184–9. Suarez gives a highly-coloured version of these proceedings, and a most catholic list of persons whose arrest was suggested, op. cit., pp. 267–74.

[37] *Commission d'Enquête,* Doc. 3386, p. 16.

[38] Art. 2 of the 'Loi Constitutionel sur les Rapports des Pouvoirs Publics' of 16–18 July 1875 provides that the President of the Republic may adjourn a session of the Chambers but not for more than a month or more than twice in each session.

[39] *Les Evénements survenus en France,* vol. I, pp. 11–13. Worry about a financial crisis if order was not rapidly restored was also a consideration in Daladier's mind. Bonnevay, op. cit., p. 194.

Lebrun was to declare many years later that he had advised Daladier to resign in order to avoid civil war.[40]

Herriot records that Daladier was most influenced in his decision by the declaration of Frot that he could not be responsible for public order. Herriot himself agreed to join the other ex-Premiers in a collective démarche to Doumergue without which the ex-President declared himself still unwilling to accept the post of Prime Minister. There was some further rioting on the 7th but with the return of Doumergue from his country retreat at Tournefeuille – soon to be the joy of cartoonists and chansonniers – the crisis was over.[41] Apart from the inclusion of Marshal Pétain and Marquet, the new Cabinet with its old familiar faces – Herriot himself, Tardieu, Sarraut, Barthou – was a curious commentary on an effort to change the regime, if such an effort had indeed been made. If Herriot was right and if behind the Parliament the attackers had aimed at the principle of universal suffrage itself they could hardly claim success.[42] Doumergue, Protestant and Freemason, was a curious saviour for the nationalist leagues to turn up.

Indeed, when Doumergue tried after a few months' time to tackle the problem of constitutional amendment on the plea that this was necessary in order to remedy the country's chronic political instability, he fell abruptly from power and returned unlamented to his rural obscurity. All that had happened was the usual shift halfway through the life of a French Chamber from administrations with a bias towards the Left to administrations with a bias towards the Right. Doumergue's immediate successor was Laval.

The political interest of the period after 6 February does not lie in the composition or programme of governments but in what was going on in opposition. On 9 February, the Communists staged a turbulent demonstration of their own; on 12 February, there was a general strike. In face of the danger, real or imaginary, of a fascist regime in France, and with Soviet policy taking a new line, the foundations of the Popular Front were gradually and painfully laid.

But what actually happened does not necessarily show what was

[40] *Les Evénements survenus en France,* vol. IV, p. 951.

[41] M. Herriot's comment on the further disturbances which he viewed from his flat high above Paris may be given: 'J'étais toutes proportions gardées, dans la situation de Néron, lors de l'incendie de Rome, avec cet avantage sur lui qu'il n'avait pas pu voir, comme moi, brûler les Vespasiennes.' Op. cit., vol. II, p. 378.

[42] Ibid., p. 376.

intended. What are the alternatives to the idea that the Sixth of February was an abortive *coup d'état*? One line of argument is that there was a plan for a *coup d'état,* but that it was to come not from the 'Leagues' but from within the system, from the younger Radicals, Frot himself, Mistler and Pierre Cot. Daladier could be presented willy-nilly as the saviour of the Republic.[43]

Even if this account was true this does not prove that Frot and his colleagues provoked the disorders in order to provide an excuse for a subsequent period of strong government. This accusation is based mainly on the rumours that we have already touched upon about efforts on the part of Frot to make contact with the Croix de Feu and also with the royalists.[44] It was alleged that Colonel de Lattre de Tassigny was a go-between here; though Frot asserted that only military matters were discussed in their meetings.[45]

The supposition put forward by M. Debu-Bridel that the failure of both the Croix de Feu and the Action Française to attempt to force the barricade may have had something to do with Frot's intrigues seems to remain pure supposition.[46] Nevertheless he suggests that Frot had a plan for an alternative government and that his names included a whole list of later collaborators: 'The whole of what later made up Vichy's general staff,' he writes, 'except for Déat and Doriot, was more or less in league with Eugène Frot.'[47]

What makes it possible to present this seemingly paradoxical version of events is a fact that we may be in danger of overlooking in view of our knowledge of later events, namely that it was by no means clear in 1934 what might turn out to be the effect of the depression on French political divisions. We must not assume that Frenchmen were all labelled in Gilbertian fashion with 'Right' and 'Left' tickets – still less that such tickets give a direct indication of the choice they would make in and after 1940.[48]

[43] Jean Galtier-Boissère, 'Histoire de la Troisième République' in *Crapouillot* (December 1935), see pp. 240–6. A plot for the dictatorship of the Left is the main theme of P. Henriot's book already referred to.

[44] Evidence of Henri de Kerillis; *Commission d'Enquête,* Doc. 3383, vol. I, pp. 303–19.

[45] Evidence of Frot, ibid., vol. II, pp. 2087–146.

[46] Debu-Bridel points out that only the Solidarité Française and the Jeunesses Patriotes made serious attempts to reach the Chamber, and alleges that in the rioting on the Place de la Concorde a number of Communists, including foreigners, took part. He quotes Admiral Darlan's view that it was the Communists who had set fire to the Ministry of Marine.

[47] Debu-Bridel, op. cit., pp. 231 ff.

[48] Frot was nominated by Pétain to membership of the Vichy 'Conseil National'

A more modest view is that put forward by M. François Goguel, namely that the objective of the organizers of the demonstration was to gain new adherents by giving a proof of their strength and at the same time to enlist 'the public discontent on behalf of the personal grievances of Chiappe. In this sense, the Sixth of February appears as a sort of vendetta directed by the former police-chief against Daladier.'[49] The difficulty here is, first, that no proof has ever been forthcoming of collusion between Chiappe and the organizers of the riots, and second, that he did not in fact get his job back – nor go to Morocco – but had to be content with a minor municipal post.[50]

Finally we come to the stock question of all students of modern revolutions and *coups d'état* – the role of the armed forces. Alexander Werth pointed out that what would have happened had a new government been set up would have depended on the line taken by the army. Without their support there would have been a general strike and civil war.[51] What would the army's attitude have been?

This can be answered on two levels. In the first place, there was the undoubted fact that the military profession as such had no reason to admire some of the attitudes encouraged by the major parties of the Left. From almost the morrow of victory the anti-militarism of the pre-1914 Left revived in full strength. Pacifism was in the ascendant and the military virtues at a severe discount. It is not surprising that the pacifist euphoria should have created a malaise in the army which felt itself divorced from the country's political institutions.[52] Alexander Werth's suggestion that one reason for Daladier's resignation was his doubt about the loyalty of the army, and that some army leaders had intervened with the President in order to persuade him to bring about the Prime Minister's withdrawal is thus on the face of it quite credible.[53] But does this mean that army personalities themselves were behind the riots?

and resigned in November 1941. After the Liberation he was declared ineligible for public office.

[49] Goguel, op. cit., vol. 2, p. 244.

[50] *Crapouillot* (December 1935), p. 244.

[51] Werth, op. cit., p. 161.

[52] Raoul Girardet, *La Société Militaire dans la France Contemporaine, 1815–1939* (1953), pp. 314–17.

[53] Werth, op. cit., p. 170.

Is there anything sinister in the fact that Marshal Lyautey was an honorary member of the Jeunesses Patriotes or in the rumours that if Daladier did not resign, the eighty-year-old Marshal would march at the head of a new demonstration? Is there more than coincidence in the wording of a proclamation of the Jeunesses Patriotes on 5 February which ran in part:

> by a real coup, even before receiving the endorsement of the Chamber, the Government is sacrificing to the injunctions of the Communists, the Prefect of Police – Jean Chiappe.
>
> 'Tomorrow, giving way to the pressure of Germany, one of the organizers of victory will be forced to quit – General Weygand.'[54]

General Weygand's own memoirs give no hint that his fate was in any way involved in the events of the Sixth of February. Indeed he professes to have only a vague impression of what actually took place, though he was as Vice-President of the 'Conseil Supérieur de la Guerre' (Commander-in-Chief designate) kept informed about the preparation of troops during the night of 6–7.[55]

On the other hand, his previous and subsequent narrative makes no secret of the intense opposition between himself on the one hand and Paul-Boncour and Daladier on the other.[56] Paul-Boncour was anyhow unpopular with the military since he had been President of the Chamber's Army Commission between 1924 and 1928 and so responsible for a bill for the organization of the nation in time of war which the General Staff thought far too Jacobin in its demands upon the country.[57] Weygand had bitterly opposed Paul-Boncour's economies at the army's expense and his willingness to consider certain measures of disarmament proposed at Geneva during the latter's tenure of the War Ministry under Herriot. He thus cannot have welcomed Paul-Boncour's reappear-

[54] *Commission d'Enquête,* Doc. 3385, p. 116. According to Suarez, Weygand's name stood high on the list of those whose arrest was contemplated during the night of 6–7 February, op. cit., p. 272.

[55] Weygand, *Mémoires,* vol. 2, *Mirages et Réalites* (1957), p. 409.

[56] Ibid., pp. 365–469 passim.

[57] Paul-Boncour, op. cit., vol. 2, ch. VI. For his account of the 6 February including the military measures taken see ibid., pp. 292–308. The military security of the capital was dealt with by direct contact between the Minister and the Military Governor of Paris. The Chief of the General Staff was not involved. M. G. Gamelin, *Servir,* vol. 2 (1946), p. 120.

ance in this post after Fabry's resignation – though Fabry himself seems to have been not altogether acceptable to Weygand either.[58] Daladier, who had himself succeeded Paul-Boncour as War Minister in December 1932 and who remained at that ministry (for part of the time as Prime Minister as well) until the fall of the Chautemps government, was even less congenial to Weygand. He seemed to Weygand to be blind to the necessity of meeting by an intensified defence effort the growing strength of Germany, now, under Hitler, scarcely bothering to conceal its determination to rearm. Instead of working out devices to meet the problem of the 'années creuses' by modifying the system of a single year's service, the Government were seemingly determined to make further economies at the army's expense and to give way to pressure from London and Washington in favour of dangerous proposals made at the Disarmament Conference. Weygand conceived it to be his duty to take his differences with the Government to the President himself, and practically defied Daladier to take the responsibility of dismissing him.

Weygand's own account does not suggest that he went beyond the limits permissible to a military adviser to a government which does not wish to take his advice. And he could point to the fact that a majority of the supreme military body, the Conseil Supérieur de la Guerre, including France's three living marshals, supported his views. We have, however, another account from General Gamelin, who had succeeded Weygand as Chief of the General Staff responsible under the Minister for the peacetime conduct of military affairs, and destined to be Weygand's successor in the position of Vice-President of the Conseil.[59] In view of the bitter rivalry between the two soldiers one ought perhaps to discount Gamelin's version to some extent. It was actually published a decade earlier than Weygand's, but Weygand does not directly contradict him though he cannot conceal his bitterness about the fact that Gamelin supported the Prime Minister's view rather than his own.

Gamelin talks of the rapid development of a personal antipathy between Weygand and Daladier, and describes how it came to a

[58] According to Henriot, he had sided with Daladier against Weygand, op. cit., p. 86. Besides, Fabry was a Joffre man and Weygand a Foch man. Suarez, op. cit., p. 190.

[59] Gamelin, op. cit., pp. 91–108.

head in December 1933 when, as Gamelin points out, Daladier had to remember that whatever his personal views on the gravity of the international situation he was a minister and a party leader, and could not therefore give way to Weygand's pressure for more drastic measures. On 17 December, the day before a crucial meeting of the Conseil Supérieur de la Guerre, Weygand openly attacked Daladier in a conversation with Gamelin and declared that if the Conseil were solid, the Government would find it very difficult to continue its policy of constant reductions in the country's military strength. 'There followed', writes Gamelin, 'a sharp criticism of our policy both in foreign and in home affairs. In short, it seemed to me that it was a question if not of a plot with political motivations, then at any rate of a trap contrived for the minister.'

The Conseil defeated the Prime Minister's project by 11 votes to 3 (including Gamelin) and with two abstentions. Daladier said that he would put the matter to the Chamber, and told Gamelin that if Weygand resigned his post he would accept his resignation and put Gamelin in his place. According to Gamelin he at once said that there could be no question of his accepting the supreme post under these conditions.

The Chamber passed Daladier's bill by 449 votes to 147, but the news of further re-armament in Germany gave rise to the hope that the Senate might turn it down. Meanwhile Gamelin felt that he had now been classified as the military man who would not lend himself to a plot against the lawful government, and that in certain circles this would count against him in his future career.

Daladier left the War Ministry on becoming Premier and Fabry's first act as his successor was to announce his intention of getting rid of Daladier's *chef de cabinet,* General Bourret. Bourret and others whom Fabry wished to replace with different officers seem to have appealed to Daladier against him. There was thus friction between Daladier and Fabry even before the latter resigned over Chiappe. In view of the previous relations between Weygand and Paul-Boncour the latter's return to the War Ministry was unlikely, as we have seen, to mollify the general and his friends.

Pétain's arrival at the War Ministry was more perhaps than a satisfaction for the old slogans of the leagues, as M. Beau de Loménie suggests.[60] It meant the burial of the Daladier bill so

[60] Beau de Loménie, op. cit., p. 27.

vehemently opposed by Weygand and the dropping of an earlier measure for the cutting down of the number of officers.[61] On the other hand, his conduct at the War Ministry disappointed those who had hoped that he would firmly insist upon the necessity for two years' service. One cannot altogether feel convinced that the appearance of Pétain on the scene was entirely fortuitous. It may only be our knowledge of his later role. But is there not something curious in the fact that an active demonstrator in the ranks of the Jeunesses Patriotes was an *inspecteur des finances,* and former *chef de cabinet* to Paul Reynaud, Du Moulin de Labarthète who was placed *en disponibilité* as a result, and that the next time he appears upon the historical scene it is as Pétain's first *chef de cabinet* in the Vichy Government?[62] In a letter to the Commission of Inquiry, Dumoulin de Labarthète wrote that the purpose was to gain access to the Chamber and to carry out 'de solides représailles (solides, mais non sanglantes) sur les élus d'un suffrage universel qui mène la France à la guerre et à la ruine'.[63]

In conclusion, one must admit that after all there may be no mystery, that the striking impression made by the map of a planned and purposeful convergence upon the Parliament buildings may be an illusion, that the rioters may have had little leadership and no defined plans for the seizure of power, that the threads that link 6 February 1934 to 10 July 1940 may be too slender to bear the weight of any theory that would seek directly to connect them. Perhaps there is nothing to it but the familiar story of a political crisis in Republican France with a little intervention this time by the mob. To the charge of 'attempted murder of the Republic' made by Léon Blum the only reasonable verdict may be the Scottish one: 'Not proven.'

Postscript. This paper was delivered as a lecture to the Cambridge University History Club on 13 February 1958. In the June 1958 number of *History Today* there appeared an article by Mr Geoffrey Warner, of Sidney Sussex College, Cambridge, entitled 'The Stavisky Affair and the Riots of February 6th, 1934': as this paper was already in final proof I was not able to profit by Mr Warner's article.

[61] Gamelin, op. cit., p. 108; Weygand, op. cit., p. 399.

[62] To this fact we owe the most revealing of Vichy memoirs; H. Du Moulin de Labarthète, *Le Temps des Illusions* (Geneva, 1946).

[63] *Commission d'Enquête,* Doc. 3383, vol. II, pp. 1836–7.

14 The Anglo-French Union Project of June 1940

The project for an 'indissoluble union' between France and Britain was presented by the Prime Minister M. Paul Reynaud to a meeting of the French Council of Ministers held at Bordeaux at 5 p.m. on Sunday, 16 June 1940. It arrrived, we are told, like a bombshell in an atmosphere little prepared to receive it.[1] The explosion it caused was a muffled one, and soon forgotten in the press of other dramatic and tragic happenings.[2] It might be held that an episode so isolated and with so little obvious effect upon the course of history is hardly worth resurrecting. But nothing in history stands quite by itself; every event has both a past and a future and this ill-fated project is no exception.[3]

[1] 'Ce projet arrivant — permettez-moi cette expression — comme une bombe dans un milieu peu preparé à le recevoir'; the words are those of President Albert Lebrun, *Rapport de la Commission sur les Evénements survenus en France de 1933 à 1945,* IV, p. 976.

[2] The present account owes much to the pioneer article of M. Léon Noël. 'Le projet d'union franco-britannique de juin 1940', in *Revue de l'histoire de la deuxiéme guerre mondiale,* no 21, (Jan, 1956). The principal British accounts are those by Sir Winston Churchill in volume 2 of his *The Second World War* (1949) and by Sir Llewellyn Woodward in his contribution to the official series of British war histories, *British Foreign Policy in the Second World War,* (London 1962). I have been able to make use of information that has come to light since M. Noël wrote, particularly on the British side. Among those who have assisted me in regard to material or by answering questions bearing on the subject, I should wish to thank the Earl of Birkenhead, Viscount Simon, Sir John Wheeler-Bennett, the comtesse Jean de Pange, Lord Ismay, Sir Desmond Morton, Mr Martin Gilbert, Lady Vansittart, Lord Salter, M. Jean Monnet, M. Charles Corbin, Lord Layton, Lord Sherfield, Mr Edward Ashcroft, Mr Randolph Churchill, the Rt Hon. Kenneth Younger, the Rt Hon. Julian Amery MP and Mr Ian Colvin. Others equally kind must perforce remain anonymous. Until the opening of the archives some details must remain obscure.

[3] The present paper is an offshoot of a wider study of the development of British foreign and imperial policy in the present century which I have undertaken under

It is necessary to recollect the extent to which the affairs of Britain and France had been jointly run on the civil side since the beginning of the war. A Supreme War Council had been set up immediately upon its outbreak and work had begun towards the establishment of machinery for joint economic planning. On 6 December 1939 this culminated in the inauguration of an Anglo-French Co-ordinating Committee. Its chairman was M. Jean Monnet who had been the principal motive force in its establishment.[4] The team of Frenchmen which M. Monnet headed in London included M. René Pleven, and the necessary link with the French Embassy was maintained by M. E. G. M. Monick who had been financial attaché there since 1934.

The work did not always go smoothly, and many difficulties were encountered, but a Franco-British trade agreement was concluded on 16 February 1940, and on 8 March the formation of an Anglo-French industrial council was announced. At the same time there were the beginnings of an effort to coordinate the resources of the two Empires and to achieve some degree of harmonization in their colonial policies. The British Colonial Secretary Mr Malcolm MacDonald had talks with his opposite number M. Georges Mandel in Paris from 16 to 18 March, and it was announced that permanent liaison machinery between the two colonial offices would be established at once.[5] This news received a general welcome in the French and British press. It may be noted that between 1920 and 1939 only one British Colonial Secretary, Mr Leo Amery, visited Paris officially, and no French Minister for the Colonies visited London.

The most important matter was however that of military security, since it was in this field that the two governments had most sharply clashed in the inter-war period. M. Paul Reynaud, who had succeeded M. Edouard Daladier as Prime Minister on 21 March, visited London on 27–8 March. After the meeting of

the auspices of the international affairs programme of the Rockefeller Foundation. I wish to take this opportunity of thanking the Foundation for their assistance and also my research assistant on the project, Mrs M. Croft (née Heeley).

[4] For a general account of general cooperation between the two countries between the beginning of the war and the French armistice see W. K. Hancock and M. M. Gowing, *The British War Economy,* (London 1949), ch. VII.

[5] See Mr MacDonald's written reply to a Parliamentary question on 21 March. 358. *HC Deb.,* 5s., cols 2125–2126.

the Supreme War Council on 28 March the following communiqué was published:

> The Government of the French Republic and His Majesty's Government in the United Kingdom and Northern Ireland mutually undertake that during the present war they will neither negotiate nor conclude an armistice or treaty of peace except by mutual agreement.
>
> They undertake not to discuss peace terms before reaching complete agreement on the conditions necessary to ensure to each of them an effective and lasting guarantee of their security.
>
> Finally they undertake to maintain, after the conclusion of peace, a community of action in all spheres for so long as may be necessary to effect the reconstruction, with the assistance of other nations of an international order which will ensure the liberty of peoples, respect for law and the maintenance of peace in Europe.[6]

At the time it was the last paragraph and the phrase 'a community of action' that attracted most attention. A correspondent of *The Times* wrote: 'Anglo-French unity has already reached a more advanced point than at any period during the last war, and what is more it is realized in both countries that this point is but the first step towards a closer and more lasting association'.[7] The French press emphasized the break that this undertaking represented with Britain's isolationist past and spoke of the communiqué as marking a turning point in the history of British foreign policy. 'England', commented M. Wladimir d'Ormesson, 'is now in Europe.'[8]

The failure of the western democracies to stand together in face

[6] *The Times*, 29 March 1940. According to M. Paul Baudouin, M. Daladier had rejected a suggestion for a joint declaration made by Mr Chamberlain at a meeting of the Supreme War Council on 19 December because of the absence of material guarantees for France's future security. P. Baudouin, *Neuf mois au gouvernement*, (Paris, 1948), p. 18. The Baudouin account, here as elsewhere, is unreliable. The idea was put up to Daladier on 19 December and he asked questions about material guarantees to which a British reply was only forthcoming on December 22. The form of declaration accepted by the French on 28 March, which did not include a precise reference to *material* (i.e. territorial) guarantees, was prepared in the Foreign Office in the interval between the two meetings, and clearly had not been rejected in principle by Daladier. Cf. Woodward, op. cit., p. 60 fn. The minutes of both meetings were captured by the Germans and published by them in 1940.

[7] Ibid.

[8] *Le Figaro*, 30 March, 1950.

of the rise of aggressive dictatorships had brought about an interest in proposals for closer unity between them even before the outbreak of hostilities.[9] Now on the margin of politics it was possible to explore these ideas in an Anglo-French context. In the spring of 1938, the Royal Institute of International Affairs (Chatham House) and the Centre d'Études de Politique Étrangère set up groups for the joint study of problems in Anglo-French relations, but their work remained incomplete and unpublished.

At the beginning of the war, Chatham House itself was divided; the major part of its staff was transferred to Oxford where, with important additional recruits, it now functioned under the aegis of the Foreign Office, with the study of the development of opinion abroad as its main task. On 19 September Professor Arnold Toynbee, who was in charge of the work, wrote to M. E. Dennery, the Secretary-General of the Centre, suggesting the need for some permanent form of union between Britain and France in order to prevent any post-war renewal of the German bid for domination. This initiative led to the creation of a group at each institution to study the problems inherent in the creation of such a union. In November 1939, M. Dennery together with M. André Siegfried and M. Robert Maheu visited Oxford in order to discuss the cooperation of the two institutions in the study of war aims.[10]

Meanwhile at Chatham House in London, the comte Jean de Pange, a member of the French group, gave a talk entitled 'The Foundations of Federalism in the old Austria-Hungary and in Franco-British Relations Today.' He pointed out that in the Supreme War Council, in the unified command of the forces in France, and in the economic machinery already in operation there existed the essential organs for a future federalist structure. He suggested as the next step a system of Parliamentary delegations – a twice yearly meeting alternately in London and Paris of forty MPs and twenty peers with forty deputies and twenty senators. Eventually, as in the old Austria-Hungary, there should be common ministers for foreign affairs, war and finance. Visible evidence of unity might be given by the immediate pooling of the two coun-

[9] M. Noël in the article referred to places emphasis on the role of Clarence Streit's *Union Now*, of which the French version *Union ou chaos* was published in May 1939. His view that it was this book which inspired M. Monick, and further that it was M. Monick who suggested the Anglo-French Union plan to M. Monnet cannot be sustained.

[10] Noël, op. cit.

tries' air forces and this might be continued in peace time. In conclusion he suggested as a common war aim the creation of a Danubian federation.[11]

The subsequent discussion by his British audience was almost exclusively concerned with this last point and the suggestion of permanent Anglo-French institutions evoked no response. Indeed at this time the subject seems to have been more alive in France. In the *Dépêche de Toulouse* of 19 March, M. Georges Scelle also took the Austro-Hungarian Monarchy as the prototype for a proposed Franco-British union which should have an inter-governmental council, inter-parliamentary delegations, a single military command, a common foreign policy, preferential arrangement in respect of citizenship and so on, and one which should eventually be open to the adherence of other states. On 24 March *Le Temps* proposed an inter-allied consultative committee to prepare questions for decision by the Supreme War Council; on 25 March, M. Pierre Cot called in *L'Œuvre* for the creation of an Inter-Allied Economic General Staff. In *L'Ordre* on 11 April, a writer suggested setting up a joint committee of the two Parliaments; in *L'Époque* on 23 April, a proposal was put forward for a Franco-British customs union.

Perhaps the general temper might be gauged from the fact that the graduating class from St Cyr in March 1940 was styled 'the Promotion of Franco-British friendship'.[12] No doubt there was in the background the rumble of criticism that the British military commitment to the common struggle was still much inferior to that of France, and there were overt expressions of alarm on the traditional French Right lest a common policy in the future might prove too lax towards Germany. But the time seemed propitious for a further examination of what might be done to strengthen the institutional ties between the two countries.

Between 9 and 12 March, Professor Arnold Toynbee of Chatham House visited Paris in company with Sir Alfred Zimmern. They were present when Senator Honnorat put forward at a lunch given by the Commissioner for Information Jean Giraudoux, and again before a wider audience at the Centre d'Études de Politique Étrangère, a definite plan for a treaty of association between the two countries to be drafted immediately for the assent of the two

[11] A version of M. de Pange's paper was published in *Le Petit Parisien* on 25 March.
[12] *The Times*, 18 March 1940.

Parliaments.[13] It was to provide for the pooling of defence and for the joint control for this purpose of the resources of the metropolitan territories of the two states, and of their non-self governing dependencies. Foreign trade should also be jointly organized with regard to credits and to the allocation of markets, and French citizens in Britain and British citizens in France should enjoy the passive rights of citizens, in the sense that they should not be treated as aliens. Finally steps should be taken towards the creation of a 'bilingual elite'. In the course of discussion at the meeting at the Centre on 11 March it was further suggested that the effort towards bilingualism should be extended beyond the elite to broader sections of the population, that the two countries should develop a common colonial policy, and that the proposed joint governmental organs should be subject to some form of Parliamentary control.

In reporting to the Foreign Office on his visit on 13 March, Professor Toynbee said that he had been very struck by the general welcome for such ideas even among Frenchmen to whom the abstract ideas of federalism made no appeal. He also noted that the group which had been working at the Centre had arrived at suggestions almost identical with those put forward by Senator Honnorat. The reasons for this favourable current of opinion were, he believed, that an announcement of such plans would act as a stimulus to French morale, and discourage the Nazis, and that such a union would provide a gravitational pull for the smaller neutrals, while discouraging the more powerful neutral states from pressing for an inconclusive peace.

Toynbee's report led to further discussions between members of his Chatham House team at Oxford and some individual officials of another wartime offshoot of the Foreign Office, the Political Intelligence Department. At the beginning of April these discussions resulted in a memorandum, for which Toynbee was presumably responsible, which took the form of a draft 'Act of Perpetual Association' between the United Kingdom and France. It proposed the establishment of common organs for the conduct of

[13] M. André Honnorat had been senator for the Basses-Alpes since 1920, and was a former Minister of Education. M. Honnorat was also the founder of the Cité universitaire of the University of Paris and was particularly hopeful that the juxtaposition of the national pavilions would lead to greater mutual understanding between students from different countries. His interest in the idea of closer association between France and Britain is perhaps seen best in this wider context.

external relations, for the development of the association of their military forces, for the joint control of the economic resources necessary for the maintenance of their common defence, and for the development of their existing close cooperation in general economic matters. The proposals for a reciprocal extension of the rights of citizenship and for the development of closer mutual understanding between the two peoples also reappeared.

The Commentary which accompanied this memorandum pointed out that the word 'Union' had been avoided because it might provoke criticism and because it would logically involve summoning constitutional conventions in the two countries.[14] What was being suggested was a treaty which might in turn give birth to some further process of integration on the analogy of the original Swiss *Eidgenossenschaft*. It was held that the proposed permanent organs could be developed from the existing wartime arrangements. The idea of passive citizenship was explained as meaning that citizens of each country would enjoy the normal rights of citizens in the other except for the right to vote; and classical parallels were invoked. Closer mutual understanding was interpreted as meaning efforts towards a measure of bilingualism.

For the first time the question of the Commonwealth was raised, and the hope expressed that Commonwealth relations might actually be strengthened by this enhancement in the position of the United Kingdom; the special interest of Canada in a rapprochement between its two mother-countries was particularly stressed.[15]

Of these proposals the one on education was the easiest to translate into action. In April, the Minister of Education Lord de la Warr visited Paris and discussed the possibilities in this field. A circular from the French Minister of Education Yvon Delbos on 20 April set out a programme of very large scale exchanges of students and schoolchildren.[16]

At about this time the British Government showed its interest

[14] This point is striking testimony to the extent to which the memorandum still bore the mark of its basically French origins. Since Britain enjoys parliamentary sovereignty a constitutional convention would have been unnecessary whatever the significance of the changes might be.

[15] The available sources give no indication that the Dominion Governments themselves were consulted about the proposals.

[16] Jean de Pange, 'De l'union franco-britannique à l'union de l'Occident', in *Revue de Paris*, (October 1945); 'A plea for Anglo-French Union', *Contemporary Review* (May 1947). 'Nous voulons', wrote Delbos, 'que notre alliance soit plus qu'une union d'intérêts, qu'elle devienne une communion profonde des âmes.'

in these themes by establishing a small committee to look into the whole question of extending the wartime cooperation between Britain and France into the post-war years. Its chairman was Lord Hankey, who had been the secretary of the Supreme War Council in the first world war, and was now Minister without Portfolio in Neville Chamberlain's war cabinet.[17] Monnet himself had discussions with Chamberlain on some form of Anglo-French union to show their common purpose as defenders of freedom and going beyond their obligations to Poland as the root of the conflict. He worked with Sir Arthur Salter (now Lord Salter), with whom he was associated on supply questions, on a paper on the subject which was however never submitted to the War Cabinet.

The evidence of official interest enabled Toynbee to write to Dennery for the last time in May in a letter still expressing support for the idea of permanent union. Toynbee informed M. Noël when the latter was writing his account of the project that it was unlikely that Churchill himself knew of the memorandum or that it could have had any influence on the eventual project. Lord Vansittart wrote to Noël that he certainly did not know about the studies that had been made and that the June proposal had a quite independent origin.[18] On the other hand the work of the Hankey Committee was of course known to the Foreign Secretary Lord Halifax who submitted to it yet another document that arose out of the Chatham House-PID discussions, a personal proposal from the pen of that fertile proponent of federal schemes, Lionel Curtis. This proposal, which was dated 22 April, departed very far from the original suggestions of Senator Honnorat which had played so great a part in directing Toynbee's thinking. Instead of providing for increasing cooperation between the two sovereign States, Curtis argued, in his familiar fashion, that all mere alliances or confederations carried within them the seeds of their own destruction. He therefore demanded a single, democratically-elected Parliament of the two countries, controlling a responsible executive which would have as its sole duty the management of defence and

[17] When Mr Churchill's war cabinet was formed on 11 May, Lord Hankey was not included but he remained in the ministry as Chancellor of the Duchy of Lancaster.

[18] Sir Robert (subsequently Lord) Vansittart was chief diplomatic adviser to the Foreign Secretary from 1938 to 1941. It is curious that neither Halifax nor Hankey should have consulted him on these proposals in view of his well-known sympathies for France.

foreign policy, and which would be empowered through a quota system to raise the necessary funds for these purposes as a first charge on the respective national incomes. He believed that opposition to the scheme could be lessened if questions of commercial policy, and so on, were left out of the scheme. On the other hand, he thought that such a union for limited purposes would attract the smaller democracies pending some future date when Germany itself might be an acceptable member, and that the Dominions might feel that it was an actual guarantee against too much intervention from London in their affairs.

But by the time this paper was sent to Lord Halifax on 14 May the campaign in the West had begun, Churchill was Prime Minister and all thoughts of a more distant future were subordinate to the needs of the desperate struggle in which Britain and France were now engaged.

By 25 May General Weygand was already warning the French Government that the armies might be unable to continue the struggle, and that it was time to sound out the British as to their attitude towards releasing France from the pledge of March 28.[19] With the surrender of the Belgian armies on 27 May the situation became still more critical; on the following day the evacuation from Dunkirk of the bulk of the British forces began, to be completed by 4 June. At a meeting of the Supreme War Council on 31 May Churchill declared that Britain would continue the fight whatever might be the outcome in France. In the House of Commons on 4 June, Churchill declared Britain's intention to carry on the struggle even if necessary from beyond the seas.[20] On 5 June in a message to the French Prime Minister sent through General Spears he made it plain that there were limitations upon the resources that Britain could throw into the battle in France.[21] By this time the absolute priority of defending the home islands was uppermost in the minds of Britain's leaders.

[19] There are useful summaries of this period with full bibliographies in chapters I and II of Robert Aron, *Histoire de Vichy*, (Paris, 1954), and in ch. IV of Paul Farmer, *Vichy, Political Dilemma* (London, 1955).

[20] 361. *H.C. Deb.*, 5 s., cols. 787–796.

[21] General Sir Edward Spears MP had arrived in France on May 25 as Churchill's personal liaison officer with the French Prime Minister. His book *Assignment to Catastrophe,* 2 vol., (London, 1954) is of the first importance for Anglo-French relations between this date and the armistice.

Reynaud now renewed the suggestion, first made by him on 29 May, that an attempt be made to regroup some French and British force in Brittany from which contact could be retained with Britain. Weygand opposed the scheme as militarily untenable, as he had before, and the discussion was inconclusive.

For the next five days the German drive continued across northern France. On 10 June, Italy entered the war and the French Government abandoned Paris and moved to Tours. That evening Reynaud telegraphed to President Roosevelt asking him to declare publicly that the United States would give the Allies aid and material support by all means short of an expeditionary war.[22] Churchill, who had been pressing for a meeting of the Supreme War Council, was now told that a meeting could be held at Briare, Weygand's military headquarters near Orléans, on the afternoon of 11 June.

The discussions at Briare on the evening of the 11th and the morning of the 12th largely turned on the continued refusal of the British to sacrifice the air squadrons required for home defence. Weygand raised the possibility of France seeking an armistice and Churchill replied: 'If it is thought best that in her agony her Army should capitulate, let there be no hesitation on our account, because whatever you may do we shall fight on for ever and ever'.[23] On the same day the Highland Division were forced into surrender at Saint-Valéry. In the evening however, General Brooke landed in France to take command of the remaining British troops 'on which the Government still hoped to build up a new expeditionary force'.[24] Next day he was informed from London that the creation of a 'Breton redoubt' was being studied.

At a meeting of the French ministers at Cangé, after Churchill's departure from France, Weygand told the Cabinet that the war was lost and advised the Government to seek an armistice. Rey-

[22] *Foreign Relations of the United States,* 1940, vol. 1, pp. 245–6. There is a detailed discussion of American policy in these days in W. L. Langer and S. E. Gleason, *The Challenge to Isolation,* (London, 1952), pp. 524–44.

[23] Churchill, op. cit., vol. 2, p. 138. The French procès-verbal of the meetings at Briare is given in Paul Reynaud, *Au coeur de la mêlée, 1930–1945,* (Paris, 1951), pp. 747–57. Churchill's frequent expressions of Britain's determination to fight on alone if need be were afterwards treated in some quarters as indicating that Britain had released France from the pledge of 28 March; there is no indication that they were ever intended to be taken in this sense. See also Woodward, op. cit., pp. 58 ff.

[24] L. F. Ellis, *The War in France and Flanders,* (London, 1953), p. 297.

naud replied that this was out of the question; if fighting could not be continued on over the whole of the metropolitan territory the Government would have to withdraw to Brittany or failing that to Algeria, or even to tropical Africa. But he agreed to inquire what the British attitude would be.[25] As Churchill's telegram to President Roosevelt after his return from Briare indicates, he was fully aware of the likelihood of a French surrender; but he urged the President to do what he could to bolster up French resistance.[26]

On 13 June, Churchill went back to France and conferred with the French ministers at Tours. He was now told that it was too late to organize the 'Breton redoubt', that resistance on French soil could not be maintained, and that Britain must face the question of France's right after all her sacrifices to enter into a separate peace.[27] 'Mr Churchill', runs the official British record, 'said that in no case would Britain waste time and energy in reproaches and recriminations. That did not mean that she would consent to action contrary to the recent agreement. The first step ought to be M. Reynaud's further message putting the present position squarely to President Roosevelt. If England won the war France would be restored to her dignity and her greatness.[28]

After this meeting the French ministers at Cangé decided to allow for a final appeal to Roosevelt before deciding upon their course of action.[29] Meanwhile the Government was to move to Bordeaux. That evening President Roosevelt sent Reynaud a mes-

[25] Much has been made of Churchill's failure to attend this meeting; but it does not appear that it was ever suggested to him. See Spears, op. cit., vol. 2, pp. 230 ff. Chautemps argues that Reynaud purposely avoided a confrontation between the British Minister and some of his own colleagues. C. Chautemps, *Cahiers secrets de l'armistice 1939–40* (Paris, 1962), pp. 130–1. Chautemps, who was deputy Prime Minister in Reynaud's cabinet, was the principal civilian advocate of seeking to make a separate peace.

[26] *Foreign Relations*, 1940, vol. 1, pp. 246–7.

[27] On the morning of 14 June Weygand was to tell General Brooke that the Breton redoubt had been agreed upon by the two governments and that his dispositions should be made accordingly. Fortified by the scepticism expressed by Generals Weygand and Georges themselves, General Brooke appealed to London and secured permission to withdraw the British forces from the allied command and to group them for re-embarkation. The remaining British forces in France were actually withdrawn on 17–18 June. Ellis, op. cit., pp. 298–305.

[28] Churchill, vol. 2, p. 161. For a French record of the meeting see Baudouin, op. cit., pp. 154–9.

[29] According to Chautemps the ministers also agreed to go to Quimper. This destination was changed for Bordeaux by Reynaud himself that night. Chautemps, op. cit., p. 137.

sage stating that American efforts to get material aid to the Allies were being redoubled, and expressing the hope that France would continue to fight even if it meant withdrawing to North Africa and the Atlantic.[30]

During the night of 13–14 June Churchill sent messages to both Reynaud and Roosevelt, still in the hope that some word might be forthcoming from Washington that would enable Reynaud to hold out against the forces making for surrender. But the next day brought no encouragement. During the morning it was made plain to Churchill by the American Ambassador Kennedy that the message to Reynaud of 13 June could not be published and that it was in no sense intended to commit the United States to the slightest military activity.[31] What Roosevelt had in mind was primarily the disposition of the French fleet. From Tours at noon on 14 June the American Deputy Ambassador telegraphed the final appeal which had been handed to him by Reynaud early that morning. It declared that France could only continue the struggle if American intervention reversed the situation by making an Allied victory certain, and that unless in the hours to come the President could give France the certainty that the United States would come into the war within a very short time the fate of the world would change.[32]

By the time the French Government arrived in Bordeaux later that day – the day on which the German army entered Paris – the dilemma before it was clear.[33] Either it must, getting what en-

[30] *Foreign Relations*, 1940, vol. 1, pp. 247–8.

[31] *Foreign Relations*, 1940, vol. 1, pp. 250–2. The Prime Minister of Canada to whom Churchill also appealed felt himself able to go further and made the following statement in the Canadian Parliament that day: 'If I know the heart of the American people as I believe I do, and as I am certain I know the heart of the Canadian people, I believe I can say to Premier Reynaud, in the hour of the agony of France, that the resources of the whole of the North American continent will be thrown into the struggle for liberty at the side of the European democracies ere this continent will see democracy itself trodden under the iron heel of Nazism.' J. W. Pickersgill, *The Mackenzie King Record*, vol. 1 (Toronto, 1960), pp. 123–4.

[32] *Foreign Relations*, 1940, vol. 1, pp. 252–3.

[33] Sir Winston himself commented on the extreme difficulty of reconstructing the events of the next three days, partly because of the difficulties of maintaining contact by telegraph with Bordeaux and the frequent use of the telephone (Churchill, vol. 2, pp. 175–6). In London and Bordeaux a great deal that mattered was decided in the course of informal conversations which went unrecorded in the press of events. Above all one must remember the speed with which events were moving and the often appalling nature of these events. Men put out of their minds whatever was not strictly relevant to their next task and later found it impossible

couragement it could from the Americans, make arrangements for the transfer of as much as possible of France's fighting strength to North Africa with a view to continuing the war there while the remaining armies in France accepted a military capitulation as the Belgians had done, or it must seek armistice terms from the Germans, if possible in agreement with the British, if not without it. If the harder line of resistance were to be taken, then it might help to rally support for it, if the British commitment to a full restoration of France's position after victory were made as formal as possible. The most formal commitment would be some form of political union entered into at once. Thus what had been considerations largely relating to 'war aims', and only secondarily to the sustaining of French morale, now became closely linked with the struggle itself.

It was for this reason that on about 11 or 12 June the idea of some kind of permanent link between the two countries re-emerged both among the French group in London and through Monnet's contacts also in British government circles. From 13 June Monnet was in consultation with Salter and with the Secretary of State for India, Leo Amery. The group also consulted Mr Lloyd George. But it must be remembered that until the very last moment the formal position was that the French Government were requesting a British release from the pledge not to make a separate peace.[34]

Churchill, who had been apprised of the idea by Vansittart, had something of this kind in mind when in his telegram to Reynaud in the early hours of 14 June he used the following words: 'We take this opportunity of proclaiming the *indissoluble union* of our

accurately to recall even their own actions and thoughts. Hence the discrepancies in both the printed accounts and in such personal testimonies as I have been able to gather.

[34] It must be remembered that at the same time the Americans were worried lest a French surrender be followed by a British one, and that just as British worries about France were to concentrate on the question of the disposition of the French fleet, so the Americans were trying to make certain that the Empire would continue resistance with the aid of the British fleet and merchant shipping. They wished some of the fleet to be based on Canada where the United States could give assistance within the limits of nonbelligerency. But nothing was offered as an inducement to Britain by the United States comparable to the Union offer from Britain to France. See on Anglo-American-Canadian relations over these days the papers of the American Ambassador at Ottawa, Jay Pierrepont Moffat, *The Moffat Papers* ed. N. H. Hooker, (Cambridge, Mass., 1956), and the *Mackenzie King Record,* vol. 1, pp. 116–28.

two peoples and of our two Empires'.[35] It was however soon clear from a message from General Spears at Bordeaux that something more far-reaching would be needed, and Vansittart was entrusted by Lord Halifax with working something out which could be put to the Cabinet.

Meanwhile the idea of a declaration of Union had also been put forward by M. Monnet directly to Neville Chamberlain through Sir Horace Wilson. Quite apart from the general hope of stimulating the French will to resist, there was one practical issue that may have given the idea a new urgency in Monnet's eyes. He was in constant communication with Arthur Purvis, the head of the Anglo-French Purchasing Board in the United States, who was trying to get authority to divert ships laden with munitions for the Allies from French to British ports. He may have felt that this would be easier if the two countries could formally be treated as one for this purpose.[36]

On 14 June there were discussions about a possible declaration of Union between Vansittart and Churchill's secretary Desmond Morton on the British side and M. Monnet and M. Pleven on the French side.[37] It may have been a product of their labours that Chamberlain brought before the War Cabinet on the morning of 15 June when he described it as a memorandum that had been shown to him and other ministers 'proposing some dramatic expression of Anglo-French unity'.[38] In his view 'the form of unity proposed – joint Parliaments and a joint Cabinet – did not seem to have been fully thought out.' The War Cabinet thought that if the French Government were to come to England frequent meetings of the Supreme War Council would suffice.[39] Sir Alexander

[35] Churchill, vol. 2, p. 165.

[36] The main contract with the Board was signed on 11 June. It was taken over by Great Britain upon France's decision to seek an armistice and the money paid by the French was refunded by Britain on 25 July. Duncan Hall, *North American Supply* (London, 1955), p. 137. The transfer agreement was signed at 3 a.m. on 17 June. P. Goodhart, *Fifty Ships that Saved the World* (London, 1965), pp. 79–80.

[37] Churchill, p. 180. Churchill is clearly mistaken in saying that General de Gaulle took part in these talks. He was not in England between 9 and 16 June.

[38] Woodward, op. cit., p. 66. There is an English text of this document headed 'Anglo-French unity', and dated 14 June in the Vansittart papers: clearly the work of the Monnet-Salter-Amery group.

[39] The idea of the French Government as such continuing the struggle from England does not actually seem to have been put forward on either side.

Cadogan, the Permanent Secretary to the Foreign Office, was also dubious about the plan.[40]

Churchill says that he first heard about a definite plan at a lunch that day with Lord Halifax, Sir Robert Vansittart, the French Ambassador M. Charles Corbin and one or two others and that his first reaction was unfavourable.[41] General Ismay, the head of the military wing of the War Cabinet secretariat, was told about it by M. Monnet that afternoon.[42] According to Churchill's account the idea was raised (as though for the first time) towards the end of a long cabinet meeting that afternoon and was to his surprise greeted with much enthusiasm.[43] If this account is correct it is curious that the matter was not raised again when the War Cabinet met the following morning, the 16th. This may have been due partly to the absence of an agreed draft to put before it and partly to the extreme urgency about the fate of the French fleet. Vansittart and his British collaborators were encouraged to go on working away at the scheme in order to produce a version which could be approved for transmission to the French Government. On 15 June, a document called 'The Declaration of Liberty' was produced and this text was shortened to 'The Declaration of Union' on the same day.

Meanwhile the atmosphere in Bordeaux was becoming increasingly unfavourable to the idea of continued resistance from outside metropolitan France, as the pressure on Reynaud from Marshal Pétain and the others grew more intense. The British began to be regarded as the principal obstacles to peace and the British representatives at Bordeaux found themselves increasingly ostracized.[44] In a discussion with the Permanent Secretary of the French Foreign Ministry, the British Ambassador Sir Ronald

[40] Woodward, loc. cit., Sir Alexander Cadogan was out of London from the afternoon of 15 June until after the cabinet meeting on the afternoon of the following day and the Foreign Office as such took no further part in the affair; nor did the French Foreign Ministry according to its permanent Secretary who was only informed about it subsequently. F. Charles-Roux, *Cinq mois tragiques aux Affaires ètrangéres* (Paris, 1950), p. 46.

[41] Churchill, vol. 2, p. 180. This statement is curious as it would suggest that Churchill had not been present at the morning cabinet, an absence for which there is no other evidence.

[42] Lord Ismay, *Memoirs* (London, 1960), p. 145. It does not appear that either Ismay or the Chiefs of Staff had cognizance of the Chamberlain proposal that morning, still less that they were consulted on its military implications.

[43] Churchill, vol. 2, pp. 180–1.

[44] There is a vivid account by one of them: Spears, op. cit., pp. 244 ff.

Campbell talked of the 'wave of anglophobia' sweeping through French political circles.[45]

At 2.30 p.m. on 15 June General Weygand arrived in Bordeaux and renewed his demand that the Government should take responsibility for seeking an armistice. At the cabinet meeting that afternoon a majority seemingly supported the view allegedly put forward by M. Chautemps that to try to find out what the Germans' terms would be would not be incompatible with France's obligations. But Reynaud by threatening resignation succeeded in having the decision postponed until the following morning.[46] A telegram sent from Bordeaux at 5 p.m. by the Deputy Ambassador of the United States, A. J. Drexel Biddle, declared that there were only two alternatives before the French Government: to sue for peace 'which would of course have to be unconditional, or to move to North Africa and continue the fight'. The decision which would be taken the next morning would depend upon the nature of the President's reply to Reynaud's final appeal.[47] Successive messages from Sir Ronald Campbell kept London informed of the deterioration of the situation. A telegram sent off at 6.5 p.m. suggested that even the surrender of the fleet might not be opposed by some of Reynaud's colleagues.[48]

Immediately after the cabinet meeting, that is to say after 7 p.m. on 15 June, Reynaud saw Spears and Campbell to inform them about the Cabinet's decision, and for a time adopted as his own the claim made by Baudouin that Churchill at Tours had already given his assent to France seeking an armistice. This was disposed of by reference to the French official transcript.[49] Reynaud then began to draft a message to Churchill following the Cabinet's decision. While engaged on this task Reynaud received a reply from Roosevelt to his appeal of the previous day.[50] He communicated its negative content to his British visitors and completed his message in the light of this knowledge.[51] At about midnight Campbell was able to telephone London and Halifax replied that he should

[45] Charles-Roux, op. cit., p. 36.

[46] Reynaud, op. cit., pp. 802–18; Baudouin, op. cit., pp. 168–71. But Chautemps himself says that he only formulated this proposition at the meeting of the following morning, 16 June, op. cit., p. 144.

[47] *Foreign Relations,* 1940, vol. 1, pp. 256–7.

[48] Woodward, op. cit., p. 64.

[49] Spears, op. cit., vol. 2, pp. 263–5.

[50] *Foreign Relations,* 1940, vol. 1, pp. 255–6.

[51] Spears, op. cit., vol. 2, pp. 265–7.

impress upon the French that they should not take a final decision until there had been the chance of a personal exchange of views with the British.[52]

Meanwhile in London Churchill had sent off two telegrams to Roosevelt. The first at 9 p.m. warned Roosevelt that Hitler might be in a position if France sought peace to extort the surrender of the fleet, and that even in Britain a pro-German government succeeding his own might make the British Isles a vassal state of the German Empire, and in that case overwhelming sea-power would rest in Hitler's hands.[53] The second at 10.45 p.m. commented on Reynaud's final appeal and said that if the United States proved unable to respond the French would seek for an armistice and that he doubted if it would then be possible to keep the French fleet out of German hands.[54] But by then it was in fact twelve hours since Roosevelt's negative reply to Reynaud had been dispatched.

Reynaud's message to Churchill was telephoned to London at 1.20 a.m. on 16 June. It asked the British to permit an approach to the Germans and Italians to discover what their conditions for an armistice would be, and said that if the British agreed to this step Reynaud was authorized to declare to the British Government that the surrender of the French fleet to Germany would be held to be an unacceptable condition. Churchill had said at Tours that the request for an armistice would be reconsidered if Roosevelt's reply were negative, and this was indeed the case. If the British refused the request it appeared likely that he (Reynaud) would have to resign. He asked for an answer later that (Sunday) morning. At 4 a.m. a further telegram was sent to London by Campbell and Spears filling in the political background.[55]

Early on 16 June Campbell and Spears again saw Reynaud and told him that though there had been no reply yet from London, there could be no doubt what the British attitude would be to the solemn engagement undertaken by France; they urged him to continue the fight from North Africa leaving Pétain, if he insisted on remaining in France, to negotiate the surrender of the forces actually on French soil.[56] Reynaud consulted the presidents of the Chamber and Senate about moving the government to North

[52] Woodward, op. cit., p. 65.
[53] *Foreign Relations*, 1940, vol. 3, p. 53.
[54] *Foreign Relations*, 1940, vol. 1, p. 257.
[55] Spears, op. cit., pp. 265–70.
[56] Ibid., pp. 278–80.

Africa, as he was constitutionally obliged to do in such an eventuality, and met his cabinet at 11. He informed his colleagues about the failure of the appeal to America. Pétain again announced his intention of resigning because of the delay in asking for an armistice. But once again Reynaud succeeded in postponing a decision until the British had been heard from, and the Cabinet was adjourned until 5 p.m.[57]

General de Gaulle, Under-Secretary for War since 5 June, had been sent to London by Reynaud on 15 June in order to seek the help in shipping that would be needed if French forces were to be transported to North Africa. He arrived in London in the early hours of Sunday, 16 June. According to his own account he was at once visited by Monnet and Corbin who informed him that unless France asked for an armistice, Churchill was ready to meet Reynaud at Concarneau on the 17th, where they could together arrange for the move to North Africa. Monnet and Corbin then went on to say that they had worked out together with Vansittart a scheme for an Anglo-French union to be put up by the British to the French Government. They argued that de Gaulle was the only person who could persuade Churchill to adopt the scheme, and that since he was due to lunch with the Prime Minister the occasion would be there.[58]

While General de Gaulle was dealing with his shipping problems, the British War Cabinet was meeting to discuss the French request to be released from their obligations.[59] It was decided to agree to this request provided that the French fleet sailed forthwith to British ports; and a telegram to this effect was sent by

[57] Reynaud, op. cit., pp. 818–9. It is not clear how far Reynaud informed his colleagues of the negative reaction from Campbell and Spears to his approach to the British Government. See Y. Bouthillier, *Le drame de Vichy*, pp. 80–83. Baudouin (like Bouthillier) says that the interval was meant to make it possible for Reynaud to keep a rendezvous with Churchill at Nantes in the early afternoon; op. cit., p. 171. There is no British source for such a proposed meeting, and it is not mentioned by Reynaud himself.

[58] Charles de Gaulle, *Mémoires de guerre*, vol. 1, (Paris, 1954), pp. 61–63. A draft believed by him to be by Monnet was also shown to Amery. In a note to Vansittart (undated) Amery expressed the view that total union was not necessary but declared himself ready to fall in with de Gaulle's preferences. (Vansittart Papers). It would appear from Amery's diary that he thought too strong a project would put off the French.

[59] It was decided that the urgency was too great for the Dominions to be consulted. J. R. M. Butler, *Grand Strategy*, vol. 2, (London, 1957), p. 203. It is worth noting that according to this official war history the Cabinet only learned of the Union scheme for the first time on the afternoon of June 16.

Churchill to Reynaud at 12.35 p.m. The Foreign Office further instructed the Ambassador in a telegram sent at 3.10 p.m. to tell Reynaud that the British Government would expect to be consulted as soon as armistice terms were received; that it was assumed that the air force would be flown to North Africa if not to England, and that the French Government was counted upon to extricate and send to England the Polish, Belgian and Czech troops in France.[60]

After the cabinet meeting was over Halifax, as he noted in his diary, was shown the latest draft of the Anglo-French union project by Vansittart and Desmond Morton.[61] An attempt to reach him before the cabinet meeting had failed. After making some minor amendments Halifax gave instructions that the document should be ready for discussion at the Cabinet that afternoon and that copies should be brought to the Carlton Club where the Churchill-de Gaulle lunch was to take place.[62]

At some point during the morning Churchill briefly saw de Gaulle.[63] De Gaulle told him that Churchill was being represented as having appeared to consent to France seeking an armistice and that the telegram about the fleet sent earlier that morning would also weaken Reynaud's effort to keep France in the war. Finally, he apparently mentioned the Union project and got a favourable response. As a result of this conversation General de Gaulle telephoned Reynaud in a conversation overheard and taken

[60] Texts in Churchill, vol. 2, pp. 181–2. Cf. Woodward, op. cit., pp. 65–6.

[61] A covering note from Vansittart said that the draft proclamation had been drawn up by De Gaulle, Monnet, Pleven, Morton and himself. They urged that it be given to de Gaulle to take back to Reynaud not later than the same evening (Vansittart Papers).

[62] The King was kept in touch with what was going on by Chamberlain.

[63] Woodward, op. cit., p. 66; Churchill, p. 182. De Gaulle himself (op. cit., pp. 63–4) talks as though he saw Churchill only at lunchtime and says that Churchill broke off the lunchtime conversation in order to send his telegram to Campbell about suspending his earlier instructions. It is difficult to see how de Gaulle's telephone call to Bordeaux could have been put through from the Carlton Club during lunch and recorded as early as 12.30 though it is M. Corbin's recollection that the call was made during lunch, and this presumably coincided with Reynaud's own recollection fifteen years later. On the other hand, M. de Pange states that his telephone call was made only at 3.15 p.m. which if correct would confirm the de Gaulle version (*Revue de Paris,* October 1945). M. de Pange, then working in an Anglo-French intelligence unit, was not in Bordeaux on that day, but as a trained historian he is likely to have checked his facts closely.

down by the French military under Weygand's instructions.[64] The extract available runs as follows:

> 12.30 p.m.
>
> De Gaulle: I have just seen Churchill. There is something stupendous in preparation affecting the entity of the two countries. Churchill proposes the establishment of a single Franco-British Government, with you, perhaps, M. le Président, as head of a Franco-British War Cabinet.

Churchill, on the urging of Halifax conveyed by Vansittart, telegraphed to Campbell to suspend delivery of the two earlier telegrams and called a Cabinet for that afternoon.[65]

During lunch de Gaulle discussed the Union project and it was arranged that de Gaulle and Corbin should be available at Downing Street where the Cabinet was called for 3 p.m.[66] Churchill and Halifax, who had received Vansittart's final draft during the last stage of lunch and had briefly conferred about it in private, now met their colleagues for the final discussion.[67] At about 3.55 they were informed that the French Cabinet was to meet at 5 p.m.[68] By 4 p.m. de Gaulle was informed that a Declaration would be forthcoming, and the following conversation took place between him and Reynaud:

> *De Gaulle*: There is going to be a sensational declaration.
> *Paul Reynaud*: Yes, but after five o'clock it will be too late.
> *De Gaulle*: I will do my best to bring it to you at once by plane.
> *Paul Reynaud*: Yes, but that will be too late. The situation has

[64] Reynaud, *In the Thick of the Fight*, (London, 1955), pp. 539–40. The intercepted telephone calls were not available to Reynaud when the French edition of his book went to press, and only appear in the English version. Previously he had referred to de Gaulle's telephone call as being made 'early in the afternoon'. Reynaud, *Au coeur de la mêlée*, p. 827. De Gaulle also stated that the arrangements for supplying the necessary shipping for the move to North Africa were completed. See A. Truchet, *L'armistice de 1940 et l'Afrique du Nord*, Paris (1955), pp. 20–21.

[65] Woodward, op. cit., p. 66; Churchill, vol. 2, p. 182.

[66] De Gaulle op. cit., p. 64. According to M. de Pange, Churchill now heard for the first time of the proposal about reciprocal citizenship, which would require legislation by the Dominion as well as the United Kingdom Parliaments, but after some discussion agreed to this as well. *Revue de Paris*, October, 1945.

[67] A document in Vansittart's papers presumably relating to the draft argues that Roosevelt's promise of help is conditional on the 'two empires' continuing the struggle.

[68] Churchill, vol. 2, pp. 182–3; Spears, op. cit, p. 289.

seriously deteriorated within the last few minutes. Unforeseen events have occurred.[69]

Presumably as a result of this further evidence of urgency an English text was telephoned to Spears who had Reynaud with him, and de Gaulle was handed a French text of the proposed Declaration to read to Reynaud himself. Churchill also came briefly on the line to commend it to Reynaud.[70]

The War Cabinet which continued in session until 6 p.m. then approved a final version of the Declaration.[71] De Gaulle left London by air for Bordeaux and it was arranged that Churchill, who had confirmed with Reynaud that they were to meet at sea off Concarneau on the following day, should take with him Mr Attlee and Sir Archibald Sinclair as leaders of the Labour and Liberal Parties, so that all parties in the State should be shown to have concurred in this momentous act.[72]

The text in question was as follows:

DECLARATION OF UNION[73]

At this most fateful moment in the history of the modern world the Governments of the United Kingdom and the French Republic make this declaration of indissoluble union and unyielding resolution in their common defence of justice and freedom

[69] Reynaud, *In the Thick of the Fight*, p. 540.

[70] The French text is printed from Reynaud's rough note in *Au coeur de la mêlée,* pp. 828–829. The Cabinet's alterations included the deletion of the draft's proposals for the creation of a single currency and the abolition of all customs barriers between two countries.

[71] At 7 p.m. Halifax saw the American Ambassador who reported sceptically: 'So the British drew up a declaration of union between France and England. All partners now, what you have is mine and what I have is yours. All damage done to France, England pays her share. A noble sentiment but just does not mean much so the armistice preparations have been held up pending finding out Reynaud's reaction to the declaration of union.' J. Kennedy to Secretary of State, 16 June, 9 p.m. *Foreign Relations,* 1940, vol. 1, pp. 259–60.

[72] Churchill, op. cit., pp. 183–6.

[73] Churchill, op. cit., pp. 183–4. There were two changes from the French draft telephoned to Bordeaux by de Gaulle. Where the final version says 'one Franco-British Union' the draft had said 'une nation franco-britannique indissoluble.' Where the final version says 'the two Parliaments will be formally associated', the original had run 'les deux Parlements seront associés.' M. Reynaud regards these changes as purely questions of form. *Au coeur de la mêlée,* p. 828 fn. It is not clear whether the latter change is what Woodward refers to when he speaks of a version saying 'the two Parliaments will unite.' According to his narrative the War Cabinet's amendments were made before de Gaulle telephoned the French version to Bordeaux, op. cit., p. 67.

against subjection to a system which reduces mankind to a life of robots and slaves.

The two Governments declare that France and Great Britain shall no longer be two nations, but one Franco-British Union.

The constitution of the Union will provide for joint organs of defence, foreign, financial and economic policies.

Every citizen of France will enjoy immediately citizenship of Great Britain; every British subject will become a citizen of France.

Both countries will share responsibility for the repair of the devastation of war, wherever it occurs in their territories, and the resources of both shall be equally, and as one, applied to that purpose.

During the war there shall be a single War Cabinet, and all the forces of Britain and France, whether on land, sea or in the air, will be placed under its direction. It will govern from wherever best it can. The two Parliaments will be formally associated. The nations of the British Empire are already forming new armies. France will keep her available forces in the field, on the sea, and in the air. The Union appeals to the United States to fortify the economic resources of the Allies, and to bring her powerful material aid to the common cause.

The Union will concentrate its whole energy against the power of the enemy, no matter where the battle may be.

And thus we shall conquer.

The opportunity to work out its full implications never arrived. The British party were already seated in their special train for Southampton when they were informed by telegram from Campbell that it was no use their coming to France as there had been a fresh cabinet crisis and no one was prepared to meet them. 'We knew then', recalls Lord Attlee, 'it was all over and Reynaud had lost. We got out of the train and drove back to Downing Street and went back to work.'[74]

The first of the morning telegrams giving the War Cabinet's conditional assent to the French request for an armistice reached Reynaud after the morning meeting of the Cabinet. Reynaud told Spears and Campbell, who transmitted the message, of a telephone conversation he had had with Churchill about a proposed meeting

[74] Francis Williams, *A Prime Minister Remembers*, (London, 1961), p. 44.

and it was arranged, as has been seen, that this should be at sea off Concarneau on the following day.[75] The telegram from the Foreign Office reached Campbell and Spears at about 4 p.m. and was taken by them to Reynaud almost at once. It was during this interview that the text of the Declaration was telephoned through and taken down. Reynaud said that he presumed that this superseded the earlier messages and both Campbell and Spears agreed that this must be presumed. After leaving Reynaud they received the message Churchill had despatched to say that the two earlier messages were to be considered cancelled.[76]

Lebrun had asked Reynaud to see him before the cabinet meeting at 5 p.m.[77] The latter was already aware of the Union offer which he had communicated to Baudouin and Bouthillier.[78] The two officials, it would appear, argued with the Prime Minister that whatever the long run attractions of the scheme, it would do nothing to solve France's immediate problem.[79]

The same arguments prevailed with the majority of the Cabinet when that body met at 5 p.m. confronted with still further messages from the military on the seriousness of the situation. Reynaud did not communicate the two telegrams since these had now been superseded. Instead he attempted to present the British offer of Union as a substitute for seeking armistice conditions. Amid considerable bitterness the partisans of the project led by Georges Mandel were overborne and Chautemps' proposition that the Germans should be approached was accepted. The cabinet meeting went on until 8 p.m. When it was over Reynaud placed his resignation in the hands of President Lebrun, and by 10 p.m. the new Cabinet of Marshal Pétain had been constituted.[80] In the further negotiations between Britain and France over the question

[75] Spears, op. cit., pp. 282–4.

[76] Ibid., pp. 289–94. Churchill says that 'suspended' would have been more accurate. Churchill, vol. 2, p. 185. Cf. Woodward, op. cit., p. 67. The telegrams were in fact transmitted to the new government of Marshal Pétain on the following day; ibid., pp. 68–9.

[77] Reynaud, *In the Thick of the Fight*, p. 540.

[78] Baudouin, op. cit., pp. 172–3.

[79] Bouthillier, op. cit., p. 27. It is fairly clear that security was very lax at the Prime Minister's headquarters and that some of the ministers at least were aware what had been going on before the Cabinet met. See, e.g. Spears, op. cit., vol. 2, p. 293.

[80] See the summary in R. Aron, *Histoire de Vichy*, pp. 49–51. For the events of the day as reported at the time to Washington, see Biddle's despatches of 16 June sent at 1 a.m. ; 4 p.m. ; 9 p.m. ; and midnight. *Foreign Relations,* 1940, vol. 1, pp. 258–62.

of the armistice conditions and the fate of the French fleet the issue of a Union was not renewed.[81] The story of the project may therefore be said to end at the moment Reynaud resigned. It played no part in the stormy relations between General de Gaulle's Free French movement and the British Government in the period between June 1940 and the liberation of France. In a speech in Oxford on 25 November 1941 de Gaulle said: 'the first result of this war, if it was won, must be the establishment of a more frank and solid collaboration than had ever yet existed between the French and the British.' But his idea on this subject fell far short of 'indissoluble union'.[82]

On 25 April 1945 it was formally announced on the British side that the proposal was not to be considered as still open.[83]

From the above account it will be seen that the project of an Anglo-French Union, although very largely French in inspiration, was translated into positive terms by British civil servants and endorsed by the War Cabinet before being submitted to the French Government. At least one French commentator later asserted that its ambiguities were precisely characteristic of its British authorship.[84]

The reasons for its failure to appeal to a majority of the French Cabinet have been satisfactorily analysed by M. Léon Noël. At this late juncture a good deal of latent anglophobia came to the fore. Some of it was because of the totalitarian and anti-democratic sympathies of some of the military and political figures concerned. Some Frenchmen cherished a particular antipathy for Churchill and his political associates arising out of the arguments in the 1930s over 'appeasement'. Old imperial rivalries also played their

[81] For these events see Woodward, op. cit., pp. 68–74.

[82] De Gaulle, op. cit., 1, p. 568. A study of Franco-British relations published in 1945 says that in 1941, General de Gaulle 'suggested' that 'he was still disposed to consider' the offer 'open.' Royal Institute of International Affairs, *France and Britain* (London, 1945), p. 56. I have found no evidence that this was the case; nor would it have chimed in well with de Gaulle's persistent efforts to maintain that his movement was a wholly independent one and in no sense an instrument of British policy.

[83] 410. *HC Deb.* 5 s., col. 840.

[84] 'La rédaction . . . est entièrement anglaise, avec ce je ne sais quoi de flou qui ne correspond pas à l'esprit français habitué à mettre les choses en équation et à les exposer avec logique.' A. Kammerer, *La vérité sur l'armistice* (Paris, 1945), p. 129. Perhaps M. Kammerer makes too little allowance for the extreme rapidity with which events were moving.

part. All these sentiments had been given free reign by Dunkirk and the feeling that Britain had managed to salvage her forces for her own defence while leaving the French alone to fight on in the common cause. Looking to the future it was widely believed that Britain also would in the end prove too weak to stand up to Hitler. If Britain came to sue for peace would she not be able to do this at the expense of French interests? Some individuals, finally, already looked beyond the immediate situation to a possible reversal of alliances such as Pierre Laval was eventually to attempt.[85]

Other opponents did not look so far forward but dismissed the idea as impracticable because of their view that resistance could not have continued from North Africa as matters then stood. Even after the war, General Weygand remained convinced that this was in fact the case.[86]

In such circumstances it is not surprising that French opponents of the project should have referred to it as merely an attempt to 'reduce France to the rank of a dominion' and that an intention to accept such a status should have been one of the charges made against Reynaud by the Vichy authorities after his fall.[87] Such arguments had at the time more appeal than the view that, given the circumstances of the two countries at the time, it was Britain that was offering to make the greater sacrifice.[88] In retrospect it was also argued in some French circles that the commitment fully to restore France's pre-war position would have been a valuable one; and that in the political uncertainties that faced France after Liberation such an association might have provided a useful element of political stability.[89]

[85] Noël, op. cit.

[86] 'Transporter en juin 1940 la lutte en Afrique du Nord eût été la perdre.' M. Weygand, *En lisant mémoires de guerre du général de Gaulle* (Paris, 1955), p. 91. Cf. *Les évenements survenus en France*, VI, p. 1848.

[87] Perhaps Churchill's own scepticism as to the likelihood of the project succeeding may be gauged from the fact that it was not even mentioned in the long telegram on the situation sent by him to the Dominion Prime Ministers at some time during the afternoon of 16 June. Text in Churchill, vol. 2, pp. 172–4.

[88] 'On pouvait penser au contraire que dans l'état de faiblesse où se trouvait encore la France, le sacrifice dans la création de l'union indissoluble venait plutôt du côté de l'Angleterre.' Lebrun, *Tèmoignage* (Paris, 1945), p. 84.

[89] 'La notion de l'autorité, si dangereusement affaiblie en France, se restaurera en s'appuyant sur le seul pays d'Europe où la tradition monarchique se maintient depuis plus de mille ans.' Le comte Jean de Pange, 'De l'union franco-britannique à l'Union de l'Occident', *Revue de Paris* (Oct. 1945).

On the British side, as we have seen, the secret was well kept, and news of the proposal was only published in the press on 18 June.[90] Meanwhile its content was becoming known to those ministers who, being outside the War Cabinet, had not been in a position to appreciate the rapid march of events on the two preceding days. Lord Simon, the Lord Chancellor, wrote to Chamberlain on 17 June:

> My Dear Neville,
>
> Hankey lunched with me and told me of the 'indissoluble union' proposal (No. 375 of yesterday). I am simply staggered by it, as I gather Hankey is also. 'No longer two nations' – what happens to the Dominions, or to the King? 'The constitution of the Union will provide for ... joint finance!' What a squabble!
>
> I well understand the need for boldness and imagination, but has the plan really been *examined,* however speedily, from the British point of view? I can appreciate that Monnet would approve it.
>
> If the French are now seeking to come to terms with the Germans in their desperate situation, how can anyone feel confident that half the 'Franco-British Union' won't want us to give in later on? Surely, surely, *our* business is with the defence of this island, and I cannot believe that a half-and-half arrangement is going to serve us *now*: nothing will serve but a *British* resolve *never* to yield. Forgive me, but I am in great perplexity and of course knew nothing whatever of this.

Hankey himself also wrote to Chamberlain on the same day:

> My Dear Neville,
>
> John Simon has sent me the enclosed letter to read. I confess I was so staggered when I heard the proposal that I broke out in a sweat.
>
> The more I reflect on the events of recent years the more clearly I realize that the French have been our evil genius from the Paris Peace Conference until today inclusive. Heaven forbid that we should tie ourselves up with them in an indissoluble union!

[90] See also Churchill's speech in the House of Commons that afternoon. 362. *HC Deb.* 5 s., col. 60.

But when I thought about it I came to the conclusion it was a tactical move in a terrible situation, with no real chance of acceptance.[91]

Whether one agrees with Hankey's view on Anglo-French relations between the wars, or believes on the contrary that a better appreciation on the British side of the two countries' mutual interests might have averted the ultimate catastrophe, there can be no doubt but that his letter, with all allowances made for the bitterness of the moment, went to the heart of the matter. If it was right for Britain at the crisis of the war's fortunes to take the tremendous risk of the unknown inherent in the merging of the two sovereignties then it must have been the case already for a long time that Britain's security was inextricably involved in preserving a balance on the continent and that this could only be done by a total commitment of her material and moral strength for European objectives. The French rejection of the offer followed by a victory achieved in partnership with the Soviet Union and the United States made it possible to regard the events of 1940 as an historical accident with no lessons for the future. And the renascent France of the post-war period could come to hold the view that for it also Anglo-French unity, while desirable, was by no means essential. The Treaty of Dunkirk in 1947 was a very modest instrument compared with what had been contemplated not only by the British Government in June 1940, but by both Governments and in influential official and unofficial circles during the preceding winter. Europe was to try to find other routes to unity, and their outcome is still in doubt.

At midnight on 16 June, Biddle reported to Washington on a meeting he had just had with Reynaud: 'While it now belongs to history Reynaud referred in glowing terms to Churchill's "reply" this afternoon. It was far-reaching in scope: it meant in reality a

[91] Both these letters are in the Papers of the first Viscount Simon. There is also a letter from the former minister, Lord Runciman, dated 2 April 1941 expressing general unease at the conduct of affairs and including the sentences:

'The suggestion of a new Anglo-French nationality is almost incredible. I wonder if he (Churchill) consulted you before putting it forward. The sacrifices we are asking.

The French bundle of errors make me terribly uncomfortable and handled badly they may be disastrous.'

It is not clear why at this late date Runciman should have been harkening back to the events of the previous June.

fusion of two great Empires. It might, he said, have marked the beginning of a United States of Europe'.[92] *Dis aliter visum.*[93]

[92] *Foreign Relations,* 1940, vol. 1, pp. 261–2.

[93] Dr David Thomson's Zaharoff Lecture for 1966 'The Proposal for Anglo-French Unity in 1940' appeared too late for me to be able to use it.

15 *The Special Relationship: an Anglo-American Myth*

The history of the eary 1960s was much influenced by the belief (of which General de Gaulle was the most eloquent exponent) that the countries of continental Europe stood in danger of the development at their expense of an Anglo-Saxon world hegemony. The French veto on Britain's entry into the European Common Market was based upon the assumption that Britain, once inside, would act as an agent of the essentially non-European policies of the United States.[1] French suspicions of Britain and America working together had of course a long pre-history.[2] And they were given particular point by the wartime experiences of General de Gaulle when he found Winston Churchill apparently acquiescing in the strongly anti-French policies of President Roosevelt and his Secretary of State.[3] But that there exists a natural harmony between Britain and the United States based upon a common linguistic and cultural heritage has long been an article of faith in continental Europe. Only the Marxists with their concentration on economic rivalries have tended to take the opposite view.

It is not difficult to challenge the belief in the closeness of the alignment between British and American policies where particular episodes are concerned. Washington did, of course, welcome the proposed British accession to the Treaty of Rome. But its reasons had little to do with the possibility of using Britain as a Trojan

[1] See, e.g., Nora Beloff, *The General Says No* (London, 1963).

[2] Writing to the Foreign Secretary from Geneva about American attempts to bring France and Germany together Viscount Cecil of Chelwood referred to the French 'Anglo-Saxon' complex as a serious obstacle. Cecil of Chelwood Papers. BM Add. 51082. Cecil to Reading, 13 September, 1932.

[3] Lord Avon's volume of memoirs, *The Reckoning* (London, 1965), provides an illuminating treatment of this theme.

horse for entering the European citadel.[4] The real nature of the two countries' political relationship demands a more critical approach than it has usually received.[5] In what sense is there a 'special relationship' between Britain and the United States from which all other countries are perforce excluded, and what is its nature?

The belief that there is such a relationship underlies two schools of historical thought upon the subject. On the one hand there are the American historians who account for the entry of the United States into two world wars by pointing to British intrigues designed to force upon the United States the burden of defending the British Empire.[6] On the other hand there are British historians who appear to justify de Gaulle's suspicions by recounting the history of Anglo-American relations in terms that suggest that a growing intimacy of action between Britain and the United States has been the great continuing theme of world history for the past sixty years.[7]

Both versions are too simple; yet both enshrine part of the truth. One must be careful not to overlook important factors in the situation simply because they are so obvious. It is for instance true that the existence of a common language has meant that ideas and myths have been able to circulate more freely between Britain and America than between most other countries.[8] Each has been prone to react rather sharply to a change of tone on the part of the other. Yet this very fact, when translated into political terms, makes the record harder rather than easier to follow. British statesmen have often been very guarded in their public utterances about the United States, concealing both their hopes and their fears just because they were worried about the reaction in the United States. And where they have wished for public declarations of unity of purpose they have found their American friends warning them

[4] See Max Beloff, *The United States and the Unity of Europe* (Washington and London, 1963).

[5] For an admirable treatment of the relations of the two countries since the end of the Second World War see Coral Bell, *The Debatable Alliance*, (London, 1964). It is not the intention of this essay to traverse the same ground.

[6] The main objective in American foreign policy since 1900 has been the preservation of the British Empire.' C. C. Tansill, *Back Door to War* (Chicago, 1952), p. 3.

[7] See, e.g. H. C. Allen, *The Anglo-American Relationship since 1783* (London, 1959).

[8] Some students would however argue that there was a relative weakening in the 1960s of the density of the contacts between the two communities as compared with earlier periods. See B. Russett, *Community and Contention; Britain and America in the Twentieth Century* (Cambridge, Mass., 1963).

against such a course, either because of domestic considerations or for fear of arousing elsewhere the kind of suspicions voiced by General de Gaulle.[9]

For this reason the public record is not a safe guide and perhaps especially not where British policies are concerned.[10] What we are concerned with here is the reflection in British governing circles of major historical developments that are themselves adequately known. For the American reaction to these developments is no great mystery.

The most important development has, of course, been the diminution of Britain's status in the world, which has been a more or less continuous process since the 1890s, even if sometimes masked by apparent accretions of territory, and the contemporaneous ascent of the United States, which at least until after the second world war was a much more uneven process. Naturally enough, as American power has increased, so has the importance of relations with Britain decreased in American eyes. In 1940 at the time of the fall of France, the fate of the British fleet was a matter of vital concern to the United States.[11] Even in the 1940s and 1950s, Britain's geographical role as a stationary aircraft-carrier was of the first importance. With the development of intercontinental missiles this has ceased to be the case. Today the transfer of the war-making potential of Britain to the side of America's enemies would not greatly add to the dangers that face the United States. For this reason by the 1960s, the whole idea of 'special relationship' was something of an irritant to Americans in official positions, in whose world-outlook the British played a fairly subordinate part.

The British response to these changes has always been a more complicated one. On the other hand there were those who were reluctant to believe the process was an inevitable one and who held that all that was needed was a proper firmness in the defence of British rights. Such arguments were at their most powerful in the

[9] See, e.g. Eisenhower's warnings to Winston Churchill in 1952 where the accent was particularly on the danger of the United States being identified with the policies of a colonialist or ex-colonialist power. Dwight D. Eisenhower, *Mandate for Change* (London, 1963), pp. 97, 249–50.

[10] The present essay relies primarily on papers not part of the public record. Public speeches, magazine articles, etc., would no doubt produce a very different impression.

[11] See above, chapter 14.

period of American isolation between the wars.[12] And they coloured British reactions to tentative American moves towards greater participation in world affairs in the 1930s.[13] Some sought to strengthen Britain's hand by exploiting her imperial or, later, her European, position. But it has been pointed out that the most remarkable thing in the whole story is that Britain never responded to the growth in American power by the classical method of organizing a counter-coalition.

It has been strongly argued that the reason was an intuitive British assumption 'that British and American interests would in the end prove complementary in the central issues of international politics'.[14] But one could put this in more modest terms by saying that as Britain's consciousness of her growing weakness increased, so did her leaders' feeling that the only way to meet the situation was to make certain of American support against her other challengers, and that it was here that the particular nature of the relations between the two societies made itself felt.

If one were to put it in the most brutal fashion possible, one would have to say that ever since the 1890s the dominant element in the British 'establishment' has known in its heart that the world order dependent on British sea-power which was the key to the unparalleled growth of the western economy in the nineteenth century could no longer be sustained by British power alone. It was therefore the intended lot of the United States, perhaps its moral duty, to take over an increasing share of this burden and to use its new strength to further Britain's original purposes.[15] In this modi-

[12] 'Time after time we have been told that, if we made this concession, or that concession, we should secure good will in America. We gave up the Anglo-Japanese Alliance. We agreed to pay our debts and we have again and again made concessions on this ground. I have never seen any permanent results follow from a policy of concession. I believe we are less popular and more abused in America than ever before, because they think us weak. The only thing that has really done any good has been the Balfour note on international debts, where we stood up to them firmly. I would refuse either to be blackmailed or browbeaten and stand absolutely to our preconcerted plan of action.' Sir Maurice Hankey to Lord Balfour, 29 June 1927. Balfour Papers. BM Add. 49704.

[13] By early 1933 there was a party 'with its representatives in the Cabinet' that believed that the record of the United States since 1920 in refusing 'all decent co-operation with the rest of the world' made any negotiations with the Americans doomed to failure. Lord Lothian to Colonel House, 13 February, 1933. Lothian Papers. (Scottish National Archives). Box 221.

[14] Bell, op. cit., p. 116.

[15] 'There have only been two long periods of world peace in history. One was created by the Roman Empire. The other was the great Pax of the nineteenth

fied form, shorn of its demonological element, the theory of the American isolationist writers was nearer the truth than the *bien-pensants* on either side of the Atlantic like to admit.[16]

It could thus be maintained that the 'special relationship' was something in which many British public figures felt the need to believe, so as to be able to argue that the displacement of power from Britain to the United States need not directly damage British interests. It was for them the test of their own ability to exercise sufficient influence at Washington, to make this interpretation of world politics plausible to themselves and others. To trace the vicissitudes of this idea would be to recount the whole story of the relations between the two countries in the present century; it is enough here to point out some of its origins and some of its consequences.

The Venezuela crisis may be taken as the moment at which Britain, alerted to the consequences of permanently alienating the United States, virtually ruled out the use of armed force as a method of settling Anglo-American disputes and appealed to the idea of belonging to a common community as the fundamental reason for so doing.[17] But this did not, of course, rule out competition in armaments nor contingency planning for their use.[18]

What is interesting is the speed with which British statesmen passed from the position of merely making certain of American

century which was created by the British navy. That is why I have always believed that the only foundation for world peace was close cooperation between the British Commonwealth and the United States for the restoration of the nineteenth-century British system operated not by Britain alone but by the whole English-speaking world.' Lord Lothian to General Smuts, 6 June 1939. Lothian Papers. Box 208.

[16] This is not to say, of course, that either Woodrow Wilson or Franklin Roosevelt were willing accomplices in a 'British plot'.

[17] Balfour declared in a speech at the height of the Venezuelan crisis that the idea of war with the United States carried with it 'something of the unnatural horror of a civil war'.

[18] See J. C. Grenville, *Lord Salisbury and Foreign Policy* (London, 1964), p. 422. There was much discussion of the problems of defending Canada in British Government circles in 1903–5. Balfour Papers. BM Add. 49707. Asquith Papers, Bodleian Library, Oxford. Dep. Asquith. 132. But some people felt from very early on that there should be no competition with the United States in armaments. Lord Selborne, the First Lord of the Admiralty, wrote to Lord Curzon on 19 April 1901, that he would never quarrel with the United States if it could be avoided. If the Americans chose they could easily afford to build a navy larger than Britain's and would probably do so. G. W. Monger, *The End of Isolation* (London, 1963), p. 72.

neutrality to that of looking for positive support.[19] But even so, in the crisis of the first world war it seemed as if the former objective might prove difficult enough of attainment.

For it is clear that in 1915 there were real fears as to how far the United States might go in defending its neutral rights against the British measures aimed against the economies of the Central Powers; and in 1916 there was very great anxiety about the possible consequences should the Germans offer, in response to American moves for mediation, to accept proposals falling far short of the Allies' minimum demands.[20] It is true that the eventual American entry into the war put an end to these particular nightmares and gave comfort to those who had placed their faith in the ultimate coming together of the two countries.[21] But it did not in fact mean any such thing when seen from Washington. The British remained baffled and saddened. For them it was axiomatic that the maintenance of Britain's world position conduced to the general welfare and they felt that all Americans should be able to perceive this obvious fact unless blinded by prejudice. It was from the contrast between the perceptions of British needs, so clear in London, so cloudy in Washington, that there were to arise all the subsequent misunderstandings and disappointments.

For the differences in perception many reasons could be adduced, particularly on the American side. Some of these were not as visible as they should have been to British leaders who found their natural American interlocutors among the educated elite of the eastern seaboard with their many personal and cultural ties with England, rather than among the preponderantly non-British elements in the country's population whose gradual ascent to poli-

[19] We must, of course, note the point recently made by an historian that the changes subsequent to 1941 must not make us forget how remote the United States felt, generally speaking, from questions of the European balance of power before 1917. G. Barraclough, *Introduction to Contemporary History* (London, 1964), pp. 106–7.

[20] In the course of Anglo-French conversations at Downing Street on 26 December 1916, Balfour pointed out the dangers inherent in President Wilson's attitude to possible peace terms, in view of the fact that 'the United States had it in their power to compel peace'. Austen Chamberlain Papers (University of Birmingham Library) AC20/76. Cf. Arthur Link, *Wilson*, vol. 5 (Princeton, 1965).

[21] When the news reached London of the United States breaking off relations with Germany, Austen Chamberlain wrote in his diary (4 February 1917): 'It is not the material support we should receive but the memory of having once at least co-operated in a great struggle that has made me long for the intervention of the United States in the war – long for it without expecting it.' Diary (University of Birmingham Library).

tical power was a principal feature of the American scene in the decades in question. It was because of the restricted social sympathies of British statesmen and diplomats that they failed to grasp the extent to which American anglophilia was part of the ideology of a class struggling to maintain its ascendency at home in the face of a series of urgent challenges from below.[22]

Indeed, the degree to which British statesmen and diplomats expected a natural sympathy for British policy to exist in the United States and equated any hostility to or criticism of Britain with treason to America and not merely to Britain can be abundantly illustrated. In the cold light of reason there was little cause for Irish-Americans, or German-Americans or refugees from Tsarism to favour American intervention on the Allied side in the first world war;[23] but it was assumed by Cecil Spring-Rice, the British Ambassador, and by other Englishmen that some especial turpitude attached to the 'hyphenated Americans',[24] while it was perfectly natural that the sympathies of New Englanders should be with old England. And while some excuse might easily be found for their spleen in the fact that Britain was fighting for her life

[22] For this interpretation of the changing climate of American opinion in the 1890s see C. Strout, *The American Image of the Old World* (New York, 1963), ch.8, 'The Old Sweet Anglo-Saxon Spell'. The course of American opinion during the Venezuela crisis should have shown where the strength of pro-English feeling lay: 'The opponents of Jingoism and of Cleveland's performance are the businessmen, clergy, professors, and the like of the eastern coast. The hatred of England in the west and south-west is rabid, bitter and ferocious, and would welcome war tomorrow. . . . The children have been taught in all the schools for twenty-five years to hate England and to believe that we can thrash her and that we did so in 1812.' E. L. Godkin to James Bryce, 9 January 1896. Bryce Papers, Bodleian Library, Oxford. USA Papers, vol. 5.

[23] The dangers to British interests arising from the importance of these voting blocs had been foreseen by some British students of affairs. Writing to Balfour's secretary, Sanders, on 27 January 1910 to suggest a compromise on Home Rule, J. L. Garvin wrote: '. . . what we most need is better relations with America. The new alliance between the Irish and the German vote is a more important thing than almost anybody here seems to realise. It is one of the greatest dangers that ever threatened the Empire.' Writing to Balfour himself on 17 October, Garvin added that through alienating the United States, Britain also ran the risk of forfeiting the sympathies of Canada. Balfour Papers. BM Add. 49745.

[24] 'If we regard ourselves as fighting for a cause bound up with the existence of America and as a people of the same blood and language and principles as the people of the United States we might be filled with rage at the thought of their attitude.' Spring-Rice to Grey, 5 December 1916. Balfour Papers. BM Add 49731. The passage is one of those omitted from the printed version of the letter in S. Gwynn, *The Letters and Friendships of Sir Cecil Spring-Rice* (London, 1929), vol. 2, pp. 357–9.

against a militarist aggressor – after all the Germans had invaded Belgium and not vice versa – the same could not be said of all such episodes.

Disputes such as those over Indian policy during the second world war and Palestine later on made the ideas of an Anglo-Saxon partnership as adumbrated in the 1890s very remote indeed when viewed from the perspective of less than fifty years later. But some of the frustrations engendered by the subsequent divergences of view are perhaps only fully understandable in the light of the hopes originally aroused: 'The danger is that we should hope too much and we should be angry because we are disappointed from too much hoping.'[25]

The notion of 'Anglo-Saxondom' has been more fully explored on the American side than on the British one and perhaps rightly so. From the time when James Russell Lowell concluded as a result of his experiences as American Minister that 'the differences between Yankees and Englishmen were "mostly superficial",[26] a whole school of American writers had dilated on the superiority of the Anglo-Saxon race and its destiny to dominate the world.[27] But there were British parallels; and the musings of Cecil Rhodes as adumbrated in the successive drafts of his will suggest a climate of opinion in which policies of Anglo-American partnership could be made acceptable.[28] It has been pointed out, for instance, that while, in general, the European press took Spain's side in the Spanish-American war, the British press supported the United States, which seemed to be fighting the kind of struggle that it had elsewhere fallen to Britain's lot to wage.[29] The words of Ambassador John Hay claiming that there was as between the United States and Britain 'a sanction like that of religion' which bound them 'in a sort of partnership in the beneficent work of mankind' were answered by Joseph Chamberlain in a number of celebrated utterances.[30]

[25] Spring-Rice to Balfour, 11 January 1917 (*not* 12 January), Gwynn, op. cit., vol. 2, p. 369.

[26] Strout, op. cit., p. 138.

[27] R. W. Leopold, *The Growth of American Foreign Policy* (New York, 1962), pp. 125–7.

[28] See generally, A. P. Thornton, *The Imperial Idea and Its Enemies* (London, 1959), and B. Semmel, *Imperialism and Social Reform* (London, 1960). For Rhodes see F. Aydelotte, *The American Rhodes Scholarships* (Princeton, 1946).

[29] Grenville, op. cit., pp. 214–15.

[30] Leopold, op. cit., pp. 206–7; J. L. Garvin, *Life of Joseph Chamberlain*, vol. 3

The vogue of such ideas was relatively short-lived, since overt racialism was compatible neither with the essential nature of American society nor with the general British outlook on world problems. Only some of its inherent ambiguities were fully explored at the time; but two of them were decisive for the failure of any fully fashioned policy of partnership to emerge. The more obvious was the ambiguity in the position of the Germans who from the point of view of an American racialist like John W. Burgess formed part of a Teutonic master-race.[31] Chamberlain in his speech at Leicester on 30 November 1899, spoke of the need for a triple alliance between 'the Teutonic race and the two branches of the Anglo-Saxon race', and attempted to bring about a British alignment with Germany. But such views, which reflected Chamberlain's concentration upon the purely commercial aspects of foreign policy, could not long survive their absence of echo at Berlin, and the clear determination of Germany to seek a position of naval strength which would inevitably threaten the established position of Britain upon which American security had also been permitted to rest. Nor were they in accordance with most British sentiment.[32] In a memorandum of November 1901 Lansdowne listed among the objections to a German alliance the risk of becoming entangled in an anti-American policy.[33]

In the long run the more important ambiguity lay in the fact that Britain's overseas dominions included not only peoples of overwhelmingly or at least predominantly British stock but also millions of Asian and African subjects of the Crown to whom the ideas of Anglo-Saxondom were clearly irrelevant. British exponents of the idea of an Anglo-Saxon brotherhood, such as Dilke, saw no future for the race in the possession of alien dependencies.[34] And Kipling's Indian sympathies did not prevent his imperial ideology from being dominated by the same reliance upon the tie of blood.[35] For Americans this aspect of Britain's imperial heritage presented even greater problems. Despite their acquisition of

(London, 1934), pp. 296–306. For this period generally see also A. E. Campbell, *Great Britain and the United States, 1895–1903* (London, 1960), and C. S. Campbell, *Anglo-American Understanding, 1898–1903* (Baltimore, 1957).

[31] Strout, op. cit., p. 141.

[32] Grenville, op. cit., pp. 128, 156, 218–3.

[33] Monger, op. cit., p. 66.

[34] Thornton, op. cit., p. 39.

[35] See J. I. M. Stewart, *Eight Modern Writers* (Oxford, 1963), p. 286.

Hawaii, Puerto Rico and the Philippines, they never seriously contemplated the idea of a long-term dependent empire for themselves, and regarded Britain's imperial possessions as an obsolete incubus. Once Britain was set on the path of developing non-white dominions which would, however slowly, come to take their place permanently beside the older members of the Commonwealth, there could be no fusion between the Commonwealth ideal and the notion of an Anglo-Saxon confederation. Different Englishmen were to face the consequences of this fact in different ways, but that they would have to face it should have been apparent from very early on.

If they chose to follow to its logical conclusion the argument that an Anglo-Saxon confederation was the safest object for Britain to aim at they might argue as Philip Kerr (later Lord Lothian) did in a paper in 1909 that this policy involved a relinquishment of British political control over India and Egypt. Its guiding concept would be the world-wide maritime supremacy of the British and American fleets able to protect territories like Australia and South Africa which, though fit for white settlement, were too thinly populated for their own defence.[36]

Americans of the more conservative school who might be attracted for internal political reasons to a policy of alignment with Britain were in an ambivalent position, since they tended to be protagonists of a policy of expansion by the United States itself and

[36] Balfour Papers. BM Add. 49797. In his life of Balfour, Mr Kenneth Young attributes this paper to him. *Arthur James Balfour* (London, 1963), pp. 281–2. This is clearly an error. It appears that Kerr, who was then in South Africa, also sent it to Theodore Roosevelt, then about to visit Europe. An introductory note marked 'not sent to Roosevelt' points out that one of Germany's objectives in building up her fleet is to be able to oppose imperial preference, and she might make this impossible if she could get the United States on her side. The advantages of the 'Anglo-Saxon Confederation' must be impressed upon Roosevelt so that he will not be tempted towards an American-German alliance. On Roosevelt's own rather ambivalent attitude to the British Empire, see 'Theodore Roosevelt and the British Empire' in Max Beloff, *The Great Powers* (London, 1959). Fears lest the United States and Germany might come together were not confined to Kerr and his friends. Admiral Fisher expressed the same anxiety in a letter to King Edward VII on 4 October 1907. Marder, *Fear God and Dread Nought*, vol. 2 (London, 1956), pp. 142–3. It was this in part, no doubt, that made Fisher, earlier rather anti-American in sentiment, a protagonist from at least 1907 of 'that great and impending bulwark against both the Yellow man *and the Slav* – "the Federation of all who speak the English tongue" '. Marder, op. cit., pp. 191, 298, 343, 346, 348, 361–2. For his earlier view: 'The Yankees are dead against us', see his letter to Rosebery, 10 May 1901, ibid., vol. I (London, 1952), pp. 188–9.

so to look upon Britain as a rival and competitor especially in the field of trade. In this way the hostility towards all forms of imperialism on ideological grounds which was the mainstay of American thinking on world affairs was fortified by a particular degree of distrust in reference to the British Empire. And such sentiments were aroused in particular whenever it seemed likely that the Empire (or later the Commonwealth) might fortify its ties by a system of preferential tariffs, common currency arrangements or other forms of commercial discrimination.[37]

In this respect the achievement by the Dominions of independence in the field of foreign and defence policy after 1919 placed the Americans in something of a dilemma. They put up some resistance to accepting the fact of the change, for instance in their strong opposition to separate representation for the Dominions in the League of Nations.

Another example was in relation to the limitation of naval armaments. The American invitations to the Washington Conference of 1921 ignored the existence of the Empire and it was Lloyd George who had to insist upon Dominion representation. 'It is amusing,' wrote one of his secretaries, 'to think that one of our underlying objects in this great Conference must be to make the United States recognize the British Empire. Their present attitude seems to be that the Dominions are nondescript appendages whose natural instincts would be to link up with the stronger English-speaking power. They have never yet even begun to grasp what the underlying sentiment of the Empire is.'[38]

But in the final treaty and in subsequent discussions, the Americans insisted on reckoning the Dominion navies in with the British. Since their ships were not under Admiralty control it could be argued that any increase in the Dominions' navies left Britain itself weaker in regard to the United States.[39] Similar problems about air forces arose at the Disarmament Conference.[40]

[37] Note the remark by Sumner Welles in 1943: 'The whole history of British Imperial Preference is a history of economic aggression' quoted by L. S. Amery, *My Political Life*, vol. III (London, 1955), p. 385.

[38] Edward Grigg to Lord Cromer, 15 November 1921. Grigg Papers (property of Mr John Grigg), General and Political. 1922. A-K. It would seem that Lloyd George was himself acting under Canadian pressure. See Roger Graham, *Arthur Meighen*, vol. II (Toronto, 1963), pp. 103–4.

[39] John Dove to Philip Kerr, 18 July 1927. Lothian Papers. Box 189.

[40] 'We can of course say that the Dominions are self-governing, that we have no control over their forces, that we are bound to them by no military treaties or

What was true of armaments was equally true of finance. Even in the second world war, British financial emissaries to Washington found most Americans unable to distinguish between the resources of the United Kingdom, which the Government could mobilize, and those of the Empire, which it could not.

Britain's relations with the United States had, of course, always been affected by the special position of Canada. As early as the beginning of the century British statesmen were facing a new worry – the 'great influx' into Canada of Americans and American capital, and were connecting with it the fact that while some Canadian statesmen were looking to ultimate independence none was seriously advocating closer ties with Britain.[41] For this reason the internal Canadian battle over 'reciprocity' was one of the neuralgic points in the whole Anglo-American complex of the pre-1914 period, though one where the United States often seemed peculiarly insensitive to British susceptibilities.[42]

It is the imperial angle which provides one persistent theme in the relations between Britain and the United States. It began with Joseph Chamberlain's campaign for tariff reform in the early years

understandings. But can we say that their aeroplanes would not be available to us in case of war? I hardly think so, and if we did, with the record of the Dominions during the last war fresh in the memory of other nations, no one would believe us. This is likely to cause grave difficulty, especially with the United States.' Anthony Eden to Sir John Simon, 1 May 1933. Simon Papers (property of Viscount Simon).

[41] See, e.g. the letter from the Governor-General, Lord Minto, to Sir Edward Hutton, 16 July 1901. Hutton Papers. BM Add. 50081. In a letter to Balfour of 25 January 1907, W. A. S. Hewins, who was looking over the prospects for imperial preference in Canada, pointed out that a substantial part of the brains and capital behind Canadian expansion was American. Balfour Papers. BM Add. 49779.

[42] On 23 July 1905, Andrew Carnegie airily wrote to Balfour pointing out that the absorption of Canada by the United States was natural and that Canadian demonstrations of loyalty to Britain merely provoked unfavourable reactions in America. For the same reaons he condemned imperial preference proposals and looked forward instead to the ultimate unity of the English-speaking world. Replying on the 28th, Balfour wrote that while he was very well disposed towards the Americans 'whom he did not in any sense regard as a foreign community', he thought them unreasonable about Canada and about preferences generally. Balfour Papers. BM Add. 49742. Balfour as Prime Minister was primarily concerned to secure American support for British policies, notably in the Far East, ibid., 49729. But he regarded himself then and later as 'a Pan-Anglican'. 'I have always held', he wrote in 1911, 'that the English-speaking peoples have traditions, interests and ideals that should unite them in common sentiments and in not inconceivable eventualities in common action.' To Philip Kerr, 18 March 1911, ibid., 49797. And indeed as early as 1898 we find him writing of himself as cherishing a 'pan-Anglo-Saxon ideal'. Letter to J. St Loe Strachey, 22 July 1898, ibid.

of the century and concluded with the rearguard action fought against the American loan of 1946 by such spiritual heirs of Chamberlain as Leo Amery and Walter Elliot. The genuine imperialist school in British twentieth-century politics with its programme of economic unity based on trade and migration, combined with common defence and foreign policies for all the self-governing members of the British family of nations, never, it is true, attained a fully commanding position in British politics. When its minority strength was sufficient for one or other aspect of its policies to get a run – as imperial preference did at Ottawa in 1932 – this aspect did so in isolation from what made up a coherent if not convincing master-plan for Britain's future. For the most part the imperialists were actually in impotent opposition.[43] But the movement lacked neither leaders nor mouthpieces, and its influence cannot altogether be set aside in discussing Anglo-American relations.[44]

In their scepticism as to the desirability and feasibility of a close Anglo-American combination the true imperialists were in line with the views of the intellectual father of the school, Lord Milner. While expressing in 1909 and 1912 his desire to keep on the best possible terms with the United States he regarded ties with that country as of quite a different order to those which could develop within Britain's 'own family' which should be of a constitutional kind.[45] Indeed, with particular reference to Canada, Milner realized that any confusion between the Empire and Anglo-Saxondom would be positively dangerous.[46]

[43] The so-called 'Suez group' of the 1950s may be regarded as the final remnant of the old imperialist school.

[44] Even more difficult was the position of 'The Round Table Group', who combined their ideal of greater Commonwealth unity with a generally 'English-speaking' orientation. For a recent assessment of Anglo-American relations from this point of view see the first Ditchley Foundation Lecture, H. V. Hodson, *The Anatomy of Anglo-American Relations*, delivered 27 April 1962. Kerr, who was one of those most preoccupied with American issues, wrote in relation to the 1927 Naval Conference: 'The English-speaking nations have either got to bring themselves under one sovereignty or they will drift into antagonism.' Kerr to Lionel Curtis, 2 September 1927. Lothian Papers. Box 189.

[45] Milner to Colonel Denison, 3 November 1909. Milner Papers, Bodleian Library, Oxford. c. LXXIV. Milner to Haslam, 24 May 1912. Ibid. Imperial Union Box.

[46] Milner to H. Bourassa, 9 October 1912. Milner Papers c. LXXIV. For Milner's not altogether consistent views on the relations between his racial and his imperial ideas, see A. M. Gollin, *Proconsul in Politics* (London, 1964), pp. 128–32, 401.

It might have been expected that this suspicion of the United States on the part of the British 'right' would have been compensated for by the traditional sympathy of the British 'left'. And there were moments when such an alignment seemed to be taking place, most strikingly perhaps in relation to the question of 'war aims' during the first world war when Woodrow Wilson was influenced by, and in turn influenced some radical elements in British politics.[47] But such episodes were on the whole exceptional. For as the British Left became increasingly identified with dogmatic socialism, so the United States as the great bulwark of capitalism increasingly appeared to be the principal obstacle to the spread of its ideals.[48]

The actual Labour governments of the period in respect of their broad attitude to Anglo-American relations showed no significant departure from their predecessors. Indeed, if anything, they attached even greater importance to attempts to secure agreement with the United States. It was Ramsay MacDonald during his second term as Prime Minister who became the first British holder of the office to visit the United States. It was under the post-war Attlee administration that the American loan, the Marshall Plan and the Atlantic Alliance together made the fortunes of Britain much more closely dependent upon the United States than in any preceding peace-time period.

Even therefore if due allowance be made for the opposition of extreme factions on both wings of British politics, it remains possible to discuss standard British attitudes towards the United States as a continuing expression of a clear majority viewpoint, first formulated by Balfour and Grey. And basically, as has been suggested, this viewpoint involved a continuous search on the British side for a 'special relationship', first, so that no British

[47] See L. W. Martin, *Peace Without Victory* (New Haven, 1958). In addition to Colonel House's information derived from the United States Embassy in London, there were also close contacts between the *New Republic* circle in New York and the British 'Union of Democratic Control' and other Liberal and Labour elements. A. J. Mayer, *Political Origins of the New Diplomacy* 1917–18 (New Haven, 1959), pp. 335–7. Wilson's biographer, Professor Arthur Link, has argued that the breakdown of the first Anglo-American alliance was due to the failure of British statesmen to understand Wilson's purposes or accept his leadership. A. Link, *President Wilson and his English Critics* (Oxford, 1959). On Anglo-American relations after the Armistice see S. Tillman, *Anglo-American Relations at the Paris Peace Conference of* 1919 (Princeton and London, 1962).

[48] See Henry Pelling, *America and the British Left* (London, 1956).

resources should be tied up in actually opposing the United States, and second so that America's power should be brought to bear wherever possible in support of British interests.

The former of these ambitions could be fulfilled without too much difficulty because it depended fundamentally on what the British were willing to surrender – their claims on the isthmus,[49] any significant naval strength in Caribbean waters,[50] the contested Canadian frontier claims. The process was virtually complete by the time of the (unratified) Arbitration Treaty of 1911, but there were still echoes of it in the celebrated destroyer-bases deal of 1940.[51]

It was more difficult to apply the notion to the general problem of sea-power and of imperial communications so long as Britain continued to regard herself as a world power. It is true that as early as 1908–9 when the Asquith Government reviewed the problem of the two-power naval standard the advice it received was to exclude the strength of the United States in assessing Britain's naval needs. The former First Lord of the Admiralty, Lord Cawdor, put the case at its simplest: 'It is incredible that we should go to war with an Anglo-Saxon power'.[52]

But the experience of the 1914–18 war itself made a difference. It was the freedom of the seas that had caused so serious an Anglo-American crisis before America herself entered the war.[53] The war ended in a wave of 'big navy' sentiment in the United States which had serious repercussions on British thinking.[54] The dis-

[49] On the significance of the Hay-Pauncefote treaty in Anglo-American relations generally see especially Grenville, op. cit., pp. 376–89.

[50] There was, of course, an element of discretion in deciding how far this retreat should be admitted. Writing to Balfour's secretary, Sandars, on 31 March 1905, Colonel Clarke (later Lord Sydenham), the secretary of the Committee of Imperial Defence, made a caveat about a proposed answer to a Parliamentary question: 'What it is best not to say is that we believe that the idea of opposing the navy of the US in the Caribbean and the North Atlantic close to its bases must be abandoned.' Balfour Papers. BM Add. 49701.

[51] See Philip Goodhart, *Fifty Ships that Saved the World* (London, 1965).

[52] Asquith Papers. Dep. Asquith. 21.

[53] For an attempt to explain to a well-wishing American friend the dangers to Britain of an indiscriminate use of the phrase 'freedom of the seas', see James Bryce's letters to Charles Eliot, 11 May, 1 July, 8 August 1915, 29 January 1916. Bryce Papers. USA 2. For Eliot's attempt to show that international guarantees would suit Britain better see his letter of 26 August 1916, ibid. USA 1. – partly printed in Henry James, *Charles Eliot* (London, 1930), vol. II, pp. 263–5.

[54] Wilson himself seems to have entertained the idea for a brief time that only the United States and Britain should be allowed to have navies at all and that

putes that ensued and the American reluctance to admit the close connection between naval armaments and the general armaments position in Europe caused intermittent and sometimes considerable friction between Britain and the United States until the London Conference of 1930, which may be taken as the point at which Britain finally acquiesced in effective American preponderance.

This naval controversy is also illuminating for a different reason. The British case for more light cruisers rested, as is well known, upon an assessment of the requirements arising from the responsibilities of patrolling the routes connecting the scattered portions of what was now an Empire *and* Commonwealth. The United States had no similar problem and tended to minimize the importance of such considerations on the British side.[55] The two countries had different geographical visions. The United States saw itself as a compact land-mass with certain outlying island bastions. The British saw their world in the form of a network of maritime communications threatened with possible strangulation by predatory powers on their flanks. The Mediterranean, the Suez Canal and its approaches, the Cape and Singapore – these were still a focus of British attention and concern whose importance Americans were slow to appreciate. Not until the great debates over strategy in the second world war was the full importance of these differences in perspective made altogether apparent.

But they had their effect even earlier. Those who during the

together they should police the world. Letters from Lord Derby to Balfour, 20 and 22 December 1918. Balfour Papers. BM Add. 49744. Later, however, he appeared to British ministers to be advocating two incompatible policies, the League of Nations and a big navy for the United States. It was thought of the greatest consequence that Britain should not be forced into naval competition with the United States, both because of the latter's greater financial strength and because the British were unwilling to assume that the United States might ever be hostile. See Walter Long to Lloyd George, 16 February 1919. Austen Chamberlain Papers. AC 25. The American repudiation of the League Covenant and the triumph of isolation naturally played into the hands of those who believed that Britain must be able to rely on herself. The idea of making Britain's future security dependent on an international organization had indeed been opposed during the war in some quarters largely on the ground that the Americans could not be relied upon. See, e.g. a minute from Hankey to Balfour of 25 May 1916, opposing the idea of a compulsory International Tribunal of Arbitration, Balfour Papers. BM Add. 49704.

[55] In a letter to General Smuts, referring to a paper by him on peace conversations, Kerr wrote that it was 'fatal to suggest that the German colonies must be retained because they are essential to British communications'. The United States would not look at that for a moment. Kerr to Smuts, 14 December 1917. Lothian Papers. Box 187.

first world war and the Peace Conference looked hopefully to Anglo-American cooperation as the basis of a future world order were dismayed at finding the Americans not merely ignorant of the problems of 'backward peoples' but convinced that the assumption of any kind of responsibility for them was 'iniquitous imperialism'.[56] In general the British were very anxious to encourage American participation in world affairs to the full extent America's revealed power justified. When it became likely that the United States would not ratify the Covenant some of those most attached to the American connection felt that Britain should also seek release from the obligations she had assumed.[57] Similarly it was felt that if Britain surrendered the Japanese alliance to Dominion and American pressure, the United States must in some form provide an alternative.[58] Only with this background in mind can one appreciate the nature of the important divergences over Far Eastern policy between Britain and the United States in the early 1930s.[59]

But the differences in outlook were not confined to naval and imperial questions. Roosevelt's indifference to Britain's need to see France restored to the position of a Great Power after the second world war was in a line with the continuous refusal of the Americans in the inter-war period to understand European attitudes to the central problem of security. The British might disagree (at times violently) with the French over methods but they did not

[56] See, e.g. Philip Kerr to Lionel Curtis, 15 October 1918, quoted in J. R. M. Butler, *Lord Lothian* (London, 1960), p. 69.

[57] See the memorandum by Philip Kerr of 14 November 1919. Lothian Papers. Box 139. The blow was greatest for those whose American friends has assured them that ratification was certain. See, e.g. Charles Eliot's letters to Bryce, 10 April, and 12 May 1919. Bryce Papers. USA 1.

[58] See Philip Kerr to John Dove, 13 July 1920. Lothian Papers. Box 186.

[59] Writing to Simon on 1 September 1934, Neville Chamberlain urged on him the danger of drifting into unfriendly relations with Japan. We should not allow ourselves to be browbeaten by the United States who would not repay us for sacrificing our interests to conciliate her. Simon Papers. 'The discussions in London are getting into difficulties', wrote Lothian, 'not because there is any fundamental difference of view but because Great Britain before she knows where she is with the United States, is doing her best to avoid taking up an attitude of hostility to Japan. The United States, on the other hand, being in a much securer position, is trying to press us to take an attitude of opposition to Japan before she has really thought out what she is going to put into the pool if Japan forces the pace.' Lothian to Colonel House, 30 November 1934. Lothian Papers. Box 221. It is significant that Canadian sympathies were with the United States on this issue. Letter from Vincent Massey to Sir Alfred Zimmern, 4 May 1934, quoted in Massey, *What's Past is Prologue: Memoirs* (London, 1963), p. 207.

dissent from their objective. And after the failure of the Anglo-American guarantee treaty itself it was the American unwillingness to under-write any other British guarantees of French and European security that conditioned the whole of Britain's European policy from 1919 to 1939.

After the Somme and Passchendaele, after the submarine warfare crisis of 1917, Britain, it was clear, could never again contemplate a full-scale battle for existence without some assurance of American support. For this support she was prepared to bid high, as the Baldwin debt settlement indicated. Britain felt obliged to surrender other guarantees of her position which were incompatible with American friendship. The Anglo-Japanese alliance was in fact terminated.[60] But caution about the 'sanctions' clauses of the Covenant and a general predisposition to limit continental commitments were also products of the priority accorded to America.[61] Unless the negative influence of American policy is properly assessed – whether exerted directly or through Canada – one cannot understand the whole complex of policy that goes under the name of 'appeasement'.[62]

It is true that the United States' attitude played rather a minor part in the final stages of this policy except in its impact upon the

[60] Walter Long, the First Lord of the Admiralty, told his colleagues on 30 May 1921, that political relations with the United States were of such transcendent importance to Britain as to outweigh all other considerations of policy. He had been assured by Admiral Sims that the American people were vehemently opposed to the alliance and that while it lasted it would be impossible to curb the American cry for increased armaments. Austen Chamberlain Papers. AC 26. Americans were prone to argue even after their repudiation of the League that the Anglo-Japanese alliance belonged to the 'old order' and could not be squared with the new theories of international relations being advocated by the British and American Governments. See Paul Cravath to Edward Grigg, 4 October 1921. Grigg Papers.

[61] The American Secretary of State Hughes worked hard to prevent ratification of the Geneva Protocol. According to the leading French historian of the period, Britain was forced to choose between the United States and the Protocol and chose the former. J. B. Duroselle, *From Wilson to Roosevelt: Foreign Policy of the United States*, 1913–45 (London, 1964), pp. 164–5.

[62] As early as 1924 Mackenzie King was pointing out to the Governor-General, Lord Byng, the importance of Canada keeping aloof from European affairs. 'If the people got the idea we were to be drawn into the European arena there would certainly be a movement to the United States which had kept out of the League of Nations on that account and political union with the U.S. might become a subject of discussion.' Mackenzie King's Diary, 1 December 1924, quoted by H. Blair Neatby, *William Lyon Mackenzie King*, vol. II. *The Lonely Heights*, 1924–32 (Toronto, 1963), p. 44.

Dominions. The reason may be that of all the British statesmen of the period, Neville Chamberlain was perhaps the least disposed to give major weight to the American factor. His position in this respect may have been affected by his deep mistrust of Cordell Hull, arising from the latter's persistent campaign against the Ottawa system of preferences. In February 1937, when Chamberlain was still at the Exchequer, he received a message from the American Secretary of the Treasury asking whether the United States could do anything to help bring about an agreement on disarmament. Chamberlain's reply (after consulting Baldwin and Eden) was that the main source of danger and fear in Europe arose from Germany's intensive military preparations, propaganda against other countries and violation of treaty obligations. Only a superiority of force could contain her.[63] Another approach from Hull made through Lord Tweedsmuir, the Governor-General of Canada, that autumn after Chamberlain became Prime Minister again, met with the answer that economic appeasement was insufficient and that Germany would have to be brought to cooperate on political issues as well.[64]

It is important to understand this background to the much-debated negative that Chamberlain returned to Roosevelt's initiative in January 1938.[65] The American offer did not alter the political and strategic facts of the situation, which was not susceptible to handling in purely economic terms. Perhaps this belief that the United States could not materially assist in 'appeasement' helps to explain the surprising decision to abandon the policy without any assurances of American support for the alternative – resistance.

For 'appeasement' was abandoned in the end without any positive guarantees of American aid, although of course in the knowledge that the nature of the enemy this time made it certain that there would be no repetition of the equivocations of the American policy of the 1914–17 period. The United States could not be

[63] See John Blum, *Years of Crisis* (From the Morgenthau Diaries, vol. I, Boston, 1959), pp. 458 ff. Cf. Lord Avon, *Facing the Dictators* (London, 1962), pp. 526–7; Ian Colvin, *Vansittart in Office* (London, 1965), pp. 140–3.

[64] Lord Tweedsmuir to Neville Chamberlain, 25 October 1937; Lord Simon to Lord Tweedsmuir, 24 November 1937 – both in the Simon Papers.

[65] It should not be thought that the answer to Roosevelt was solely the work of Chamberlain. Roosevelt's message kept the Foreign Policy Committee of the Cabinet 'busy all week.' Sir John Simon's diary in the Simon Papers.

genuinely neutral in a Hitler war. Whether British statesmen felt in their hearts that sooner or later the United States would be drawn into it is something we are not entitled to say. Only after June 1940 was Britain's resistance obviously predicated on the expectation of ultimate American involvement. Anything else would have been quixotic in the extreme.

But although Churchill might play upon the sentiment of the 'special relationship' in public utterances, once it became safe to do so without causing embarrassment to Britain's friends in America, he did not rely upon it. He was fully aware of the seniority of the Americans in the partnership and in spite of some gestures of defiance, in the main acted upon this awareness.[66]

In the post-war world, America's growing determination to 'contain' communism coincided with the British desire to perpetuate American involvement in the defence of areas considered vital from the British point of view. Few incidents in the transfer of responsibility were as clear-cut as that involving the enunciation of the 'Truman doctrine' when Britain confessed herself unable any longer to uphold the independence of Greece and Turkey. And this perhaps prevented a full understanding on the British side of the sacrifices that had to be made in return for this aid.

The new nature of the 'special relationship' was most fully illustrated in relation to the 'Suez' affair of 1956: and this not so much in John Foster Dulles' cavalier disregard for his allies' interests, and the deception practised upon them in leading them to expect that the Americans would give an active support to their claims which they clearly had no intention of providing, but rather in the nature of the opposition to, and subsequent criticism of the British Prime Minister. If one discounts the hysteria of the Left, the real charge brought against Sir Anthony Eden (now Lord Avon) was precisely the fact that he had kept America in the dark about his plans, and undertaken military measures without mak-

[66] Churchill's inevitable concessions to the Americans contrast with the ability of Asquith in the First World War to resist suggestions that Irish policy might be modified, or the blockade, 'black list' and censorship measures seriously weakened in order to conciliate American opinion. See the Foreign Office memorandum of 30 October 1916. Asquith Papers. Dep. Asquith 130. and other documents in this collection. Some concessions were of course made: 'on 22 July 1915, the Cabinet decided that in view of the situation between the United States and Great Britain, no American ships outward bound from Europe, although presumably carrying German goods, should be stopped by our cordon.' Cecil to Balfour, 12 April 1916. Balfour Papers. BM Add. 49738.

ing certain of American backing should the country run into difficulties. The Americans were not blamed for frustrating a British action so much as the British Prime Minister was blamed for embarking upon an action which the Americans would wish to frustrate. And it is significant that criticism of this kind was most strongly echoed in Whitehall itself, the repository, as it were, of the central traditions of British external policy.[67]

It is therefore likely that when the records of Mr Harold Macmillan's premiership come to be revealed we shall find the re-establishment of confidential relations with the United States as having been his prime objective. The 'Nassau Agreement' of December 1962 may fittingly illustrate the nature and limitations of his success. It could be represented, as it was by its British negotiators, as a perpetuation and recognition of Britain's survival as a Great Power, based upon the availability to it of an independent nuclear deterrent; it could be represented by their critics, and notably by General de Gaulle, as the acceptance by Britain of a degree of dependence upon the United States amounting virtually to satellite status.

What is not clear is the extent to which earlier British statesmen and the exponents of their policies perceived the rate at which the process of the substitution of American for British power and influence was in fact taking place. One is inclined to think that for the most part they were largely unconscious of it.[68] The equilibrium had been shaken but sooner or later it would be re-established. The concessions that had to be made did not detract from Britain's essential requirements, or were paradoxically justified by the need to hold the Commonwealth together.[69] For one must not forget that after 1930 at any rate the relationship did not any

[67] Here time had made a difference; in 1911 the Foreign Office was described as so anti-American as to make it always difficult to discuss American affairs with its senior officials. George Young to James Bryce, 13 July 1911. Bryce Papers. USA 32.

[68] There were exceptions. In reference to the Kellogg Pact negotiations Gilbert Murray wrote to an American friend: 'The world is now engaged on an exceedingly delicate process, inevitable whether we regret it or rejoice in it, viz. the transference of wealth, power and leadership from Great Britain to America. It wants a very high standard of public duty and intelligence on your part and on ours to conduct that proceeding of the dethronement of Great Britain and the enthronement of the United States without war.' Gilbert Murray to S. Levinson, 3 December 1928. Gilbert Murray Papers. Bodleian Library.

[69] It was argued by some that the Locarno Pact would have the effect of destroying the diplomatic unity of the British Empire and so would drive the Dominions

longer concern London and Washington alone. First, and inevitably, Canada, and later Australia and New Zealand sought their own national security in relations with the United States of a kind that did not necessarily strengthen the Commonwealth tie, although for Britain to have opposed their intentions would certainly have weakened it. We have still to estimate the role of Mr Mackenzie King either as an honest broker between the British and the Americans, or as a grave-digger of Empire, according to one's point of view.[70]

The degree to which Britain had no choice was something which many either failed to understand, or if they understood it, kept to themselves. The faith in the 'special relationship' that seemed to remove Anglo-American relations from the ordinary sordid calculations of power politics may have prevented proper assessments of the situation from being made, or from being made known to the public when they were. The difficulty was that the 'special relationship' was not an illusion – if it had been there would have been no problem. On the personal level, whether between government servants or between the armed services – truer here of army than navy, one must add – a very high degree of intimate contact between individuals was possible in wartime;[71] and from its existence both nations, but especially the British, reaped handsome dividends.

Nevertheless, as we have noted, such personal contacts, including the numerous Anglo-American marriages, as might have fitted British statesmen or diplomats for working closely with the United States were limited by their narrow social range.[72] Business ties were largely those with the financial world of New York, which

into the arms of the United States. Kerr to Lloyd George, 13 November 1925. Lothian Papers. Box 188.

[70] This brokerage began before Mackenzie King's premiership. Theodore Roosevelt tried to use him as a channel to the British Government in 1908 over the question of reaching a joint policy on Japanese immigration on the Pacific Coast in respect of which the Anglo-Japanese alliance put difficulties in the way of a direct approach to London. R. MacGregor Dawson, *William Lyon Mackenzie King: a Political Biography*, vol. I (London and Toronto, 1958), pp. 152–61.

[71] On sentiment towards Britain in the American navy see Goodhart, op. cit., pp. 64–5.

[72] Before the First World War the contacts at the highest level were few indeed. In 1917, the American Ambassador, W. H. Page, pointed out to Balfour that neither he himself since 1875 (when he was only twenty-seven years old), nor Grey, Lloyd George, Asquith, nor any other important British statesman had visited the United States. Kenneth Young, op. cit., p. 379.

was itself a very suspect factor in American life.[73] The two trade-union movements were separated from each other by profound differences in structure and spirit.[74]

Much has been made of the alleged failure of the British at large to give serious attention to American history or the American polity. Such criticism was in part exaggerated; at no time was the British reader at a loss for serious studies of important aspects of the American scene. No other European country was served so well. Indeed, the real damage may have been done when conscious efforts were made to improve things. For the introduction of American history and politics into school and university courses was inevitably the work of enthusiasts who, being animated by laudable desire to improve the two countries' relations, were prone to underestimate the importance of the real differences between them. Similarly the visiting American professors who after the first world war, and even more during and after the second, found their way to British shores were unlikely to represent those elements in the American population which it was very important for the British to get to know, just because they repudiated so many aspects of the British view of the world.

Even in peacetime, such personal relationships played an important part as those who took part in the transactions of the Marshall plan era are always willing to testify. It was a rare Frenchman who, like Jean Monnet, could acquire a similar place for himself in the counsels of official Washington. But great democracies are not deflected from their national policies by personal friendships. Individuals might help to dissipate genuine ignorance or genuine misunderstandings – a Lothian, a Halifax might make relations easier, a Brand or a Purvis break through jungles of red-tape that might have strangled lesser men. On the other side, it was no doubt helpful when a John Winant and a Harry Hopkins replaced Joseph Kennedy as the authorized interpreters of the British scene to Roosevelt. But the great issues were decided as a result of factors outside their control.

The very understanding which some Americans showed for the British predicament in wartime may have contributed to difficul-

[73] The utility and limitations of such contacts are well brought out in Sir Harold Nicolson *Dwight Morrow* (London 1935).

[74] When Ramsay MacDonald visited the United States in 1927 he felt no strong bond of sympathy with the American trade-union hierarchy: 'he was amazed at their conservatism and affluence.' Massey, op. cit., p. 150.

ties later on, when the essential national interests of the two sides came into play. For it may have been hard for people to believe that this cordial intimacy in the common task did not preclude an ultimate divergence in aims. The abrupt termination of lend-lease, the exclusion of Britain from further cooperation in the exploitation of what were largely British discoveries in the field of nuclear energy are two examples of the way in which reliance upon assumptions of a common purpose was falsified in the event.[75]

Even if one discounts the personal factor there remains the need to study in much greater detail than has been done the changing scope and nature of the official representation of each country in the other during the whole of the period. Apart from the special case of Bryce, it was the first world war which saw the first serious attention paid to the question of who should be British Ambassador in Washington, and what additional representation was required. And for some time afterwards, it was still felt that the post could not be treated on ordinary lines. If, following Lord Reading's successful mission, a diplomatist were appointed, it was felt that Britain would lose the exceptional position she had acquired at Washington.[76] Almost everyone of note in British political life was canvassed for the post – Speaker Lowther, Lord Finley, Devonshire, Lord Salisbury, Lord Crewe, Balfour, Austen Chamberlain and the Canadian, Sir Robert Borden – before the choice fell on Sir Auckland Geddes.[77] Subsequently, the view that a public figure was required was abandoned,[78] and until Lord Lothian's appointment in 1939 the Washington Embassy was again filled by the normal operation of the diplomatic *cursus honorum*.[79]

[75] To adduce the latter example does not mean to exculpate the British from all blame for developments that were to be costly for both countries. While they had a comfortable lead in the field the British were slow to enter into full partnership with the Americans; had they done so it is possible, though not certain, that they might have got arrangements for cooperation which would have stood them in good stead later on. See M. M. Gowing, *Britain and Atomic Energy*, 1939–1945, (London, 1965). Cf. Sir John Wheeler-Bennett, *John Anderson, Viscount Waverley* (London, 1962), pp. 332 ff.

[76] Drummond to Balfour, 23 January 1919. Lothian Papers. Box 141.

[77] Balfour Papers. BM Add. 49734.

[78] In 1929, MacDonald thought of sending Gilbert Murray, but it was felt that his own visit to the United States had so changed the situation that a diplomat would serve Britain's purposes better. Murray to Miss I. Munro, 25 October 1929. Gilbert Murray Papers.

[79] No ex-Ambassador to the United States had until 1961 filled the post of

Under Lothian and Halifax the Washington Embassy became, as it has remained, and for obvious reasons, rather a microcosm of Whitehall overseas than an ordinary diplomatic post.[80] But the issue as between a 'political' and a 'career' ambassador remains interestingly unsettled.[81]

To sum up: the 'special relationship' is a fact, but a fact of a rather peculiar kind; for myths are also facts. It would be interesting, however, to inquire further into why it has been found psychologically so necessary to dress up in this way a perfectly honourable relationship as though national self-interest were something which should play no part in this branch of international politics. One possible reason is to be found in a remark by A. V. Dicey, whose own legal and constitutional interests may explain his advocacy of an Anglo-American *entente cordiale* and his tendency to talk of the Americans as though they were just another branch of the English people.

> Both branches of the English people [he wrote in 1911] seem to me to act somewhat better in foreign affairs than do Continental States. But to make up for this both branches of the English people persuade themselves that they pursue in international affairs far more disinterested principles than they in reality act up to.[82]

In this respect the lapse of over half a century has brought about little change.[83]

Permanent Secretary at the Foreign Office.

[80] On the background to this see Max Beloff, *New Dimensions in Foreign Policy* (London, 1961).

[81] One also needs a study of the effectiveness of the representation of the British Press in the United States. A beginning has been made in *The History of The Times*, vol. IV (London, 1952), chapter 9.

[82] A. V. Dicey to James Bryce, 23 March 1911. Bryce Papers. English Papers. Vol. 3.

[83] The themes sketched out in this essay are being further developed in my book, *Imperial Sunset* of which volume I was published in 1969, in London and 1970 in New York. I would like to thank the Rockefeller Foundation for its hospitality at the Villa Serbelloni where this essay was written and my research assistant Mrs M. Croft (née Heeley).

16 Reflections on Intervention

Very little is new in the practice of international relations. The twentieth century's record of wars, revolutions, and institutions has added little that has not long been familiar to the international system. 'Intervention' – the attempt by one state to affect the internal structure and external behaviour of other states through various degrees of coercion – is not new to the system in this decade or even in this century, but has been a commonplace in the history of world politics. The United States' action in Vietnam in 1968, for example, is no more original than the Soviet Union's intervention in Eastern Europe at the end of the second world war.

The indignation which the *modern* practice of intervention arouses in the United States is not due to the novelty of the event, but is the result of a still powerful Wilsonian political philosophy. Woodrow Wilson and his successors saw the world as ideally composed of equally sovereign and impermeable political communities, and they based the League of Nations and the United Nations on the assumption that the real coincided with the ideal. This idealistic misconception, moreover, was the result of a confusion of the subject matter of modern international law – self-contained sovereign states – and the subject matter of international relations, which is a far more complex thing. The contemporary resentment toward interventionist politics, then, is not so much due to the novelty of intervention, for intervention is a recurrent feature of the system, but is a result of the aberrations implicit in Wilsonianism.

It is plain that intervention, loaded with normative overtones and plagued by misconceptions, is a difficult concept to analyze. However, our understanding of the concept may be advanced by using the relatively wide frame of reference of our historical experience. This will enable us better to comprehend intervention as it is presently practised by the United States and the Soviet Union, and determine its relationship to the contemporary international system.

Several observations about intervention may be drawn from Western historical experience. First, it is obvious from Western history that intervention is one of the instruments employed in *any* competitive international system. Thucydides begins his history of the Peloponnesian War with an account of Corcyra's attempt to reinstall the exiled oligarchs of Epidamnus and the counter-intervention of the Corinthians. It is indeed clear that the Greek city-states exploited civil strife to further their own policies, and that political parties existed within the Greek states which depended, or came to depend, upon 'foreign' support.

Secondly, intervention clearly becomes more frequent, as well as more dangerous, when party divisions or religious-ideological passions give states automatic access to the sympathies of partisan groups in other states. The struggle between Guelph and Ghibelline in Florence is the most celebrated example of a political-party division that led to intervention. Both the invasion of the Spanish Armada, which followed thirty years of almost unceasing intervention by Catholic Spain in the internal affairs of England, Scotland, and France, and the Thirty Years War, when both Catholic and Protestant forces engaged in intervention in Germany on a scale unequalled until the wars of the French Revolution, are notable examples of intervention caused by religious passions.

Thirdly, the wars of the Revolutionary period, the subsequent interventions of the anti-revolutionary Holy Alliance, and Russian intervention during the revolutions of 1848 make it plain that basic *social* and *political* differences among members of a multi-state system result in conflicts of ideologies and interests. This will, in turn, give rise to the politics of intervention. Thus, Russian policy between 1815 and 1914 was influenced by the desire to defend the monarchical principle as much as by any other aspect of the 'national interest.' British policy under Palmerston, and more circumspectly later on, was similarly influenced by ideology and national interest. The British desire that constitutional government should prevail, for example, influenced British policy toward the Iberian peninsula and Italy, and affected both of their subsequent political histories. Differing ideological ambitions, as much as any other aspect of their national interest, influenced both Russian and British policies toward intervention throughout the nineteenth century, and gave rise to the politics of competitive interference that characterized Anglo-Russian relations.

This may seem to be labouring a fairly obvious point, but it is necessary to do so to make it clear that when Lenin, Trotsky, and their successors appealed to the working classes of the capitalist countries or the subjects of colonial empires to rise and overthrow their rulers, and then provided them with money and armed force, they were doing nothing that governments before them had not done. Indeed, the Kaiser's government had intervened in Russia's affairs with the financial assistance it had given the Bolsheviks. And while it is true that the motives of the British, French, and Americans in the 'intervention' in Russia, 1918–20, were more confused than Soviet historiography allows, it cannot be denied that the Allied forces were trying to bring about a regime in Russia that they preferred to that of the Bolsheviks. And yet at the same time the Bolsheviks were trying to overthrow the Allied governments.

Soviet intervention in the inter-war years, 1919–39, was distinguished by two characteristics. First, intervention was justified by stressing the unity of the international revolutionary movement – the proletarian brotherhood was supposedly ignorant of national differences. Secondly, the formal separation of the Russian Communist Party and the Soviet State meant that intervention could almost always be disavowed when it inconvenienced conventional Soviet diplomacy. Organized by the Comintern or the result of bilateral arrangements between the Russian and foreign parties, Communist intervention was formally autonomous from activities of the State, although this autonomy was of course a fiction. It was as though the Papal Curia of the sixteenth century had been situated in Madrid, with Philip II the mere executor of Papal instructions. But this dissociation of intervention from the State did not depend on the particular nature of the Soviet Regime or even its ideology. 'Volunteers' have always been one of the ways in which a state can intervene in the affairs of another; Italy and Germany were to use this cover for their own intervention in the Spanish Civil War, just as Russia's intervention was largely camouflaged by the so-called International Brigade.

The United States, on the other hand, intervened or attempted to intervene in the internal affairs of other states under the guise of the slogan, 'making the world safe for democracy.' This period of American intervention ended with Wilson's defeat by the isolationists, and the United States for the remainder of the inter-war period abstained from positive intervention in a manner unusual among great powers and inconsistent with the positive American ideology. American non-recognition of the Soviet Union until 1933, however, was a form of *negative* or *passive* intervention, hopefully leading to the downfall of the Soviet regime. (Indeed, non-recognition is always a form of intervention and is rightly treated as a hostile act by any *de facto* government against which it is applied.) Nevertheless, it can be said that until 1941 there was a pronounced contrast between the Soviet and American attitudes on intervention.[1]

Since the American and Soviet involvement in the second world war and its immediate aftermath, intervention by the two superpowers has been continuous and has included all conceivable modes of intervention – direct military assault, economic aid, subversion, propaganda, non-recognition, and the expression of moral support and sympathy. Each country has, from the viewpoint of its own interests and ideological commitments, blamed the other for intervening in the affairs of other states. Soviet intervention, in American eyes, has used illegitimate pressures to install and maintain regimes that free people would repudiate. From the Soviet viewpoint, the United States has been and is acting to retard the progressive course of history and to support or reinstall 'reactionary' regimes.

The two superpowers, thus, are doing no more than their predecessors throughout history, and if we accept their respective concepts of intervention and views of their own interests, each of them is basically correct in accusing the other of intervention. Each is correct in the sense that is unreasonable in the light of history to suggest that a great power should not use its strength to further the cause it has at heart; indeed, the West is unlikely to cease intervening to promote the kind of society and governmental structure that it believes to be good.

[1] For an analysis of their respective policies see my chapter, 'L'URSS et L'Europe', in vol. III, part 2 (Milan, 1964), and 'Le rôle des Etats Unis dans la Politique Européenne', in vol. III, part 3 (Milan, 1967), of Max Beloff, Pierre Renouvin, Franz Schnabel, and Franco Valsecchi (eds), *L'Europe du XIX et du XX siècle*.

The nations of the West have not reacted similarly to each and every example of intervention, even when the intervention was intended to promote their own interests. For example, a great many people in Europe supported the Marshall plan as a form of anti-Communist intervention but deplore the United States' intervention in Vietnam, despite the fact that this latter effort is inspired by a similar desire to check the expansion of Communism. Ideology, by and large, has ceased to serve as a sufficient cover for intervention.

Modern nations tend to judge the practice of intervention by an examination of the means and ends, apparently disregarding ideological sympathies. The West has, for example, ruled out assassination as a method of intervention, although assassination was used during the Counter Reformation in Europe and is now used in some Asian countries. More rigorous criteria of suitable means may be established. For example, one may rule out all armed force and recognize the legitimacy only of peaceful intervention through propaganda or financial assistance, or anything other than overt propaganda may be regarded as over the permissible threshold.

More importantly, however, modern nations have distinguished between genuine national interest and ideological ambitions in formulating and judging interventionist policies. The criterion used to determine national interest in an interventionist situation is *comparative risk:* it is unwise to intervene where the risk involved entails an even greater evil than the one the intervention is intended to prevent. For example, the West's fear that a general war would escalate into a nuclear war caused it not to intervene to assist the Hungarian uprising of 1956 (although it would be hard to think of a case where the other arguments for intervening were more compelling).[2]

Comparative risk or its synonym, *limited liability,* also suggests that there is little reason to intervene in a situation where the likelihood of success is marginal and the penalties for failure fall on one's allies. The Bay of Pigs fiasco is the best example of this. The limited liability criterion avoids all the apparatus of deception involved in the Bay of Pigs affair and judges the debacle for the failure it was. Intervention, like most other political actions, takes on a very different appearance according to whether it succeeds or fails, for victors write history. This line of argument is bound

[2] This essay was written before the Soviet invasion of Czechoslovakia in 1968.

to be regarded in some quarters as an example of old-world cynicism, but it should not be abandoned on that account – much of the harm in the world has been done by the high-minded.

According to this criterion (comparative risk or limited liability), rapid and easy success *alone* justifies intervention. If this is an acceptable statement, then we may have established a standard which will help us to judge the legitimacy of future cases of intervention.

The contemporary struggle between Sino-Soviet Communism and Western-style democracy is restrained by this recognition that intervention is justified only where it is successful, although Vietnam may prove to be the exception to this observation. In the immediate post-revolutionary period, the Russians believed that revolutionary conditions existed in the principal European countries and that only a limited amount of agitation was needed to overturn the balance in the Communists' favour. But experience and defeat in Hungary and Germany convinced them that this was not the case. Similarly, the almost total annihilation of the original Chinese Communist Party by Chiang Kai-shek in 1927 gave the Russians second thoughts about the malleability of Asian nationalism and the practicality of intervening in Asian politics.

As a result of these lessons, the Soviet Union throughout the 1930s observed the rule of *limited liability* in intervening in other states' affairs – Soviet participation in the Spanish Civil War is the best example. Two objectives came to be given priority over the Trotskyite belief in the expansion of the area of Sovietization: (1) maintaining the strength of the Soviet fatherland; and (2) preserving Russian control of the European Left. In view of these objectives and their reluctance to over-commit themselves in the affairs of other states, the Soviets emphasized indirect struggle against their rivals within the Spanish Republic rather than risking a direct confrontation with the Fascists.

After the second world war Stalin took an almost precisely opposite line to that of Lenin's expansionism after the first world war, and he limited intervention to those countries where the Red Army was in a position to intervene directly. Indeed, Stalin made it plain in his controversy with the Yugoslavs that the military position of the Red Army was the test of a 'revolutionary situation'.

The lessons for the Soviets in Asia and Africa, on the other hand, were less clear-cut and resulted in some miscalculation. The opportunities for intervention on the Communist side in China from 1945 to 1949 were underestimated rather than overestimated, and recent Russian and Chinese experience in Africa and Indonesia has indicated the difficulty of making prudent calculations in an unfamiliar environment.

The United States has had a more difficult task in formulating a coherent theory of intervention than the Soviets. The guidance provided by American political science and sociology has been scarcely more helpful than the Russians' crude Marxism, but Russia at least learned from her failures in the 1920s while America was misled by the famous success of the Marshall plan. Why?

The United States' task in positive intervention is more difficult than the Marxists' because the Americans would like to see a highly complex social order universally established. But the United States has not yet successfully discovered, either in theoretical *or* empirical studies, the preconditions of this desired social order (at least not to our intellectual satisfaction). A self-sustaining democratic form of government, able to meet popular demands for rising levels of consumption without the use of more than a minimum of social coercion, is not a norm of human organization, as Wilsonian theory seems to have assumed, but a highly artificial and fragile growth, easy to destroy and very hard to foster. The Russians, on the other hand, have been more easily satisfied with the destruction of regimes, a task not very hard to achieve except in the highly organized industrial states of the West or in Japan. The Soviet alternative to the Western democratic order – a totalitarian form of government able to supplement a high degree of coercion with limitless resources in propaganda – is both easy to install and easy to sustain once in power; no modern totalitarian regime has ever been destroyed except by all-out war.

Latter-day Wilsonians, including Franklin Delano Roosevelt, his advisers, and many recent American policy-makers, have had a world-view which proclaimed the democratic system as the norm. Since all departures from the norm required particular explanations, these latter-day Wilsonians saw most of the world outside Europe as a vast area under alien colonial rule, which they assumed prevented the emergence of full-fledged democratic societies. They believed that if the European overseas empires were swept away, the new nations would emerge in full glory and happily cooperate in the *Pax Americana.* Let the American Prince kiss the Sleeping Beauty, the spell would be removed, and the enchanted palace come to life again.

It was not like that at all. American action, some of which could certainly be called intervention, encouraged the empires to disappear, but the Sleeping Beauty has not shown much appreciation of the lavish gifts with which the American Prince endowed her. For the truth is that it was not the empires that caused the weakness of the ex-colonial peoples as much as their weakness that had given rise to the empires.

America most needs a coherent theory of intervention in dealing with the underdeveloped world. Some of the newly-emergent peoples, if given support against other nations' intervention and if *not* propelled along unfamiliar paths in the name of development, democracy, or the latest fad from outside, may be able to build viable institutions. Some may even be able to build systems not too remote from the democratic model. For many, however, the post-imperial phase will merely be a transition to a new colonial era.

It is endlessly surprising, as one studies American experience over the last two decades in Asia and Africa, to see how little the United States has learned from the object lesson provided by Latin America. After nearly a century and one-half of independence, Latin America still shows little if any sign of following the North American model, and United States' intervention in Latin America since the Platt Amendment has almost always proved counterproductive. If it has proved impossible to manipulate a civilization with many of the same roots in European pioneering as the United States itself, how can one expect that Asian and African societies, belonging to quite different cultures, will respond to American guidance?

The Vietnam conflict best illustrates how little the Americans have learned from their experience in Latin America and how badly they need to formulate a theory of intervention compatible with the facts of modern international development. The idea that the Americans are conferring self-determination on the South Vietnamese by limiting the intervention from the North is plausible only by postulating two separate countries, one of which is being defended against the other. In fact the separation is a purely artificial one, and the intellectual case for American intervention is weakened, not strengthened, by pointing to it. Moreover, the

United States, lacking a plausible rationale for intervention in Vietnam, is making its task harder by claiming to be encouraging the South Vietnamese to take responsibility for their own affairs, and at the same time propelling them toward democratic forms of government for which they have no experience and for which they require a strong American presence. And yet the US constantly argues that its ultimate purpose is to return home, unmindful of what it is creating in South Vietnam. Limited-liability intervention and half-hearted imperialism, practised together, are not kindnesses to the countries in which they are practised.

The United States is not alone in making this kind of error and in paying for it, although the toll in Vietnam is exceptionally heavy. Britain, in her post-imperial phase, is also propelled into follies by similar intellectual inadequacies. For example, Britain rejected the Rhodesian Government's 'unilateral declaration of independence' because of the belief that post-colonial rule must be based on the enfranchizement of the indigenous majority and not prolong settler control. But Britain was not in a position to intervene by force, and so chose economic sanctions as the appropriate form of intervention. As a result, all Rhodesians have felt some hardship, Britain has suffered grievous economic losses, and the black population of Rhodesia is no nearer exercising political power than formerly. This demonstrates that unless intervention has a built-in likelihood of success, it is to be discouraged because it is more likely to increase than limit the sum of human suffering.

This does not mean that the United States is not entitled by historical precedent to intervene in a foreign civil war to prevent the victory of the side it believes inimical to its interests, or opposed to the cause it claims to represent. No doubt, with sufficient exertion of military strength, America can force a victory for its allies. But to prevent future Vietnams, the United States must clearly see the difference between the idealism of Woodrow Wilson and his successors and the realism required for limited liability intervention.

The West may put a high value on racial equality and democratic liberties and be prepared to make supreme sacrifices for them, but there must be some restraint on what may be imposed upon other peoples for ideal ends. The Soviets long ago ceased intervening to pursue purely ideological goals. As for the West, anyone who says 'better dead than Red' has all one's admiration, but this is not a choice one can make for others.

If we are to consider intervention seriously we must first clear our minds of cant, and above all, of the rhetoric of Wilsonian idealism. Too many people act from a strictly moral position before they have either ascertained the relevant facts or acquired the intellectual apparatus to deal with them.

The United States may be doomed to continue intervening abroad by virtue of its size, interests, and vulnerability, but if it is to avoid humiliations it badly needs a theory of intervention. The Russians and Chinese on the basis of recent experience must also re-examine their respective concepts of intervention; they at least seem aware of the inadequacy of their present theories. The West, unfortunately, does not seem aware of the need for a coherent concept of intervention, much less seem capable of criticizing Soviet and Chinese concepts. Our business must be to understand intervention in the context of modern international politics. This is likely to be difficult enough.

IV THE RUSSIANS

17 *Soviet Studies – and Russian Reality*

Western tourists have various motives for going to Russia; for many of them, especially among the younger generation, it is just one more country in which to have a good time. The great hopes and fears that Russia evoked a generation ago are forgotten. For them, Lenin's mausoleum is no different from Napoleon's tomb at the Invalides – another sight to tick off in the guidebook. My own difficulty in making the best of a first brief visit was of quite another kind.

For twenty years Russian history and politics have been a major academic interest of mine. The difficulty was to avoid seeing in everything a confirmation of the theories I had drawn from my reading. It was as though a zoologist specializing on the larger mammals from inside the four walls of his study had suddenly gone outside and found himself face to face with a live elephant.

For me the great fact of modern history had been the Russian Revolution; it was the Soviet system and its ideological base that seemed the proper object of study and reflection. And having originally approached the subject as a branch of international relations, I had become bemused by the Soviet-American rivalry and by the current tendency to see the Soviet Union and the United States as having much in common. I somehow assumed that the Soviet Union was another great power like the United States, but distinguished from it through its economic and political system and because of the domination of a different set of governing ideas.

This illusion was strengthened by the frequent public juxtaposition of similar military apparatuses and of such prestige-symbols as earth satellites and cosmonauts, by the assertions of Soviet leaders that their aim was to overtake and outstrip the

United States in material wealth and by the readiness of western experts to accept these claims in principle and to differ from the Soviet leaders (and from each other) only as to the time-scale to be applied to the process.

I was aware, of course, that this goal was still some way ahead – I was prepared to make every allowance for Russia's enormous losses during the war; I am ready to accept the assurances of Russians and of more frequent visitors that in respect of the material conditions of life and of personal security things are much better than they were even a few years ago; the fact remains that no one who actually experiences the Russian reality can accept the frame of reference within which such discussions are normally conducted.

Whatever may be said about armaments or the space race, these are the only levels at which a Soviet-American comparison is meaningful. The inhabitants of Moscow and Leningrad neither live the lives, nor can conceive of the aspirations or frustrations, of the inhabitants of Washington or New York any more than of London or Paris or Milan or São Paulo. It is not just the absence of material goods or of variety in what there is – whether it be foodstuffs, clothing, vehicles, consumer goods, medicines – or the fact of a housing shortage so persistent that the very notion of privacy may become obsolete; what is much more striking than the ubiquitous drabness is the lack of the sense of movement and bustle, of construction and renewal, of experiment and adventure that gives its peculiar quality to twentieth-century metropolitan existence.

Moscow, for all its six million inhabitants and new residential quarters, remains an overgrown sprawling village nestling up against an historic fortress; Leningrad, gradually recovering from the unparalleled horrors of its three-year siege, remains a breathtakingly beautiful memorial to that Europe of the age of absolutism which it was the mission of the Russian revolutionaries to sweep away once and for all.

It is not without significance that by far the most impressive examples of human ingenuity and effort to be seen by the visitor to Russia are the magnificently restored summer palaces of the Tsars, wantonly and systematically looted and destroyed by the Nazis and now patiently room by room, being recreated by Russian craftsmanship. They are monuments to Russian endurance as well as to Russian skill; but can they compensate for the apparent lack of any creative urge in a modern idiom? What can one deduce from the priority thus given to resurrection over construction?

Nor is it the case that all intellectual and spiritual striving is sublimated in some vast revolutionary urge. A few tired slogans on walls or building sites are all that attest the fact that one is present at the building of what was once heralded as a 'new civilization'. A tourist could easily forget that this is not just Russia but *Soviet* Russia. I now think that we shall have to consider the Russian present in quite a different way. We may have to demote the Revolution to yet another 'Time of Troubles'. I suggest that what we face today is what we have had ever since the Princes of Muscovy first created the nucleus of the modern Russian State; a divorce, such as exists nowhere to the same extent, between the State on the one hand and the people, society, on the other.

For political reasons the Russian State has always found it necessary to impose upon the Russian people an economic system that was bound to frustrate and bring to nought its inherent capacities. Once it was serfdom, then the curious amalgam of state and private exploitation that passed for capitalism in Russia: now it is the top-heavy bureaucracy of public ownership and central planning with its low productivity and widespread waste of human and material resources.

In order to achieve at least a degree of acquiescence in these conditions, the Russian State has always striven to isolate its subjects from the outside world, so that they should neither be aware of their own full potentialities nor be excited by new ideas of foreign provenance. To do this in earlier centuries proved hard enough – from the Decembrists to the Bolsheviks, foreign stimuli were inherent in every attempt to change the shape of things. To do it now in our world of instant communications is a feat more considerable than anything else achieved by the regime.

Indeed, for the reflective tourist the other thing that makes Moscow and Leningrad so different from the other metropolitan centres of the mid-twentieth century is precisely this sense of being cut off from the normal intercourse of men and nations. Access to world newspapers and magazines, to the international literature of the paperback – these are things the traveller takes for granted until he finds that he, like his hosts, must do without them.

The Soviet State is devoted to the manipulation of society for its own purposes, not with the good opinion of foreigners nor with domestic popularity – and this is traditional also. Would not a more sensitive organism feel a double sense of shame in reintroducing the special shops at which for foreign (capitalist) currencies goods can be obtained which for the Soviet citizen are unobtainable altogether or obtainable only at much higher prices? And we at least know, as the ordinary Russians do not, of the Canadian wheat purchases that explain the preoccupation with obtaining foreign currency.

As for the Russian people, their strength lies in being able to live with any system, to display in the face of the irritations and deprivations of daily life the same patience that brought them through the greater ordeals of war. For the visitor contrasting the unfailing warmth and kindness of ordinary people with the impervious rigidity of the authorities, historical and literary memories take charge again. The recognition of continuities does not preclude the possibility of change; but it is most unlikely to take the form of an approximation of the Soviet system to that of the industrial West. It needed the shock of actual contact to show me how far apart the two worlds still remain.

18 Before the Flood

Between ourselves and the reality of pre-revolutionary Russia there exists the barrier of the Revolution itself. In the intervening half-century much has been written to explain its coming, and most of the historical writing about the preceding century has been devoted to an attempt to explain it. Furthermore although much documentation has been produced at various times during the Soviet period, historians have perforce been influenced by both the self-exculpatory memoirs of exiles and the hagiography of Soviet writers. It is only with distance that it becomes possible to see clearly and to describe without passion an important era in the history of an important section of the human race.

Professor Seton-Watson has worked long and conscientiously to provide for the English-reading public a full-scale narrative of Russian history between the accession of Alexander I and the abdication of Nicholas II.[1] It is designed on comprehensive lines and makes few concessions to those more interested in opinion than in facts. Economic life is dealt with in statistical detail; changing administrative structures are painstakingly analysed; diplomatic relations set out in such detail that occasionally they go rather beyond what the historian of Russia as such might need; military campaigns are not ignored. A particular strength of the book lies in its full and on the whole sympathetic treatment of the non-Russian minorities – we are never allowed to forget that it is an empire we are dealing with.[2] Although due attention is paid to the contribution to the story of notable individuals, full personal portraits are rarely attempted; anecdotes are kept to the minimum;

[1] Hugh Seton-Watson, *The Russian Empire 1801–1917* (Oxford, 1967).

[2] Perhaps I may be pardoned for the comment that neither the text nor the ample bibliography pays enough attention to Russian Jewry, which was not simply the victim of pogroms.

the tone is austere and scholarly; the author's own judgements and interpretations are largely reserved for a brief epilogue. Yet Professor Seton-Watson has a point of view and it is both personal and typical of the present moment with our obsession with 'development' and 'modernization'.

Previous historians have approached Russia with the yardstick of the West or of some idealized vision of Slavdom. Professor Seton-Watson is familiar with the new worlds of Asia, Africa and Latin America where the Russian experience is being lived through again in part and where the Russian example is part of the mental background of many leading figures. Not many people have the capacity to look at Russia from this point of view, and fewer have the detailed knowledge to make much of this approach. It is one of Professor Seton-Watson's strengths that he knows a good deal about Japan and we are often asked to contemplate the successes of Japan (for instance, in agriculture and in the diffusion of education) with the failures of Imperial Russia. The comparison is both apt and illuminating. Such assets make up for what is perhaps a weakness in Professor Seton-Watson's equipment for his task, namely the fact that he clearly finds the subject as such unattractive. He admits to the glories of Russian achievements in literature and the arts, and to the devotion and self-sacrifice of men on both sides of the political divide; but fundamentally it is clear that Professor Seton-Watson finds Russia and the Russians unsympathetic. He is no Russophobe but he is no Russophile either.

In part this may be the inevitable reaction of a Western scholar whose professional interests oblige him to take cognizance of the Soviet regime which controls the sources upon which he would wish to work, and the writings and contacts of the Russian scholars with whom he would wish to collaborate. A government capable of bottomless mendacity, and historians who have no choice but to bend to its directives – these are unlikely to stimulate warm feelings among foreign intellectuals. It must be hard to give adequate expression to the denial of individual freedoms by Tsarist governments and yet retain clearly in one's mind the fact that for most of the period, and especially in its final years, Tsarism was in the sense of limitations placed upon intellectual and cultural pursuits almost infinitely 'liberal' compared not merely with Stalinist Russia but even with the Russia of the 'thaw'.

Because Professor Seton-Watson deliberately breaks off his narrative at the February Revolution, one may inadvertently overlook what seems to me the most striking feature of the book, namely its contribution towards understanding what came afterwards. The Revolution, the Civil War and collectivization – events which like those of the Tsarist era itself already belong to the historical past – seem more and more to resemble a mighty flood. As the waters recede, much destruction is visible, some rivers even have changed their course, but the great landmarks remain.

What are the characteristics of modern Russian history as Professor Seton-Watson teaches us to see them? First of all, the fact that modernization and developments were imposed from above and did not for the most part, as in the West, develop from the autonomous action and private strivings of social groups or individual leaders of them. Peter the Great made so rapid and successful a beginning that by the end of the eighteenth century the gap between Russia and the West was much lower than it had been. During and after the Napoleonic wars Russia was as 'European' as she has ever been. Then with the freezing of social advance and intellectual development after the failure of the Decembrists – the last Russians in whom Professor Seton-Watson sees some hope of development along Western lines – the gap widens again. Another movement to catch up is precipitated by the defeat in the Crimea – but the era of reforms is stultified by its bureaucratic and non-popular character and renewed reaction follows. In the final decades of the Empire, important progress is made on the economic front; but the unwillingness of the regime to allow this to have its natural consequences in the spreading of social and political responsibilities even among the loyal, together with the relentless policy of Russification of the subject peoples – even when like the Finns or Armenians they had no other reason to be anti-Russian – render this progress inadequate to stand the strains of large-scale war and catastrophic loss.

In the second place, just because modernization and development came from above its incidence was unequal. Industry went ahead – with foreign capital to help – while agriculture languished. The peasant problem remained the Achilles' heel of the regime, and the peasants in uniform who were the soldiers in the mutinies in the capital in February 1917 were the essential element in the fall of the regime.

In the third place, because of the fears of the monarchy and its servants, the development of education was deliberately stultified so far as the masses were concerned. The resulting gap between the intelligentsia and the masses explains both the impossibility of a gradual and constitutional approach to modifying the regime and the particular nature of the revolutionary movements themselves. It is because of the nature of educational development as well as because of the lop-sided economic structure that Russia did not develop a true bourgeoisie, and without a bourgeoisie there could be no liberalism. An intelligentsia is not a substitute. The educational reforms of the final decade of Tsardom were again too late.

Finally, and most significant of all, for reasons that take one away back beyond 1801 to the very origins of the Russian State – religious as well as secular – the Russians had no deep sense for the importance to any but the most primitive of societies of the idea of law and of legal and constitutional processes. The autocracy and its servants – even those with some genuinely progressive and intelligent ideas like Stolypin – could never allow themselves to be bound by constitutional requirements, not even by those of their own devising. But the opposition also was much more influenced by ideas of translating the general will into action than by the Western view of the State as holding a balance between conflicting but legitimate interests. Again, there was in the Russian intelligentsia a commitment to equality so extreme that absolutism itself was generally preferred to any kind of oligarchy – and it is hard to see how in a developing society there can be any progress towards democracy without an oligarchic stage such as every successful Western society has known at one time or another.

It is of course much more difficult to write about the history of the Soviet period than about its predecessor; the materials are lacking to a much greater degree. Thus Professor Seton-Watson can show how within a bureaucratic regime like that of Tsardom, individual ministries may come to stand for different policies and the interests of different social groups that have no more public forum for pressing their views. It is obvious that much could be written in terms of the different outlooks of the Ministry of Finance and the Ministry of the Interior; and it may be that one day one will be able to make some progress towards understanding Soviet (or even contemporary Chinese) history in these terms. But what is clear is that all the main distinguishing features of Tsardom survive often in an exaggerated form to characterize its successor. We have had fifty more years of development and modernization 'from above' – with political and military rather than welfare considerations as the dominant ones. We have had the same preference for industry over agriculture, and the same consequential exploitation of the peasant. Russification has not ceased to be the ideal even if it is a secular rather than a revealed religion that is to provide its content.

There is one important difference – the general acceptance of the need for a much more equal access to education at all levels. Thus the Soviet regime has carried out what was inherent in the policies the final years of Tsardom; though we clearly can never know whether the process would have been allowed to work itself out. The gap between an educated élite and the masses which Professor Seton-Watson sees as the main weakness of all developing societies is largely absent from the Soviet scene. But for this achievement a high price has been paid; not only is there no bourgeoisie to carry the seeds of liberalism, there is not even an intelligentsia, except in the official Soviet mythology. For an intelligentsia cannot function if its livelihood and means of expression are wholly within the control of the State. The price of equality has been enforced uniformity and the price in human terms – perhaps in social terms – is high as well.

A book with so much work and thought packed into it will answer different questions for different people. It has answered a personal one for me. Between the wars I knew in London an elderly and scholarly gentleman, one Jacob Schapiro, a Jew from the Baltic provinces, who had been a member of the Second Duma. He devoted his exile to assembling and studying a library of English history and literature with a strong concentration on the seventeenth century and its constitutional problems. (Many of his books which I ultimately acquired line the room I write in.) I used to wonder vaguely why a Russian should be so interested in this particular period of the history of a foreign country, but I was too young and shy ever to inquire. I now realize that the reason must have been precisely that which underlies Professor Seton-Watson's treatment of his theme. It was just this element – the serious and earnest examination of points of legal and constitutional development, a serious concern over individual liberties and the means for safeguarding them – that was lacking on both sides of the great conflict between Tsardom and the Revolution. This concern has been a central element – the central element – in English history. It may explain why an earlier and more famous Russian constitutionalist Paul Vinogradov spent his years of exile on studying even earlier periods of English history. What would have distressed both Vinogradov and Jacob Schapiro would have been to see how little we seem to care about such things in England – how the cruder concepts of egalitarianism and statism have become the fashionable thing among our own intelligentsia. It is all very fine to make fun of the Whig interpretation of history – it was having no Whigs that was Russia's essential tragedy.

19 A Neglected Historical Novel

For those of us who live in the provinces, perhaps even for megalopolitans too, the most civilized and civilizing institution in London is the London Library. Not the least among its amenities is the half-shelf in the entrance hall of newly acquired or recently returned novels. It introduces into one's reading, all too likely to be disciplined by the demands of one's 'subject', an agreeable element of chance. It was there that I picked up *The Witnesses.*[1] It had been published a year or more back, and absence abroad must have made me miss the original reviews, since its title awakened no echo in my mind. Indeed I am in retrospect a little surprised that I added it to my load, since on the whole I dislike 'historical novels' finding the real past sufficiently difficult to penetrate without having to worry about imaginary ones. And here clearly was an historical novel – a long and elaborate work spanning the history of Russia between 1903 and 1917 – and making precisely the demands upon the reader which theoretically I most dislike. However for some reason I allowed myself to borrow it and to begin it, and since then it has become for me one of those books which form a reference point in one's thinking. It would be absurd to say that it has altered my view of the Russian Revolution; but it has provided some concrete symbols for what would otherwise be a mere abstract set of convictions. I feel now that I know better – just as I knew better after reading *Dr Zhivago* – why it is that despite all the cruelties, injustices, and even cynicism of Tsardom, despite the horrors and futilities of the long-drawn-out agony of the war, the Bolshevik Revolution (Lenin's *putsch* in October, so often confounded with the earlier democratic revolution of 1917)

[1] M. W. Waring, *The Witnesses* (London, 1967).

remains for me the major catastrophe of our century, the breach in the walls of civilization through which so many other bands of barbarians have poured or are pouring.

For as *The Witnesses* makes plain, not by argument but by the selection and narration of incident, what was defeated in the Revolution was not the old regime which had already collapsed but the ordinary instincts of decent humanity expressing themselves in rating human relations and human loyalties higher than ideological conviction and in an affection for the beauties and decencies of ordinary material life which meant nothing to the revolutionaries and which their victory manifestly destroyed. And it is significant that the pseudonymous author – it seems to be accepted that it is a woman writing – passed her childhood in St Petersburg, that always beautiful and now sad city, which is the background, lovingly and carefully delineated, for much of the narrative.

We are told that twenty years of research including interviews with the 'witnesses' of the title have gone into the making of this book; and one can well believe that if the final outcome haunts the reader the writing of it must have been a painful if cathartic experience. To spend twenty years in the company of Lenin and his friends even if only in imagination, to see them take power amid the orgy of brutality and contempt even for their own associates and accomplices that marked the initiation of Bolshevik rule – this cannot be anything but a spiritual ordeal. To some Miss Waring's prose style might seem formal and mannered; but one feels that something like it was necessary if the horror of the subject-matter so gently introduced with its scenes of rural Russia and high spirits among political exiles was not to overwhelm the writer and her readers with the pity of it all.

Miss Waring is explicit about the way in which she has treated the technical as opposed to the emotional problems of writing a novel of this kind. This is always the question of whether dealing with great events one should bring in, and therefore add a fictional dimension to, personages known to us from the history books themselves. Miss Waring assures us that all her characters whether 'historical' or not are drawn from life, and that what is told about them could be documented. But in the case of principal figures, the members of a Russian aristocratic family seen through the eyes of a young American woman who marries the heir, and the illegitimate son of another such family who forms the link with the revolutionary movements with whom he is implicated through sympathy, and involved through his relationship with Lenin's putative mistress Inessa Armand,[2] but who can never sink his identity in the cause and is ultimately its unknowing victim, the actual prototypes have been concealed. Lenin himself and others of the revolutionary circles – Trotsky, Plekhanov, Martov, Bonch-Bruevich, Kerensky, and Inessa herself – appear under the names they used in the underground or in emigration. Other historical figures appear under different names from those they bore, or must be regarded as merely convenient inventions for the purpose of the narrative.

This device enables the author to enjoy a freedom which she would not otherwise possess; incidents of the biographies of the real historical personages are altered or transmuted or ascribed to dates which do not accord with those of formal biographical record. And this naturally provided a handle for those for whom the whole purport or meaning of the book was an offence. It was easy for the *Sunday Times* reviewer (himself coyly using initials) to point out that the character who stands for Trotsky is not the historian's Trotsky. But the criticism itself is beside the point.

If one wants to know what people actually did and said, the circumstances of their lives, then the biographer and the historian reign supreme. And even here one needs to take care when the Russian Revolution is the subject. Was it not that eminent scholar the present Master of Balliol who contrived to write a life of Lenin in which the Bolsheviks manage to win the Civil War without the help of Trotsky?

What the novelists can offer – and perhaps only through the novelist's techniques – is some sense of what the instruments of the historical process were like as men, and what the historical process felt like to its victims. Perhaps the latter is the more important; and perhaps the historian is driven to accept the views of which Mr E. H. Carr has been so vigorous an exponent, that history deals only with success, with the movements that proved on the right side in the historical dialetic, and that the defeated, the victims belong to 'the dust-bin of history' (to use a favourite Lenin phrase). The novelist is not bound by such a view and for Miss Waring the weaknesses that led to defeat are as worthy of interpretation as the qualities that made for victory – the humourless

[2] See Bertram D. Wolfe, 'Lenin and Inessa Armand', *Encounter*, February 1964.

self-dedication, the utter repudiation of ordinary moral values, and the commitment to total violence and total duplicity that were the key elements in Bolshevik tactics.

It may be that no one was duped by Inessa's use of her sexual attractiveness to provide financial assistance for the underground in exactly the way in which it happens to the central figure in the novel; but the willingness to manipulate human beings and cast them aside once their task had been performed was an element – and an important element – in the revolutionary mentality. And it was not only human instinct and emotion that was manipulated; it was idealism as well. The true idealists of Revolution, all those who could not subject themselves totally to the will of Lenin and his clique – the favourite Soviet term of abuse is in order here – were cast aside and eventually in most cases 'eliminated'.

But of course, the main reason why the book angered several reviewers has nothing to do with the liberties taken with historical data. There is after all a deeply seated view (and one not restricted to 'the Left' – the *New Statesman's* review was actually one of the most perceptive) that the Russian Revolution was 'a good thing', or (in the Master of Balliol's words) 'a movement of the poor and oppressed of the earth who have successfully risen against the great and the powerful.' And any other vision – even in fiction – must be stamped upon.

The writer in *The Times Literary Supplement* is, as one might expect, the most vocal on this score. Unblushingly asserting that the Bolsheviks got no money from the Germans (for the purity of Lenin is not to be called into question), he goes on to accuse Miss Waring quite untruly of minimizing Russia's war-time losses – one suspects some skipping of the text – and then comes to the heart of the matter. Although Miss Waring can be praised for her portraits of members of the old society – 'their devotion to the regiment, their worship of the Tsar, their ludicrous [*sic*] patriotism' – she ignores the masses. She seems to believe that 'the revolution was brought off by a handful of unscrupulous engimatic conspirators.' And this is intolerable. In the words of Mr Irving Wardle (writing in the *Observer*), Miss Waring 'conveys practically nothing of the undercurrents of the Revolution.' Or as the *Methodist Recorder* observes: 'Scant reference is made to the poor. Who suffered under the Tsarist régime? Why was revolution inevitable? . . .'

One has a feeling as one reads comment of this kind (and even some of the more favourable reviews are disquieting in their unwillingness to accept the book not just as a novel but as an historical novel) one is tempted to question not Miss Waring's handling of the literary problem but the function and fate of historical research itself.

> The strikes, the 1905 revolution, the creation of the Douma are almost ignored and it is strange that when an authentic prime minister was assassinated in 1911, M. W. Waring should invent the murder of a Minister of the Interior at about the same time.

So the *Oxford Mail*. In fact, there are good reasons in a novel both for the transposition of the assassination and for not dwelling on historical events, well known in themselves and unrelated to the novel's theme. What is wrong is the obvious failure to grasp the light which both increasing distance and the fruits of scholarship have now thrown on the Revolution, its causes and consequences. Professor Leonard Schapiro, Professor Hugh Seton-Watson, Mr George Katkov, not to speak of the many American scholars in the field, might just as well not have written for all their impact on Miss Waring's critics.

Clemenceau's demand that the French Revolution should be taken *en bloc* has always been regarded as a political statement, not as an historical assessment. The French Revolution was a multiple event and so also was the Russian Revolution. To say that it was 'inevitable' is to say nothing, unless one means that it would have been an almost unimaginable anomaly if the structure of the Tsarist régime, even in its post-1905 version, could have remained unchanged into the last third of the twentieth century. It could also be argued that changes were more likely to take a revolutionary than an evolutionary form. But to say that this means that it was inevitable in some sense or other than the sense in which all history is inevitable because it has happened – that the outcome of the Revolution could only be the seizure of power by the kind of man Lenin was, and the establishment of the kind of regime Lenin established – is to make the kind of judgement that makes historical writing quite otiose.

Could one seriously imagine that Miss Waring, brought up in Tsarist Russia and devoting herself to its study, was ignorant of the fact that there was much poverty and oppression? If one did the text itself would disabuse one; the sketches of domestic servants and their role – particularly that of the female domestic servants – would be enough to disabuse one about social relations being of some wholly idyllic kind. (*Pace The Times Literary Supplement,* Miss Waring's peasants are not just 'a contented bunch who dress up exotically on St Vladimir's day.') She has not found it necessary to dwell on the social background any more than on the political or military background because for her purpose it can be taken for granted.

Her question is – as the historian's should be – granted a revolution, why *this* kind of revolution? And that is important. Mr E. H. Carr is right. The victors survive; the vanquished disappear. What has followed in Russia and the rest of the world has followed from that fact. We suffer in our perspectives from the shadow that Stalin casts across the earlier history of Soviet Russia. Yet is there anything that Stalin did, any element of cruelty or treachery in his behaviour (or in that of his successors) that could not be paralleled from Lenin's own behaviour or deduced from Lenin's own morality? Miss Waring's creatures of the old society, good or evil – and her half-hearted, perplexed, anxious, foredoomed hero – are not just victims of an abstract historical process. The murder of the defeated officer cadets, the last defenders not of Tsardom but of the Republican Provisional Government be it noted, was not just part of 'the dialectic'; and Miss Waring is right to evoke the physical nausea induced by these events if only to de-glamorize the Ché Guevaras and the other progeny of Leninism.

Men (and even some women, like Inessa Armand) consciously let loose upon the world once more an alternative method of settling political debate – terror. In so doing they did not give a new heaven to the poor and oppressed but postponed their emancipation. Having no moral sensibility, the Russian Bolsheviks cauterized the sensibility of others. The world has been a poorer and bleaker and more dangerous place because Lenin lived. Historians have given the material for forming our judgement on this man; but perhaps only fiction can help us to understand the history. Pasternak is a finer writer than Miss Waring; but those whom *Dr Zhivago* made uncomfortable are uncomfortable with *The Witnesses,* and that too is praise.

V THE JEWISH PREDICAMENT

20 *Zionism as Nationalism*

Nationalism, one of the prime motors of modern world politics, has a large literature devoted to it; Zionism also has not lacked attention. But on the whole there has been little systematic effort to treat Zionism as primarily a remarkable example of nationalism and as subject to the same historical laws as the nationalisms of other peoples.

Now that nationalism is evolving – embraced by formerly acquiescent peoples but repudiated at least in principle by some of the more advanced in favour of wider groupings – it may be appropriate to ask whether Zionism is likely to be affected by the same changes in the environment as have brought about this more general evolution. Just as the early exponents of Zionism were on the whole more impressed by the uniqueness of what they were doing than by the parallels that could be drawn between what Jews were demanding in the secular sphere and what other submerged nationalities – Czechs, Poles, Ukrainians – were claiming, so now the important contest over the future relations between Israel and the Diaspora is fought out in exclusively Jewish terms. Who, for instance, thinks of comparing it with what is being said about the likely effect on the French Canadian diaspora should Quebec achieve independence of the rest of Canada?

There are a number of reasons why there should have been this inward-lookingness in the past. Zionism has largely interested either Jews (both pro- and anti-Zionist) or anti-semites. Both groups have been more concerned to manipulate opinion than to establish a logical framework for their views.

Historically, too, Jewish nationalism in the Diaspora tended to meet with the hostility of the other new nationalisms whose objectives could in part only be obtained (especially in the economic sphere) at the expense of the Jewish communities in the countries concerned. In virtually all the nationalisms of Eastern Europe there was some degree of anti-semitism; and it is therefore not surprising that the Jews should not have felt disposed to compare their case with those of their neighbours. What happened in German-occupied territory in Eastern Europe during the second world war and what evidence there is of popular, as opposed to officially sponsored, anti-semitism in the Soviet Union today suggests that this factor remains of importance.

It is even more understandable, and perhaps in the upshot more disastrous, that the Zionist pioneers failed to take into account the inevitable growth of an Arab nationalism following in the path of European nationalism, and made no serious plans to come to terms with it. It may be that reconciliation of the objectives of the two movements was inconceivable; the point is that few people seem to have thought that this would matter in the end.

The only way in which it would be seriously possible to argue that the fate of other nationalisms was, and is, irrelevant would be to take a strictly religious view of the Zionist mission. If one says that the Jews are strictly a religious community and a people only in virtue of their religious adherence, that their return to the land of their ancestors is merely a fulfilment of the Divine purpose as set forth in the prophets, then of course all parallels break down; for no other nationalism has been justified exclusively in this way, although nearly all nationalisms have had some religious component and in most of them, except perhaps in Africa, it has been rather important.

But it is difficult to argue this case in the face of the facts. In the first place, the most strictly religious leaders of Diaspora Jewry at the time of the birth of Zionism tended to regard it as a departure from, rather than the fulfilment of, the Divine will. The return to Zion could not, they held, be brought about by the methods of secular politics. In the second place, the majority of the population of the State of Israel itself (which is the fulfilment of Zionism) would not accept a religious definition of their national status, and is indeed rather determined in its secular approach to the whole question. For most Israelis, the State of Israel is a state like any other – and to the extent that it is not like any other, this is not for religious reasons but because of its two-way relationship with a Diaspora much larger than itself.

It would therefore seem more in accordance with fact to accept the view that Zionism is best understood as one of the many examples of the way in which the changed circumstances of the nineteenth and twentieth centuries – changes in economic and social life – impelled successive groups to seek a status of political independence.

What distinguished Zionism was that the group concerned was defined by a religion and by a culture largely conditioned by that religion and the experience its followers had undergone, that the ethnic factor was derived from the religious rather than being based on actual or presumed biological inheritance, that a common vernacular either did not exist, or where it existed (as in the case of Yiddish) was repudiated for nation-building purposes, that the land where it was hoped to build the State was only to the most limited extent inhabited by members of the group concerned.

All these distinguishing features, except the last, were shared with other groups moving at the same time towards national self-consciousness and national self-assertiveness.

The interplay of land and religion would seem to indicate the way in which the case of Zionism today differentiates it from other nationalisms which have achieved their immediate objectives, and which are groping towards methods of transcending them through wider groupings which far better reflect the imperative demands of modern technology and modern economic life.

The land is of course too small for the people if we take all those whom the world agrees to count as Jews. Even if the idea that the rest of mandatory Palestine is an irrendenta – and the idea is a dangerous one and easily exploited by Israel's enemies – it would still be too small. Other nations have irredenta – and these are also dangerous. Many of them have a diaspora outside their borders. But the difference in most cases is that while they may have useful links with their Diaspora they scarcely expect its long-term survival.

The traditional answer of Zionism has been that the Jew of the Diaspora would gain in stature and self-respect from the existence of the Jewish State and from its achievements. It would prove that the occupational and temperamental peculiarities of the Jews of the Diaspora were due to the disabilities imposed upon them and not to their inherent inability to be farmers, soldiers, sailors or anything else that the rest of the world thought honourable. One would suppose that in the Western world this belief has been justified. (The Communist case is too twisted by ideology for the test to be a fair one.)

But it is not clear what kind of Jews these Jews were going to be; how were they to be defined? Clearly not as resident aliens of the countries in which they lived. Indeed the Balfour Declaration pledged the contrary. As a religious group? – but most Zionists rejected such a definition and indeed in countries where religion was increasingly a private rather than a public matter it would be hard to ascribe meaning to such a definition.

We thus come to a situation in which one purpose of Zionism as a movement is to endow the Jews of the Diaspora with a secular culture of which the foundation must clearly be the modern Hebrew of Israel and its literature. To do this for those who regard themselves as destined to live in Israel is an obvious measure of precaution; but for those who propose to spend their lives in Western Europe or North America or Australia, can it really be successful in the long run?

Israel is, of course, for its size a country of quite extraordinarily rich cultural achievement – but can Israeli secular culture continue to compete with the inevitably richer and more varied cultures of the United States, England or France in the case of people living in those countries and following the same occupations, and exercising the same political rights as other citizens? What would be the effect on the nationalism of even the most tolerant host-nation of a rival *secular* culture in its midst? Again one has to argue for a special providence – and if one comes as close as that to the religious argument, why not accept it altogether?

What of the main mutation of nationalism away from the nation-state as the ultimate goal? Almost everywhere one sees movements, however embryonic, unsuccessful and even in part fraudulent, whose purpose is to create larger unions superseding the nation-state. With the exception of the Commonwealth, which might once have provided a supranational grouping for Israel to belong to, but is now itself under strains that make this impossible, these have all got a geographical basis – though ethnic, religious and ideological arguments may all be used to fortify the case. Clearly the same consideration must apply to the area of which Israel is geographically and historically a focal point.

It is a tragic irony that the Jews in Israel have achieved the dignity of a nation-state at the very moment when the entire concept of the nation-state has become outmoded. It seems hard to believe that some degree of unity in the Middle East can forever be postponed by the greed and folly of Arab rulers. But such unity achieved without Israel would put Israel in mortal peril. In other words, there must be a sufficient mutation in all the nationalisms of the Middle East, including Israeli nationalism, to make some supranational institutions possible; the alternative – as other groupings are realizing – is subordination to external forces. How Zionism can adapt itself to this new task upon which, in the long run, the survival of its existing achievements are based is by every standard the major question now confronting the Jewish people.

21 Rootless Cosmopolitans?

Why, one begins by asking, should there be an 'Association of Jewish Graduates' in London? It suggests a desire to emphasize or promote the quality of Jewishness in its members alongside the more tangible fact of possessing a university degree. Unlike most societies that Jews may belong to – philanthropic organizations of various kinds – the Association seems to have no purpose other than the common pursuit of heterogeneous activities of a cultural or recreational kind. Why, then, as Jews?

Such Jewish self-consciousness and the questions to which it gives rise has clearly been a consequence of the emancipation of European Jewry and did not exist before it. Before the emancipation, Jews throughout the world were members of a religious group whose close-knit communal existence was the product at once of the commandments of their religion and of the pressures of the external world. It was a position the easier to maintain and understand because most of the societies in which they lived were societies of a similar kind. The antithesis of Jew was not Englishman, Frenchman, Turk or Arab, but Christian or Moslem.

It was thus easy for Jew and non-Jew to have a quite concrete picture of Jews and Jewishness. When Shylock says: 'I will buy with you, sell with you, talk with you, walk with you, and so following; but I will not eat with you, drink with you, nor pray with you', he is defining Jewishness in a perfectly understandable way in which the key word is 'pray' as Shakespeare well understood when he made Shylock explain his grudge against Antonio by saying 'he hates our *sacred* nation.'

After emancipation it was more difficult to explain a community of this kind or indeed to maintain it in face of the increasing secularization of the host society, and out of this in turn came the modern search for a Jewish identity. For someone like myself, brought up as a Jew but in an atmosphere wholly penetrated by the values of the Enlightenment, not being (in Zangwill's phrase) 'a child of the ghetto' nor even strictly speaking a grandchild, but perhaps rather a great-grandchild, the question must always be a difficult one. For those of my age-group the Hitler holocaust cut across and perhaps deflected what would otherwise have been a further mental evolution. If six million die because they are Jews, one cannot say that Jewishness is not a serious matter. The final destruction of the ghetto must have a special significance for the great-grandchildren too.

But to be sensitized to a subject does not necessarily enable one to speak of it from the inside; it is easier to begin from the outside and with the obvious fact that one thing all Jews have in common and in abundance is enemies. Let me begin with a phrase that encapsulates the charges made against Jews in modern secular societies: *rootless cosmopolitans*. This is not of course the only way of defining the hostility felt towards Jews; anti-semitism is almost as complicated a concept as Jewishness itself. There are its religious roots, and on the other hand, the secular antithesis of religious anti-semitism – the hostile reaction of those who having urged Jewish emancipation in the name of an anti-clerical secularism objected to the fact that the Jews refused then and there to merge with the rest of the community.[1] We thus get a secular (or 'Left') anti-semitism which has its progeny in Marx and the Marxists who regard the Jewish fact and the anti-Jewish fact as deriving from economic and social conditions, destined to disappear as these conditions are transformed by the proletarian revolution.

Marxist anti-semitism is only partially the result of an intellectual position; it is linked up with popular anti-semitism, with the undoubted fact that in peasant societies Jews could be regarded as part of the exploiting class and the easiest target among exploiters. Popular anti-semitism should present no great intellectual problem today, since we see the similar antipathy manifested by newly enfranchized Africans against the Asians in their midst. But this illuminating parallel is often glossed over, for while the Western liberal intelligentsia was not expected to sympathize with the Ukrainian or Polish perpetrators of pogroms, Africans have to be handled more gently. Finally, of course, and cutting

[1] See the pioneer study by Arthur Hertzberg, *The French Enlightenment and the Jews* (New York, 1968).

across all these intellectual constructions there is the deep psychological tendency among human beings to reject and fear the unfamiliar and alien and make it the focus of their buried mistrusts and violent antipathies.[2]

In this context the recent Soviet synonym for Jews, 'Rootless Cosmopolitans', becomes something worth following up, and this for practical reasons as well. Very large numbers of Jews are directly affected by Soviet anti-semitism; the Jews of the Soviet Union itself, the survivors of the holocaust in Soviet-controlled Eastern Europe – as recent events in Poland remind us all too clearly – and also if one believes that Soviet hostility towards Israel is an aspect of Soviet anti-semitism as well as of Soviet *realpolitik* – the Jews of Israel itself.

The phrase can also, however, serve as a clue to our proper search, that for a Jewish identity. In one of his essays Isaac Deutscher records his emotions when Ben Gurion gave vent to his feelings about non-Zionist Jews: 'They have no roots, they are *rootless cosmopolitans* – there can be nothing worse than that.'[3] Despite Ben Gurion's rather lame effort to unsay the phrase, there is clearly within Jewish thought an echo of the most pervasive form of modern anti-semitism. Deutscher's own perception of this fact is both very moving and very revealing. For he himself represented one way of trying to escape from the dilemma of Jewish self-consciousness, acceptance of the Marxist analysis and devotion to the fulfilment of the Marxist vision. Not for nothing was the Jew, Lev Davidovitch Bronstein, the subject of his *magnum opus*.

His commitment to Marxism made it necessary to interpret the events of his own life-time in order to fit a rigid doctrine from which they appeared to be escaping. The ingenuity he showed in doing so may perhaps without fancy be ascribed to his early training in Talmudical studies. But it is an ingenuity which is in the end unconvincing. In one essay, for instance, Deutscher maintains that the Nazi phenomenon in no way invalidated classical Marxism:

> Nazism [Deutscher writes] was nothing but the defence of the old order against Communism. The Nazis themselves felt that

[2] The work soon to be published of Professor Norman Cohn and his colleagues at the University of Sussex should throw much light on this.

[3] Isaac Deutscher, *The Non-Jewish Jew and other Essays* (Oxford, 1968), p. 92.

> their role consisted in this; the whole of German society saw them in this role; and European Jewry has paid the price for the survival of capitalism, for the success of capitalism in defending itself against a socialist revolution.

No serious student of nazism not blinkered by ideology could possibly accept this view. Nazism did not regard itself as defending capitalism; its ideological roots went back to a time before there was any question of a communist revolution; nor was Germany threatened by one in 1933. If Deutscher had to ignore facts in order to retain his Marxism where Germany is concerned, it was even harder in the case of Russia itself. How did anti-semitism survive when the Communist revolution had taken place; why was anti-semitism used and usable as a weapon by the Soviet regime as by its Tsarist predecessor? To explain this once more by 'survivals of capitalism' is to strain dogma to breaking-point.

But when Deutscher was not involved in these ideological gymnastics, his genuinely cosmopolitan culture overlaid upon his basic Jewish experience enabled him to see – as his essay on Marc Chagall bears witness – how much the contribution of Jews to contemporary civilization has been the result of their cosmopolitanism itself – the result of their having lived and worked on the margin of two societies from one of which they had not fully escaped and the other of which they could not fully enter: Spinoza, Heine, Marx, Rosa Luxemburg, Trotsky, Freud. But could not exactly the same generalization be made about those whose energies were directed back towards their own people – Herzl, Weizmann, Jabotinsky?

Deutscher does not consider this possibility because he regards the achievement of the latter group as something alien to him. While he found some aspects of Israel exhilarating, others repelled him. His Marxist categories obtrude themselves again. Just as in feudal and post-feudal Europe, Jews were objectively exploiters whatever their own vision, so an Israel dependent upon the capitalist West is objectively 'neo-colonialist', and rightly suspect to the Arab masses. The renewal of the Jewish presence in Israel could only be justified if Jews and Arabs worked hand in hand to create 'a socialist Middle East.' What happened after 1945 to the hopeful Jewish Communists of Eastern Europe is not alluded to.

Deutscher's talents were remarkable, but his dilemma was that of other Jewish Marxists. It was possible to solve it only by postulating a society in which rootlessness would not matter and cosmopolitanism be a matter of course: not the bourgeois society that first attracted men like Herzl and then was found wanting by them, but the socialist society of the future. Communism and Zionism have their origins in the same experience. After their initial divergence they come together in the latest victims of antisemitism where Jews in Poland who have for two decades collaborated in 'building a socialist society', and who have repudiated any Jewish allegiance or even interest, are branded as 'Zionists' and agents of an international conspiracy. 'Rootless cosmopolitans' and 'Zionists' come to stand for the same thing.

It is too soon fully to understand this tragic paradox. What we do know is that in the Revolution in Russia, its humanitarian side was overborne by its doctrinaire and totalitarian side largely because the latter could exploit the enormous and primitive strength of Russian nationalism. The bureaucratic elite which rules Russia has always been alert to elements which were unassimilated or even unassimilable. Any element in the Soviet world which has any external point of reference is automatically suspect. Jews cannot be totally indifferent to the existence of Israel, however tenuous their Jewish links, and this is sufficient. In Poland, a national state rather than a multi-national empire, we may be witnessing the final extrusion of the last non-national element; though Soviet pressure may have counted for something. In the Czech case, the sympathies for Israel expressed in part no doubt an anti-Soviet reaction; and the consequences for the few remaining Jews in that country were fully predictable.

But the way things have gone in the Soviet Union and its satellites does not mean that the more human aspects of the Russian Revolution could not have a practical outcome in more favourable circumstances; and the place where they have been put into practice is Israel. In the Israeli *kibbutz* we have the implementation of ideas which spring wholly from the Russian revolutionary movement and which are not at all 'Jewish' in their essence, since Jewish experience in the diaspora had perforce been individualist rather than collectivist. It is true that some have sought to minimize the ideology behind the *kibbutz* movement and to treat it as a practical response to frontier conditions of settlement; but a recent brilliant description of *kibbutz* life shows that this is not the case, and that the movement remains very bound up with the ideological debate out of which it grew and which is still continuing.[4] Israelis, or at least some Israelis, might be seen as cosmopolitans who have found roots. Are we then to say that from the socialist point of view at any rate, there is no escaping the need for a defined territorial base if a people is to flourish and give the best of which it is capable? Was Ben Gurion, after all, right to align himself intellectually with Khrushchev?

Not only Jews are concerned with the answer to this question. All diasporas are today in danger. In Europe the Nazi destruction of the Jewish diaspora was followed by the enforced repatriation of the German diaspora; in Africa, and perhaps one day in the Caribbean, the Indians; over much of South East Asia, the Chinese; and white men wherever coloured men are in control – all these seem threatened and vulnerable. Capitalism and imperialism were great mixers up of peoples: slaves, indentured workers, migrants, technicians, managers, *entrepreneurs*. Modern collectivism is reversing this trend.[5]

It is a curious as well as an awesome thought. Just when man's links with the soil have become less urgent, when peasant agriculture is becoming a declining element in the world economy, a philsophy based upon closeness to the ancestral soil has become increasingly dominant. To Israel's enemies, she represents the last belated wave of the migration that accompanied imperial expansion, an unnatural European infiltration on to Asian soil, with no more right to international recognition than Mr Ian Smith's White Rhodesians. To her friends Israel is the natural place for Jews to put down their roots since it was from this soil that they were first torn.

For Israelis the question does not arise. They are clearly not 'rootless cosmopolitans' but citizens of a defined country with no more need to justify themselves than the English in England. Any Jew – and here Ben Gurion is right – can solve the dilemma of emancipation by opting for Israel. But this does not mean that Israel has no problems of a conceptual kind. For if the logic of

[4] I follow the argument set forth by Professor Dorothea Crook of the Hebrew University of Jerusalem in her 'Rationalism Triumphant: an essay on the Kibbutzim of Israel' in *Politics and Experience*: *Essays presented to Michael Oakeshott*, ed. P. King and B. C. Parekh (Cambridge, 1968).

This dimension of the problem is not taken account of in Mr E. J. B. Rose's report on British Race Relations, *Colour and Citizenship* (London, 1969).

modern territorial nationalism is accepted, the Arabs must either be assimilated or they must leave. And what was difficult within the pre-1967 frontiers is more difficult now.

It is generally admitted that the Zionist solution as adumbrated by its pioneers did not adequately take account of the difficulties posed by the fact that Palestine was not an empty land. Why this should have been so is easy enough to see. In the world of late imperialism in which both Zionism and its Marxist antithesis flourished, people thought of relations between more- and less-advanced peoples in quite a different way from what is now normal. It would have taken rare perceptiveness to see that the explosively vital Jewish communities of Europe could ever come to be equated with the oppressed Arab subjects of the Turks. The Arab national movement in as far as it existed at that time was hardly present in the consciousness of those involved in the ideological ferment of Eastern and Central Europe. Now that the problem has arisen, there is one familiar way of cutting the Gordian knot. It is to deny that Jewish history can or should be subjected to this kind of analysis at all. The denial can take two forms – traditional (or religious) and modern (or secular) – though if the traditional one had not existed it is hard to believe that the modern version would have come into being.

The traditional view would be that the return to Zion is a fulfilment of prophecy and a manifestation of the will of God. (That some traditionalists argue that the time for the return was not ripe, that the Messiah is yet to come, or that the State of Israel is not the proper embodiment of his kingdom is interesting but irrelevant.) If you accept the religious view, little more need be said. Faith requires no support from argument.

But Israel is a secular State and what we commonly get is a more subtle version of the argument, again emphasizing the uniqueness of the Jewish experience but seeing it as manifest in that experience and not as deducible from scripture.

Of this view, the most eloquent exponent has been the Israeli historian, Jacob Talmon.

> No historian [he writes] can be a complete rationalist. He must be something of a poet, he must have a little of the philosopher, and he must be touched a bit by some kind of mysticism.... The Jewish historian becomes a kind of martyr in his perman-

ent and anguished intimacy with the mystery of Jewish martyrdom and survival. Whether he be Orthodox in belief or has discarded all religious practice, he cannot help but be sustained by a faith which can be neither proved nor disproved.

What we have here is an affirmation of the belief that the survival of Jewry is an essential part of human history as a whole, that its embodiment in the State of Israel is not just adding one more to the list of modern nation-states but has a meaning that extends beyond its frontiers:

> I believe that notwithstanding all the vexations and entanglements caused by emergency and inescapable necessity – all so reminiscent incidentally of the times of Ezra and Nehemiah – Israel will one day be spiritually significant and in conjunction with the Jewish diaspora, spiritually effective in the world. History would somehow make no sense otherwise.[6]

To say that History makes no sense unless this or that happens is to make a claim which transcends the study of history itself; and that is why I regard this approach as the 'modern' equivalent of the earlier religious one. Nevertheless, Professor Talmon would, as his own writings show, fully admit that the role of Jewry in the last two centuries is only understandable in the light of the general social evolution of the West, and it is not difficult to show which forms of society and of the State have allowed the Jew in the Diaspora to pursue with the minimum of friction his own individual and communal life.

As Jews saw it, the problem was what minimum of separation from the host society was sufficient to ensure the survival of the community if pressure were removed through emancipation; for the non-Jew, what derogations from total identification with the host-nation were implied in Jewishness. When Zionism came into the debate it was widely assumed that its success would itself improve the conditions of Jews in the Diaspora because it would show that Jews were as capable of full nationhood as any other

[6] J. L. Talmon, *The Unique and the Universal* (London, 1965), pp. 89–90. Nor need one be an Israeli or even a Jew to see the extraordinary story of the rebirth of the Jewish State in this light. As Soustelle writes: 'L'Etat national d'Israel, limité à une fraction de la terre qui le vit naître, est aussi le témoin d'une réalité qui le dépasse, le porteur d'un message de portée universelle, tel que l'ont voulu ses sages et ses prophètes.' Jacques Soustelle, *La Longue Marche d'Israël* (Paris, 1968), p. 326.

people, and relieve them of the charge of being merely parasitic upon other communities owing to the forced maldistribution of their economic energies.

For these reasons the idea of Jewish nationhood had an attraction for some Jews even if they had no particular interest in the possible content of Jewish existence in a Jewish State. Professor Talmon rightly says of the late Sir Lewis Namier that his 'essays of wrath and pride on the Jewish question are among the most moving of all Jewish writings.' He adds, however, the fact that 'for all its intensity, Namier's Zionism had little concern with Judaism. He knew no Hebrew Literature, he hated the Jewish religion, especially the religious parties in Zionism . . . Namier's Zionism was political, untouched by any cultural Ahad-Haamism.' And in this he was not unique.

But although it is true that Israel and its achievements, particularly, the world being what it is, its military achievements, have raised Jewish prestige in the abstract, the concrete problems indicated by the phrase 'rootless cosmopolitans' are if anything harder to solve. For there is now the new possibility that the interests of Israel as a state may at any time clash with those of the state of which a Jew is a citizen. How does he resolve this clash?

Where Britain was concerned, the circumstances at the time of the Balfour declaration appeared to preclude such a dilemma; on the contrary many people regarded the National Home as an element in the future security of the British Empire. A British Jew could easily and without self-consciousness espouse the Zionist cause, as did the first Lord Nathan, and yet feel quite free to play an active role in Britain's own political life. But Nathan was to live long enough to serve in a government which under the guidance of an anti-semitic foreign secretary – Bevin's anti-Zionism certainly ended up as anti-semitism – came to the very brink of war with the fledgling Israeli State.[7]

Or, one may ask what were the sentiments of M. Michel Debré, grandson of a Chief Rabbi of France, when General de Gaulle imperilled Israeli's very existence by an arms embargo. It is at these moments that the degree of assimilation of a particular Jew cannot always be relied upon. Sometimes it can be – and one thinks of some Jewish Labour MPs both in the Bevin era and in 1956 – but

[7] See Richard Meinertzhagen, *Middle East Diary* (London, 1959) and H. Montgomery Hyde, *Strong for Service: the Life of Lord Nathan of Churt* (London, 1968).

sometimes it cannot. And one remembers the way in which M. Raymond Aron turned upon General de Gaulle after the latter uttered his famous phrase about the Jewish people *'sûr de lui-même et dominateur'* and the explanation he gave of how an attack of this kind moved him to recognize the unity of the Jewish theme in history, though not to identify himself with a purely territorialist solution.[8]

When Edwin Montagu and other leading members of Anglo-Jewry opposed the Balfour declaration on the ground that it could not but prejudice the situation of their community because, to the religious and cultural differentiation which a Liberal State could tolerate, it would add a political dimension which it would find harder to swallow, they were right for the wrong reasons. It was not the existence of an alternative home for Jews nor even of a political expression for the Jewish people that would matter but this possibility of a concrete conflict between that expression and the land of which an individual Jew might be citizen, between Israel and Britain as it turned out.

The problem cannot be dismissed by saying that it is not unique, that there are for instance Irishmen or Poles in the United States or Ukrainians in Canada the cause of whose homeland may run across the policies of the land of their adoption. For in these cases there is no obstacle to ultimate assimilation including biological assimilation. In the Jewish case it is indeed possible, and the absence of a clearly marked religious element in the life of a particular Jewish community makes it more likely. If the Soviet Government had been able to surmount its fears and treat Judaism with the same mixture of pressure and contemptuous toleration as it has the adherents of the Orthodox Church for most of the time, Jewish consciousness might well be on the way to extinction. But the existence of Israel has made up to some extent for the declining force of the religious factor. Israel has done the opposite of what Ben Gurion expected. It has not ended the problem of 'rootless cosmopolitans' – this would only happen if all Jews went to Israel and stayed there. It has given the enemies of Jewry a new weapon, and the opponents of assimilation a new means of defence. Israel needs the Diaspora – and work for Israel (even simple fund-raising) is a new and important aspect of Jewishness.

[8] See Raymond Aron, *De Gaulle, Israël et les Juifs* (Paris, 1968), and his article in *Encounter*, 'The General and the Jews', June, 1968.

It is certainly much more important than Jewish secular culture. What we really mean when we talk of the cultural achievements of the Jews of the Diaspora is rather what Deutscher meant when he wrote about the contributions of those thinkers and artists who stood on the margins of two worlds and were therefore untrammelled by either. It is the juxtaposition of the Jewish element with some major cultural tradition of another people that produces most of the names that adorn the honours lists of Jewish encyclopaedias. Even so, the phenomenon is likely to decline in importance as natural science becomes the most prominent element in contemporary culture. Seeing that natural science is in essence universalist and non-national, it makes little sense to claim as a Jewish achievement the work of a physicist or chemist or biologist who happens to be a Jew. He gets no advantage from being a 'rootless cosmopolitan' while the artist may.

The Jewish condition will thus continue except in the event of another holocaust. Little can usefully be said about the future except that the condition is bound to be an uncomfortable one. The examples of distress within such tolerant, liberal, and humane societies as those of Britain and France are evidence enough. Things will be better or worse according to the temper of the host societies from time to time. What people will tolerate in conditions of security they will not accept in conditions of insecurity, rapid change, or violent upheaval. For this reason, I get a sense of anguish when I see so many Jewish names among the lists of student rioters in this country or America. The risks for a non-Jew of destroying the foundation of bourgeois liberal societies are perhaps measurable; for the Jew, they are not. The turning of the American Negro against his American liberal patrons, many of them Jewish, is something that should be taken to heart. Revolutionary activity by Jews is playing with fire – the anti-Israel and hence anti-semitic direction taken by the German SDS should bring the point home. The Jews who play along with such movements do not know what they do, and their elders and betters have given them all too little guidance.

Jews are stuck with being Jews. However wholehearted their identification with the countries of which they are citizens, the element of distinctiveness remains so long as a Jewish ancestry can be traced. It is a mutation from an originally religious distinctiveness which nowadays takes on new and secular forms. It has been given a powerful new drive by the living miracle of Israel. The most that Jews can do for themselves is to work to build up – or to preserve where it exists – the kind of society in which this distinctiveness does not count too harshly against those who bear its mark. Why this particular burden should have befallen the Jews is something which individuals can attribute to the will of God or the legacy of history, as their own particular faith (or lack of it) dictates.

22 *The Israel-Haters*

In an essay entitled 'Rootless Cosmopolitans?' in the November issue of *Encounter,* itself based on an address given to the London Association of Jewish Graduates, I used in relation to the German SDS – the extreme left-wing student movement – the phrase 'anti-Israel and hence anti-semitic direction' to describe its line. It has been suggested that this phrase risks being misunderstood in a wider context, and that it ought perhaps to be explained.

When I say 'anti-Israel and hence anti-semitic' I do not of course imply that every individual who takes a pro-Arab and anti-Israeli attitude towards the Middle East problem is necessarily motivated by anti-semitic feelings, overt or suppressed. I am, of course, well aware that there are people in this country – some of them personal friends – who regret the establishment of the State of Israel because of its effect upon Britain's own position in the Middle East, or who deplore some aspects of Israel's more recent policies, and who would yet never countenance any discrimination against Jewish citizens of the United Kingdom and would indeed actively oppose any manifestation of an anti-semitic kind. Some of them might be classed in the language of an earlier period as judaeophiles.

What I have in mind is not something relating to individuals but to the inexorable pressure of political considerations which may get people into positions which they would not otherwise have reached.

It is of course clearly so in East Europe. Vulgar anti-semitism is obviously incompatible with Marxism. But in a totalitarian country any particular connection that a group of citizens may have with the external world is inevitably suspect. Zionism which suggested that Soviet Jews might be affected by a movement external to Soviet Communism was anathema to the Soviet authorities long before the political re-entry of the Russians into the Middle East, and their espousal of Arab nationalism as the most effective local force for their own purposes.

Many Jews in the Soviet Union may have been unaffected by these developments; some clearly were not. Some feeling that the fate of Israel was their affair obviously existed in some quarters. And the same has been true of Jews elsewhere in Communist Europe. The only certain way of making sure that Zionism or even pro-Israel sentiment was thoroughly stamped upon was not to try to draw distinctions between Jews who cherished pro-Israeli sentiment however vague and Jews in general.

It was easier in many countries to mobilize for this purpose the traditional themes of anti-semitic propaganda than to excite active support for, or interest in, the Arab cause. It thus became and is impossible to distinguish the repression of Zionist tendencies from anti-semitism in word and deed.

In the democracies of the West the issues are more complicated. For those who believe that the interests of their own countries or of the West at large are to seek an accommodation with Arab nationalism at the expense of Israel, their Jewish citizens are an obstacle. In my *Encounter* article I referred to the particular tragic case of General de Gaulle. Despite the traditional anti-semitism of the milieu from which he came, I know of no evidence that he was himself anti-semitic; Jews played, not unnaturally, a considerable role in his original movement.

It was only when having settled the Algerian war and having decided to attempt to rebuild France's position and prestige in the Arab world that he found it necessary to cut back on France's commitments to Israel, and discovered that the threat to Israel's security which this posed produced hostile reactions in France, and of course most obviously among French Jews. It was resentment at not getting his own way without opposition in a matter of foreign policy that led him to the almost overtly anti-semitic tone of some of his utterances around the time of the six-day war.

In the United States the problem did not arise to any great extent until the second world war. The United States, having no special interests in the Middle East other than those generated by some missionary and educational activities, American Jews were free to follow a Zionist line and to bring pro-Zionist pressures to bear upon the Government, without singling themselves out among other pressure groups. But once the United States did come to feel that it had important economic and security interests in the area, a school of policy-makers was bound to come into existence which thought that these could best be forwarded by taking a pro-Arab line.

They naturally and inevitably resented the political pressures upon American statesmen inherent in the existence of an identifiable 'Jewish vote'. And from resenting the political influences of Jewish voters or writers to feeling that the Jews themselves were a nuisance, was not a long step.

In Britain a similar development went back much further, indeed to the time of the Balfour Declaration itself. But there have been differences. The 'Jewish vote' is not a factor of significance in Britain. But American criticism of British policies in the Middle East could be put down to the American 'Jewish vote' and produce some of the same reactions among pro-Arab British officials, as among their American counterparts later on. The failure to carry the Government with them as far as they would like has naturally produced among pro-Arab British politicians a search for the reasons for this fact.

The existence of Jewish members of Parliament in particular on the Labour side has been advanced as one reason. Influence on the press through Jewish advertisers or because of the existence of quite a contingent of Jewish journalists can be alleged as another.

The argument then is that if it were not for the existence of British Jewry – for sympathies with Israel are, of course, not confined to Jews who call themselves Zionists or take an active part in Zionist work – Britain would have followed a policy of enlightened self-interest which would have placed her unequivocally on the Arab side.

Finally, there are those who regard the cause of the Palestinian Arabs as simply part of the world-wide anti-colonialist, anti-imperialist struggle and who resent the attitude of the great majority of Jews who do not see the issue in these simple terms. So the Jewish community comes to be at once an obstacle to a British policy of *realpolitik* and an obstacle to one particular aspect of a worldwide revolutionary movement.

But if one arrives by whatever route at the view that one's own political objectives are being frustrated by the existence of a Jewish community in one's own country then what begins as being anti-

Israeli cannot but end up as a campaign against Jewish influence in public life: and from that to conventional anti-semitism is a relatively short step, even though some to their credit will never take it.

VI EMPIRE AND AFTER

23 *English as a 'Second Language'*

No country is so likely to give a visitor the sense of the ultimate unpredictability of historical events as Israel. The more familiar one is with the amalgam of biblical prophecy, Central European nationalism, Western European democracy, and Eastern European socialism, that fused into the Zionist ideology, the greater must seem the contrast between the reality of today and the dreams of the precursors. But equally, wherever one touches a problem of today one sooner or later comes up against ideological considerations that lead one very far afield – back from the strip of Mediterranean coastline and the bleak Judean hills to the forests, steppes, and small towns of Eastern Europe, to those Jewish communities of the 'pale of settlement' which have no longer any physical existence but whose moral legacy is unexhausted.

Let me look then, at one problem which is not peculiar to Israel; it is common to all those countries in Asia and Africa now for the first time after centuries seeking to make their own way in the world. For although the international problems that face Israel would seem to derive largely from its position as an outpost of European settlement and civilization in a non-European and hostile environment, its external problems are in many cases much more like those of 'developing countries' everywhere. Indeed, since more than half of its Jewish population now originates not from Europe at all but from the Middle East and North Africa, the similarity of Israel's problems to those of other Asian or of African countries is much more evident than when the State was founded. The particular problem I would select is the problem of 'communication'.

Language presents all states other than the possessors of one of the world-languages with special difficulties. On the one hand the native language must serve as the vehicle of domestic intercourse, of general culture, and (in the case of a newly enfranchised people) of national cohesion. Indeed the nineteenth-century history of European nationalisms sometimes reads like a branch of philology. On the other hand, if the citizens of such a country are to take a full part in the advancement of knowledge, in the dissemination of techniques – if the country is not to be excluded from the mainstream of the world's economic, political and cultural life, if indeed independence is to be more than an objective in itself, but is to be an actual instrument of growth – then its elite must have a very far-reaching knowledge of at least one of the world-languages.

Sometimes the choice of language for this purpose may be automatic, and the consequences of the choice obvious. Tunis must choose French and Nigeria, English. Often the choice will be obvious, but its implications less so. For a country may have the problem of making one of many indigenous languages into a truly national one, while retaining a world-language for use in its international intercourse. And this dual task is one that will confront India, for instance, for more than one generation. The ideology of nation-building and the demands of efficiency may pull in opposite directions.

In Israel, the problems which face India, Nigeria, and many other new states can be seen on a miniature scale, but in a condition of probably unrivalled complexity. The first choice – that Hebrew should be the country's vernacular and the teaching language in educational institutions at every level including the highest – was reached at an early stage in the creation of the 'National Home', and long antedates the state itself. The reasons are easy enough to understand; the effort involved may not always be appreciated. For Hebrew had to be re-created as a spoken language from classical or medieval models, and then taught to immigrants speaking a wide variety of other languages. It is true that the majority would have had some acquaintance with Hebrew through the scriptures and the liturgy; still there were differences in pronunciation to contend with. (And if there ever were to be a massive immigration from the Soviet Union – where the traditions of the synagogue have been interrupted – even this preliminary conditioning would be absent.) For immigrants from Europe their original vernacular – German or Russian, Polish or Rumanian, even Yiddish – was in basic structure no preparation for a highly inflected Semitic language. For the large numbers of Arabic speakers among the newer immigrants, Hebrew is linguistically speaking less remote; but with the much lower degree of literacy prevailing among them, the mental effort of learning a foreign language and becoming literate in it cannot be much less.

But these are matters of which time will take care. Children will learn their Hebrew if not at their mother's knee at least at their school desk. As new generations of the Israeli-born reach manhood (and as the sources of available immigrants run dry), Israel will become a Hebrew-speaking nation. It is emphatically not a piece of antiquarian nonsense like making the countrymen of Oscar Wilde and Bernard Shaw talk Erse; it is a national policy boldly executed and, in the long run, certain of success. But meanwhile what of external communication? What of the second language?

The choice of English for this purpose has been as wholehearted as that of Hebrew, but a less obvious one. If one were to go by the world-languages naturally accessible to important elements in the population, Russian, German, and even French would seem to have at least equal claims. The Russian Revolution ruled out the first; Hitler made the second unthinkable; French came too late (despite the impressive battle for 'French culture' which France is waging in Israel today). The thirty years of British rule, the fact that the Jewries of the English-speaking world are the only ones in a position to give Israel serious help – these have settled the issue for good. The older inhabitants of Tel Aviv, Haifa, and Jerusalem may still read Russian or German; the window on the world for the young Israeli is the English language. It is no disrespect to the young Israeli to say that it is often and perhaps increasingly a rather misty window: if he will nearly always know some English, he will rarely know enough for its use not to be a barrier between himself and what it is he wants to know and express.

English is universally taught but taught at a level dictated by the fact of its universality. What Israel needs at any rate for the immediate future is a relatively large number of people who read English with facility, write it correctly, and speak it with sufficient ease for the need to use it not to be embarrassing. It needs to replace among its officials and diplomats, its doctors and engineers, its teachers and experts, not only the present generation of English-speakers, but also those who still speak Russian and German, who provide alternative means of contact with the external world. And it needs to do even more than this if it is to fulfil its plans for becoming an advanced industrial economy, or of continuing to act as a magnet for students, particularly of technical subjects from other developing countries, notably in Africa. It may be picturesque for a Kenyan to be learning Hebrew as part of his studies at Jerusalem; I find it hard to persuade myself that it is useful.

To achieve this objective of teaching English to a high level for those who need it most (from the point of view of their future contribution to the national well-being), one would ideally have to begin by selecting who they were to be, by seeing that their curricula were designed to give the place to English that would be necessary, and by seeing that they were taught by teachers of whose own qualifications in the subject there could be no doubt.

Not only does this not happen, but even to visualize it happening would involve challenging some of the basic assumptions upon which the country's educational system is based. If one excludes the *Kibbutzim* which jealously maintain an educational system of their own for all up to the age of eighteen – not, by general consent, at a very highly academic level – the basic educational system for the Jewish population of the country is a uniform one. (There are some concessions to the religiously orthodox which need not concern us here.) It consists of compulsory elementary education for eight years (ages six to fourteen) followed by four years secondary education in different types of institutions roughly corresponding to the familiar British divisions at this level. Something like eighty per cent of children do in fact go on to some kind of secondary school. English is taught from the age of eleven in all schools, and is a compulsory subject in the school-leaving examination – roughly corresponding to the French *baccalauréat* which is itself the passport for admission to university.

To see what this apparently privileged position of English in fact denotes, one must dig a little deeper. In the first place, all streaming in elementary education is ruled out. All children must begin English at eleven, and none may begin it earlier. The child whose parents may talk English, or who may have had other special opportunities, is taught alongside and at the pace of the child who may have serious difficulties even with Hebrew. Furthermore, the dominant educational theory of the country is a child-centred theory of the American type: 'happiness' and 'good adjustment', rather than specialized knowledge or skills. School hours are short and tasks light. Not much English will be acquired at this stage.

But when one comes to secondary education, the case is not much better. It is true that there is some 'streaming' in that only a proportion go to the *lycée*-type or grammar-school-type institution. But even here the possibilities of getting a first-rate grounding in English are not very good either. Not only is four years a very short period for secondary education anywhere, but the Israeli child is expected to cram into it not only a full secular education but a full 'Jewish education' as well. While the elementary school of ability is given too little to do, the secondary school-child is overburdened. Furthermore, 'Jewish' education – taken in a national rather than a religious sense – has the disadvantage of being very much detached from other subjects, and of contributing little to their understanding. However nationalistic the teaching of French or English history may be, a French or English child cannot help but pick up some idea of European or even world history from the way in which the fortunes of his own country have been interwoven with wider movements. But the contacts of the Jewish people as an organized community with world history have always been peripheral, almost accidental. The other nations have mainly appeared on the scene as persecutors.[1]

Furthermore, even in the upper grades, teaching must normally be left to teachers who have themselves often learned English only as a foreign language, not even as a *first* foreign language. One thus gets a neglect of diction and pronunciation which are anyhow perhaps particularly difficult for ears used to a Semitic language. Whatever the paper accomplishments of the pupils may be, I cannot help feeling that the spoken language is drifting away from its original basis. (Nor, in passing, is the University likely to help with this difficulty, since there 'American English' and 'South African English' add their quota of aural confusion.)

The problem of English at the secondary-school level has become a subject of public controversy largely because English ranks with mathematics as the subject in the school-leaving examination

[1] Similarly, in Western Europe, Latin has been regarded both as a training for the mind and as a preparation for the study of modern languages. Latin does not figure in the curriculum of any Israeli school. Instead one has *Talmud* – no doubt as is claimed, an excellent training for the logical faculties, but hardly a help to the learning of English.

which produces the highest rate of failures. Since there is a presumption among Jewish parents that all their offspring should be capable of entering a university, this creates some bitterness, and an inevitable pressure for lowering the standards. One demand is that the schools should drop any attempt to teach English literature, and confine themselves to inculcating everyday speech on the lines of commercial language-schools in other countries. Israeli children may need to use English for certain practical purposes (so it is argued) but their cultural needs can be met in their own language. Shakespeare is superfluous. Other educationists argue that to meet this demand would further emphasize the nationalist self-sufficiency of Israeli youth. English for them is not a matter of business or official correspondence but the only link with the European culture to which European Jewry contributed so much (and from which they do not want the Jews of Israel to feel wholly alien). And so in what can only be described as a new 'Battle of the Books', parents and teachers clash over whether *Macbeth* and *Julius Caesar* are suitable set-books for Israeli examinations.

It is at the university level that the most serious difficulties begin. Israel – like many other countries – is bent on large-scale university expansion based upon the view that all who achieve the minimum school-leaving qualification are entitled to a higher education.[2]

Whether the resources being put into university expansion would not be better devoted, at least in part, to upgrading the standard of secondary education is a question important for Israel as well as for other developing countries. What is clear is that at the university level – except in purely Jewish subjects – a reasonable level of English (or of some other world-language) is essential. It is absurd to expect more than a minute fraction of the world's scientific and learned books and periodical articles to be translated into a language spoken by a bare two million people. On the other hand the number of translations that have been produced, and the fact that the university has adopted the American pattern of 'courses' and 'course credits' means that the unambitious student

[2] In addition to the two old-established institutions, the Haifa '*Technion*' (now developing courses in the social sciences and the humanities along MIT or Cal. Tech. lines) and the Hebrew University of Jerusalem, Tel Aviv has one secular and one religious university of its own as well as branches of some of the Jerusalem faculties. There is further talk of a university at Beersheba to assist in the economic and cultural development of 'the South', Israel's American-style 'frontier'.

can get by with a minimum of English. As the American-trained head of the economics department remarked to me, 'At one time there were no economics textbooks in Hebrew. All the students in the department required English. Now three of the staple textbooks have been translated and students are beginning to believe they can scrape through on these alone. . . .' In a university where so many students are only part-time, necessarily devoting much of the day to earning their keep, and where there is an overall 'wastage rate' of as high as thirty per cent, it needs a very strong will or some very marked incentive for the student not to be seduced by short cuts of this kind.

It is doubtful if the university has fully taken into account the importance of this factor in relation to its changing social and linguistic intake. It does of course provide courses in the use of English as an auxiliary language; but as elsewhere such courses are regarded as less important than those offered by the 'English department' proper which is more concerned with keeping the students in touch with the latest fashions in literary criticism in England and America. Furthermore the time taken up with English diminishes the student's chances of learning French or German as well. It could be argued that the rather large production of English 'majors' might go some way to helping the schools in their difficulties. But with the low salaries and low prestige of schoolteaching as a profession the hope may be an over-optimistic one.

For those for whom English is to remain simply an auxiliary – for the future active elite of the country – the final hope is a period of study abroad. But even here problems remain. In the first place, a country which almost by definition has a standing balance-of-payments problem is unlikely to encourage many to go abroad. In the second place, the weakness in linguistic training before the student comes to Britain or America means that much of his precious two years will be taken up with acquiring a standard of English – 'active' as opposed to 'passive' English – sufficient for him to take full advantage of the facilities for study now at last available.

One suspects that such problems vary with different branches of learning. In the pure sciences where symbolic language prevails the problems may be fairly easily surmountable. The Weizmann Institute is clearly an integral part of what one could almost call the 'scientific international' of our era; and only the different beauties of the site remind one that one is at Rehovoth, Israel, and not at Berkeley, California, or Cambridge, England. But in the humanities or the social sciences an institution such as the Hebrew University which works in a language not used elsewhere must have very special difficulties both in maintaining academic standards and in keeping its students' eyes fixed upon a wider world. It is a sad commentary on our times that the only Israeli institution which *does* demand a high standard of English as a *prerequisite* for entry is – the National Defence College!

Indeed it is no surprise that a country beset with enemies (and with a unique task in nation-building still before it) should find it hard to strike a balance between its need for special skills and social-democratic ideals of equality. I can understand resistance to any ideas of educational selection or streaming where this would in effect amount to separating children according to the countries of origin of their families – a near *apartheid*. But this response is no final answer.

Whether other developing countries have solved 'second language' problems more successfully is something one would like to know. I find that an ambitious attempt has been made by Turkey, which has created alongside the University of Ankara a new Institution planned on a very grandiose scale – the Middle East Technical University; here not only technical subjects but also the social sciences are actually taught *in English*. I was told by Turks of their hope that this would enable the Institution to attract students from other developing countries – and of its present students about ten per cent are indeed non-Turkish. But Turkey, too, is a country where both democratic and nationalistic sentiments run deep. The original arrangement was that a good knowledge of English was a precondition of admission to the METU, but its people began to complain that since English was taught only at a minority of good *lycées,* this effectively closed the door of the University to all but the children of the wealthy. So the provision has been abandoned; and the University itself offers a first year devoted to the study of English before ordinary studies are embarked upon. It is too early to say how successful this is likely to be. Turkish is as remote from English as is Hebrew – though I suspect (as an amateur) that the phonetic difficulties are rather slighter. But this may well not be the terminal point. And I can visualize political parties in a few years' time insisting that to teach in a foreign language is a mark of 'national inferiority', and on converting the University into a purely Turkish institution.

Indeed, when one confronts governments of new states declaring their intention to embark upon programmes of economic growth and of social development, one good touchstone of their seriousness of purpose might, I suggest, be the question: 'And what are you doing about the teaching of a world-language – English? French? . . . Russian? . . .'

24 The Crisis in Parliamentary Democracy[1]

Last year ceremonies were held in London to mark the 700th anniversary of the calling of a gathering of representative figures in the England of the time by Simon de Montfort, the temporarily successful leader of a movement of rebellion against the policies of King Henry III. In the view of some people that assembly deserves to rank as the first English Parliament, and therefore as the ancestor of parliamentary government all over the world. A hundred or even fifty years ago an anniversary of this kind would have been celebrated without the undertone of irony which must have been detectable on this occasion except among the most complacent.

A hundred years ago, even fifty years ago, there were certainly many fewer parliaments in the world than there are today: but where they existed they were prized indications of political maturity and where they did not exist because of tyrannical or foreign rule, it was the first objective of all forward-looking people to bring them into existence. It was assumed by men of all nations that parliamentary government – whether of the strictly parliamentary or of the presidential variety – was the essential element in civilized government. Unless there was a forum where the public issues of the time could be ventilated and debated in freedom by the chosen representatives of the people, nothing effective could be done. Even the incidentals and excesses of the system made an appeal where it was absent. I remember an elderly diplomat telling me many years ago of his apprenticeship as a younger member of the British Embassy in St Petersburg. While there he acquired the acquaintance of a young Russian lady, no doubt in order to assist him in learning the language. On one occasion his lady friend went

[1] Lecture at South African Institute of Race Relations, August 1966.

off on a trip to Paris. When she returned our diplomat asked her how she had enjoyed it. 'Wonderful,' she replied, 'in France there is real freedom, not like our poor Russia.' 'Well, what was it about freedom in France that most impressed you?' 'Oh, I went to the Chamber and there I saw the deputies shouting and throwing ink-wells at each other . . . that is freedom.'

Today we would be less confident than we would have been that all problems are solved when a newly enfranchized country adopts a parliamentary system and even in countries where it has been long established, even in Britain itself, few would now claim that it is performing satisfactorily. The situation we confront is not so much that of a direct challenge to parliamentary democracy – the fashionable anti-parliamentary ideologies of the inter-war decades have largely gone down to oblivion with the regimes that embodied them. What we have now is something rather different – a fairly general and widespread lip-service to parliamentary institutions and a fairly general disregard of some of their fundamentals.

Last autumn, for instance, I took part at Geneva in an international symposium held to mark the opening of the new reference library of the International Parliamentary Union – an old-established body which took its origins in that optimistic period of the world's history when it seemed reasonable to believe that if parliamentarians of different countries got to know each other, world peace would be the more secure. I would not wish to disparage the international benefits provided by this and similar institutions. But one could not help reflecting, as one looked round the delegates from the different continents and countries who were taking part in the discussion of parliaments and their work, how different were the realities that the label covered. The subject of Parliament and the Executive was dealt with in two papers, one by myself and one by the Vice-President of the Polish Parliament. Now it is clear that whatever functions the Polish Parliament may perform in relation to the Executive it does not stand in the same relationship to it as does a parliament in a multi-party system. In Poland both the Executive and Parliament are emanations of the single ruling party and the control on both is an external one: therefore the Executive not being dependent upon Parliament cannot in the last resort be regarded as subject to its control.

I have no doubt but that the representatives of Poland and of other one-party states whether communist or not were perfectly genuine in their belief that their own parliaments had a constructive part to play in their countries' affairs but their presence was sufficient to remind one that the mere existence of a parliament does not itself imply that we are in the presence of a parliamentary democracy. And discussions of what might be called parliamentary mechanics – the detail of procedure and organization – although of great interest to the specialist are not really more than peripheral to the major issues.

One way of explaining the reasons why parliamentary institutions have been such fragile growths in countries where they have been implanted on the British or other Western European models is to postulate that some peoples or even races are more fitted to enjoy them than others; and that other peoples somehow lack the necessary qualities of temperament required to make them operate successfully. Like most other arguments based on alleged racial characteristics, this one is nonsense. A more sophisticated version would point to the peculiar difficulties which parliamentary democracy faces in countries with a relatively small educated elite, deep national or tribal divisions, a vast discrepancy between the problems that they face and the resources available with which to face them. And it is of course true that where things are in this state, parliamentary government is not normally likely to flourish: but equally government of any kind runs up against fairly considerable difficulties. Parliamentary democracy like all other systems of government except the most primitive presupposes an existing social and political community – it cannot of itself create one though it can in some circumstances provide a powerful adjunct to efforts in that direction.

The trouble with all such explanations that begin from the particular circumstances of the case is that they are too easy – they assume that parliamentary democracy is a norm, and that departures from it require justification or explanation. It would, in my view, be more illuminating to start at the other end and to say quite frankly that parliamentary democracy is so extremely difficult a system of running a country's affairs that any long period of success deserves to be treated as a miracle for which thanks are due, and which itself requires explanation.

The reasons why it is so difficult are only to be discovered by looking at it from the inside: when we have established what these reasons are, we can consider what phenomena peculiar to our own times have brought about a situation in which it is permissible to talk of 'crisis', and whether these are common to all countries or relate especially to certain types of society. To do this properly would demand both time and knowledge neither of which are at my command. I can only put before you for discussion some crude and dangerous generalizations – all generalizations are dangerous and crude – and hope that they will suggest the kind of attitude we ought to take up where these problems impinge upon practice.

To begin with then, it is possible that the term 'parliamentary democracy' itself contains ideas that are themselves irreconcilable, that what we are dealing with here is actually a concealed contradiction – that perhaps we can have parliament or we can have democracy, but not both. Now this is a view with a very respectable ancestry – it is set out in classic pages in the writings of Rousseau. More important still, it would seem to correspond to much of the actual evolution of institutions. Democracy is after all a relatively simple term and means or should mean nothing more nor less than the rule of the majority. As Rousseau saw, this was not a difficult thing to achieve in a city-state or mountain canton where the whole citizenry could be got together for the general will to be established. The danger of any democracy that individual rights of the minority will be ignored – which is also a danger to the majority, since there may be no one left to warn it that it is making mistakes – this danger could under conditions of direct democracy be regarded as relatively remote on the grounds that people would hardly devote themselves to oppressing neighbours with whom they were in ordinary social relations from day to day. They might – and some small communities do offer rather frightening examples of majority tyranny – but on the whole this was not too much of a risk to run.

The traditions of direct democracy have little direct application to modern societies where the political unit however small is hardly ever small enough for it to function as a single deliberative body. But this tradition had left important traces in modern thinking and practice. There is for instance its modern equivalent, the plebiscite or referendum. On the face of it, this would seem a very valuable device for ascertaining the will of the majority without the distortion of intermediaries, but I hardly need to remind you of the powerful evidence that leads one to believe that under conditions where the resources of the mass-media are unequally distributed, the ability to manipulate such expressions of opinion on the part of the unscrupulous is so considerable that the use of referenda except in countries of exceptional social and political maturity like Switzerland is on the whole a demagogic rather than a democratic practice.

More important still is the connection between the idea of direct democracy and the modern one-party or would-be one-party states. In many – not quite all of these – the single party or would-be single party rests its claims not on the fact that its leaders have been chosen by the people or by a majority of them, but on its assertion that it embodies in some mystical sense the will of the true majority, the only majority that counts – the proletariat, the master-race, the genuine nationalists, or whatever the acceptable criterion may be. Such a ruling group may or may not think it worth while to fortify their claim to embody the majority's rule and hence the democratic principle, by plebiscites, one-candidate elections or similar means. The important thing is that by assuming that it is the voice of democracy it is enabled to proceed to take steps to make certain that this view cannot be challenged. It is the most familiar of all governing techniques and we hardly need to go into it further.

But the legacies of direct democracy concern us less in the present context than the fact that the conditions for its actual operation have scarcely existed in modern times, and that democrats despite Rousseau have had to fall back upon representative institutions to do the job, in other words, upon parliament. Now it is clear that there was from the beginning the likelihood of trouble and misunderstanding arising here because parliaments are not in their origins nor in their essence necessarily democratic institutions. Indeed in England, where Parliament had after all existed for some five centuries when Rousseau wrote, and elsewhere in Europe where it had existed in the past but had by now largely been eliminated, its functions were certainly not democratic in the sense of it being an agency through which the majority of the people chose their rulers or decided what policies those rulers should pursue. Indeed the question of who should rule was not really relevant, since all parliaments had come into existence in circumstances where the question of who should rule was decided by the hereditary principle, by prescription, and not by election.

Parliaments were necessary to the medieval monarchs not to tell them what to do but as a method of making certain that in what they wanted to do they could rely upon the cooperation of those of their subjects whose cooperation mattered, the powerful and the rich. Representation went to those who counted – the right to be represented arose from the rights of property not from the rights of persons. Parliament was thus the scene of what amounted to a search for bargains, or in modern jargon of consensus, between the Crown and its servants on the one hand and the representatives of the effective social and economic forces in the country on the other. Out of such bargains grew the structure first of the rights of property and gradually as society became increasingly individualistic in structure and temper, of individual rights as well. By the time Rousseau wrote, in England, though in England almost alone, the balance had shifted to such an extent that the effective rulers of the country, the ministers of the Crown, required parliamentary support if they were to hold office and carry out the tasks assigned to them. You had therefore, in England, something very like parliamentary government; you clearly did not have parliamentary democracy.

Now one ought not to despise parliamentary government as such. It has a great many virtues quite irrespective of whether it is democratic or not. It makes it possible for the merits of rival courses of action to be freely canvassed and hence makes error less likely and easier to remedy if once embarked upon: it makes it possible for individuals and groups to alternate in power without recourse to violence: it promotes effective government by the opportunities it offers for criticism and opposition. Provided that it is a genuine parliamentary system, that is to say that within the enfranchized classes there is full freedom of expression and full opportunities for registering dissent, it can be highly productive. It may well produce less injustice and be more effective for good than certain forms of plebiscitarian democracy.

But it has of course one obvious flaw and that is to say that the very degree to which it is responsive to the enfranchized classes is likely to make it insufficiently responsive to the needs of those who do not qualify for representation. A single absolute ruler who regards all his subjects as equally unentitled to a say in public affairs may well do greater justice to the majority of his people than if he is bound by a constitution to pay particular attention to the wishes of a section of them only. This argument which has found classic exposition from some imperial pro-consuls, British and other, may be a cover for oppression but it is not necessarily or intrinsically false. But this is not the aspect of the argument that need detain us.

What we do know is that where parliaments with a limited non-democratic franchise have existed, and been the focus of power, they have tended to ignore or over-ride the interests of those unrepresented as compared with those classes which have the vote and supply the representatives. This is true even in the context of a traditional society where one might expect that injustice would be prevented by the mollifying effect of propinquity – the history of the enclosure movement in England will serve as evidence for the vanity of such hopes. But it is of course even more true that discrimination will prevail when the population legislated for is to a large extent removed geographically and isolated socially from the politically enfranchized classes. The history of the English working class under the unreformed Parliament or of the French working class under the July monarchy are instances familiar enough to need no comment.

However high-minded the leaders of the politically emancipated classes may be, however desirous they may be of doing the best for the entire society and not just for those they represent, the evidence of history – from many different countries and in many different contexts – is quite overwhelming; those who have no vote in a parliamentary system will always find the prevailing legislation biased against their interests. There is no sphere in which it is more true to say that out of sight is out of mind.

This was the logic of the English parliamentary reformers, the logic of the framers of the fifteenth amendment to the Constitution of the United States, it was the logic of the suffragettes; it is within its limits one of the truisms of political history.

It is of course in many circumstances a very uncomfortable doctrine. And from the time when it began to be given currency attempts were made to get round it. In the eighteenth century, it became one of the grievances of the American colonists that imperial legislation was favourable to the interests of the mother-country rather than to those of the colonies, because the colonies were unrepresented in the imperial legislature, the Parliament at Westminster. In reply, recourse was had to what became known as the doctrine of virtual representation. It was propounded that in a parliamentary system it did not matter whether you were represented by representatives chosen by yourself provided that there was someone available to put your point of view, and that in the British Parliament there were enough members whose connections with the colonies gave them an insight into the colonies' needs and interests for the case not to go by default. It was pointed out too that the same thing applied to Britain where many cities newly grown to importance were unrepresented and had to rely on the fact that their views would be made known through the activities of the representatives of those cities which did send members to Parliament. The argument was repudiated by the Americans and there being no way by which they could participate in the making of imperial policy they preferred instead to opt out of the system altogether. Their hostility to the notion of virtual representation had not prevented the adoption of devices that savour of it, of having in some parliaments persons nominated to represent special interests or particular communities – while in a transitional system, this may be better than nothing, it is in no sense a substitute for the real thing. No parliamentary system can do justice to all unless all are represented.

A marginal case here is that of female suffrage where it has been demanded on the same grounds as those that support the claims of other excluded groups and where the same arguments about virtual representation have also been used to deny it to them. My own impression would be that unlike excluded classes, women as such have not once enfranchized made much direct use of the suffrage to press matters of particular interest to them, nor have they shown any marked tendency to vote for members of their own sex. It would also be hard to have to exclude Switzerland from one's list of parliamentary democracies. I shall therefore leave it as being an open question though not in many countries an actual one.

It is at this point where the extension of representation to its natural limits is concerned that parliamentary government and democracy join hands: indeed one would normally mean by parliamentary democracy a system in which parliamentary government operates within a regime of universal suffrage. This is often equated with the slogan 'one man – one vote'. But the two things are not in fact identical. All may have the vote and be represented and yet all votes may not weigh equally in determining representation. Indeed one could normally only guarantee that this is so in a system where the electoral arrangements were those of absolute proportional representation: and there are well-known arguments which have tended to make people fight shy of this formula. Strict democratic theory may demand it: the theory of parliamentary democracy does not.

For when we have brought together the idea of parliamentary government with the democratic idea we have still not solved the problems of either. For we have still to show that the virtues of parliamentary government are compatible with the extension of the franchise to the democratic limit. And it is here that the real difficulties begin. The fact is that quite irrespective of democracy, parliamentary government is a difficult form of government to operate because it runs counter to many profound human instincts. It demands of those who wield power that they should be prepared to surrender it, that they should not use their position as would be more natural to consolidate it, that the majority of the day should be ready to protect the rights of the minority in and out of Parliament to the point where they are able to use those rights in order to become a majority in their turn. It is remarkable that this has ever been the case – anywhere. But one can say in part explanation that it has been easier under a restricted franchise than under a democratic one – because a party that feels that the scales are weighted against it may always threaten to call the unenfranchized to its aid. Indeed in such a system, the unenfranchized have a bargaining value of their own which to some extent though always insufficiently may compensate them for the exclusion practised against them.

This argument does not apply where there is a permanent consensus arising from racial or other forms of objective discrimination to rule out in principle further extensions of the franchise. In such situations the political arena and the contestants may be considered as defined once and for all. And some of the arguments about the difficulties of parliamentary democracy may be held also to apply to situations of this kind.

In these parliamentary democracies upon which I now propose to concentrate the difficulties of self-denial are considerable. However liberal in spirit a political party may be it is of its nature that it should try to extend its authority and its appeal. All political parties in multi-party or bi-party systems are always trying to become single parties unless the society is so pluralistic and the parties so clearly based upon group-appeals that their possible expansion has inherent limitations. But in a relatively homogeneous society such as that of the United Kingdom, major parties will be expected to contest all parliamentary seats and one party could therefore in theory – and with only a slight plurality in the votes cast – win all of them.

In mature democracies the exploitation of the majority party's position will be limited by the expectations of fairness that exist in the electorate and whose violation would redound to the party's disadvantage. So no tampering with the ballot boxes. Similarly it will refrain from taking advantage of its control of the parliamentary timetable to deny opportunities of expression to the opposition. It will not in ordinary circumstances resort to censorship or other methods of curtailing the expression of hostile opinion. Rather more difficult has been the problem of the new mass-media, broadcasting and television. Here a government has to walk very warily indeed in order to tread the path between denying itself a legitimate means of communication with the citizens and hence an instrument of government and the temptation to give itself an advantage as against opposition parties in the provisions for its more general use. There are some instructive contrasts here between recent experiences in the United Kingdom and France.

But in fact the favourable conditions for the exercise of self-restraint that are essential if parliamentary democracy is to give satisfaction are rare and are probably becoming rarer. I have already indicated that difficulties and temptations confront even parties that are liberal in spirit. I must now define this term in the argument since I would contend that only liberal parties can really work the system. I do not of course mean that parties must profess the principles of particular schools of liberalism that have flourished from time to time, still less that they must call themselves Liberal. What I mean is that they must be restricted in their ambitions to objectives that are obtainable by piecemeal improvement in social arrangements so that the dialogue between parties becomes essentially one over means rather than over ends. They must be prepared to accept defeat, and even the reversal of measures they have put through during periods of power – though a parliamentary democracy in which this was frequently necessary would be showing danger signs. In a healthy system governments should be able to build upon the work of their predecessors.

If parties go further than this and regard themselves as the vehicles for making large-scale and irreversible changes in the pattern of society, if they claim for themselves and for the sections of the community they happen to represent some superior position as against the rest of society, if their aim is to manipulate society as a whole to the likeness of some ideal image, then their ability to play according to the rules of parliamentary democracy will probably come into question as soon as they run up against serious opposition whether in the legislature or outside. Comprehensive social planning – as distinct from indicative planning on the purely economic front – is difficult to reconcile with parliamentary democracy and those who favour it are normally impatient with the restraints of a parliamentary system whose procedures are devised in order to secure maximum consent and hence are bound to involve an element of delay.

If parliamentary democracies incorporating parties of this kind have hitherto survived though not without strain, the reason is that normally even such parties will have to generalize their appeal for electoral purposes and so dilute if not their doctrines at any rate their practice. The electoral system is thus or can be a limitation. But one should not draw too much comfort from this fact because the 'appeal to the centre' has two disadvantages also. It will tend to reduce the degree of clarity and definition with which particular programmes are propounded and explained and so minimize the educational function that electoral competition within a parliamentary democracy should always perform. Secondly, by appearing to deprive the victors of the full fruits of victory, it may make the more militant members of the majority either break away and form a more extreme party with some possible consequences for the stability of the system, or even more dangerously lead them to despair of the parliamentary system altogether and to seek their goals through direct action of one kind and another.

But this picture is as I have said only applicable to societies where a high degree of homogeneity permits some approximation to the model of rival parties fishing for votes in the same pool. Outside Western Europe and parts of North America, Australia and New Zealand there are hardly any countries to which this model applies. Japan, Israel and one or two Latin American countries about finish the list. Leaving aside those countries that have formally abandoned parliamentary democracy altogether, one can see two other basic models.

You may have a system in which one principal party is dominant and in which opposition parties although not seriously impeded in their activities are too divided to be serious competitors for power. The obvious example is India where one might paradoxically suggest that the condition of its success in running parliamentary institutions has been the dominant position of Congress as the essential unifying factor. Whether a transition to a multi-party system in the full sense is possible must certainly be regarded as at best doubtful.

The other Indian parties are for the most part like the bulk of parties in the new states what one might call 'objective' parties, meaning by this that adherence to them is dictated not by subjective ideological preference or programmatic considerations such as might lead an Englishman to vote Tory or Labour or an American to vote Republican or Democrat but by objective factors. In other words with such parties once one knows some basic fact about an individual, his race, religion, language, tribe or whatever it may be, one can determine with a high degree of certainty the political party to which he will be likely to adhere. And where this model prevails the odds against parliamentary democracy working are always very high indeed.

I do not mean by this to be forced into the position of saying that no pluralist society can ever work a parliamentary democracy: federal institutions, the restrictions of a rigid constitution and other devices may overcome the inherent difficulties. But if parties which are essentially 'objective' in their foundation assume the privileges that should accrue only to 'subjective' parties, if they believe that majority positions give them the right to promote the interests of their own group at the expense of other groups, it is very difficult indeed to avoid breakdown.

Once again one must be on one's guard against over-simplification and against drawing over-rigid distinctions between the two situations or the two types of party. Most 'subjective' parties have an objective element – in the United States, income-level which is an objective factor is a probable though imperfect guide to political allegiance: and this is even true in many Western European countries. But occupational and similar distinctions in modern industrial societies with a rather high degree of social mobility are less rigid a divisive factor than the factors of race, language and religion particularly where as so often all these coincide. A brown-skinned Moslem Malay seeking advancement through the use of English and a yellow-skinned Confucian Chinese demanding the right to advancement through the medium of his own inherited language and culture will not easily cross-vote unless very special conditions are deliberately created as they were in Malaya.

But this is not the end of the matter. In a relatively static society where change where it takes place at all is little affected by the actions of government, parliamentary democracy even with the handicaps imposed by pluralism may well survive them given the appropriate leadership. But in the modern world, this is rarely if ever the actual situation. Instead there is a high propensity towards change and change of a kind that is almost certain to affect unequally the different groups within a society – particularly if their geographical and occupational distribution is different to begin with. And government's task has come to be both the regulation of change and in many cases its actual sponsorship. In such circumstances more will be asked of government than it can ordinarily perform and all systems will fall short of the expectations that are aroused. To the frustrations thus created parliamentary democracies in pluralist societies are particularly vulnerable. It remains for us to see why this should necessarily be the case.

In the first place, economic development which is at the heart of the changes in most countries cannot proceed evenly: it is bound to help some areas and some classes of society more rapidly and more effectively than others and where these areas and classes correspond to the divisions between the groups of which the society is composed it will be difficult to persuade those who get less that they are not being discriminated against. In the second place, since the basic planning decisions will be made by governments, the question of the composition of the Government and indeed of the civil service also will become more distinctly a political question than in more static societies: the possibilities of discrimination will mean more.

This means in fact that it is only possible to progress if one has a virtually permanent coalition – either by means of a single party open to all groups, in which case the political struggle becomes an internal one in the party concerned – or through a coalition of separate parties. The healthy alternation of government and opposition which parliamentary democracy inherited from pre-democratic forms of parliamentary government can rarely work in such conditions. Ceylon is the only new state which has seen an alternation in governments through the working of the electoral process – and then by no means smoothly or without violence and other departures from the democratic norm.

The fact that this may be an indispensable condition of the survival of any kind of democracy at all – the only alternative to direct military rule under one guise or another such as is now the norm all the way across the map from Pakistan to Ghana – does not mean that there are no disadvantages. Without an opposition, government becomes lax: quite apart from policy issues the guarantees against corruption and maladministration are weakened and any questioning of what purports to be an expression of the national unity all too easily incurs the stigma of treason.

An additional strain has been put upon the system in most of the new states by the division of the major industrial powers into conflicting blocs. The dominant ideology and indeed almost the only ideological factor of unity in most 'new States' is nationalism – in the sense of looking forward to nationhood rather than cherishing what exists. The other favourite word, 'socialism' in its African or Asian context is, when looked into more closely, only another way of putting the same thing. The support that regimes have among the local-based elites is for development, but development on an autonomous basis. Few of these countries have the resources both to assist in a rapid rise in the standard of living and to accumulate capital for development in the future. In some of them – in India the most important of them – it is all one can do by running hard to stay on the same spot. Therefore those who actually face the problems – statesmen in power and civil servants – acutely recognize the need for foreign assistance. But such assistance can hardly be unconditional, and even the most tenuous strings – actual or implied – will be repudiated as betraying the national principle. They echo Cavour's *Italia fara da se* but have no Piedmont to begin with. Much of the history of recent years in many of the new states is a commentary on these simple propositions.

It is nevertheless wrong to assume that because the crisis assumes too acute a form in the new states that it is their crisis. Many of the phenomena I have alluded to can be found if in a less acute form in long-established parliamentary democracies. Economic interdependence is not confined to new states: what about Britain and the gnomes of Zurich? The whole notion of over-all economic planning demands thought and action on a time-scale that has little in common with the conventional notions about the swing of the political pendulum, and calls the value of alternating governments into question. Powerful ideologies dividing society into in and out groups can easily arise even in highly developed states.

As against this, the countries that form the second model, that have tried to by-pass parliamentary democracy and have opted for the overall control of a self-selecting and self-perpetuating elite which is what we mean by totalitarianism, Soviet or other, have run into problems no less daunting than those of the democracies. So there is no easy way out: we know where that particular blind-alley leads – back to some attempts to rediscover the basic elements in parliamentary government itself.

So, the crisis is with us – where do we go from here? I do not think that this is a meaningful question. The answers must differ from country to country. It may well be that there are institutional devices that could still be devised to meet some of the quite genuine problems I have been attempting to suggest. The resources of constitutionalism have not yet been exhausted. All one can say is that from the past two centuries of development, there are certain lessons that can be learned. In the first place, parliamentary institutions are in themselves the major safeguard yet contrived against the tendencies towards absolutism and the temptations towards corruption always and everywhere inherent in the act of governing. The second is that whatever the precise nature of the franchise or the precise composition of representative bodies and their relations to the executive branch, it is never to be forgotten that the essential reason for representation is to prevent the neglect of the interests of fundamental elements in the society concerned. The final justification for parliamentary democracy is to be found in the medieval tag: *Quod omnes tangit, ab omnibus approbetur.* There is no higher wisdom.

25 *Apartheid Policy under New Leadership*

After the shock of the murder of Dr Verwoerd had given way to speculation about the future, and after the inevitability of Dr Vorster's accession to power had been accepted, it was natural that the regime's opponents should try to find some crumb of comfort in an outlook bleaker than ever; and not surprisingly they found it in the argument familiar to students of politics, that power and responsibility are mellowing factors in a statesman's make-up.

Dr Verwoerd, it was argued, had developed from the narrow extremism of his past into a statesman with a subtle comprehension of the limitations of circumstance, and with some sensitivity to world opinion – why should his successor not develop in the same way? Whether this optimism is justified, whether the new Prime Minister has it in him to modify the crude racialism and antipathy to democratic ways that led to his wartime internment as a Nazi sympathizer, whether he has ambitions and understanding beyond the scope of mere policy – these are questions to which time will give the answer. But important as they are, they are not the questions that should most concern the world outside South Africa.

South African policy in world affairs is not determined by individual preferences, except within the narrowest of margins; it is the product of an interaction between the ideology of the ruling party – and the Nationalist Party is a ruling party in the full sense and not simply a competitor for power in an open arena – and the situation in which the country finds itself.

It is perhaps because a phenomenon of this kind is more usually found in connection with fully fledged totalitarian societies – a category to which South Africa does not as yet belong – that it has been possible to make the kind of miscalculation that helps to explain the failure of the Rhodesia sanctions to make an impact within the time scale thought adequate by the British Government. The idea that the South African Government would allow the Smith regime to be crushed by economic action could not seriously be entertained by anyone with the least understanding of the South African scene; indeed it is for South Africa a matter of national rather than governmental policy, since support for the Smith regime by now extends well beyond the ranks of the Government's committed supporters.

The fact that the necessary support has so far been given without overt recognition of the Smith regime is a testimony to Dr Verwoerd's diplomatic finesse – let us see what international law will do for us before we flout it openly! – and does not indicate any underlying questioning of the main issue. Mr Smith and his colleagues are to South Africa what William the Silent and the Dutch rebels against Philip of Spain were to Queen Elizabeth I – the outer bulwark of a small country determined to avoid for as long as possible direct confrontation with a clearly superior enemy.

From South Africa's point of view the enemy is almost everywhere – communist, democratic, liberal – the labels make little difference; from China to Peru the world repudiates the idea of a political system based upon the assumption of the natural superiority of a single race, and so long as this repudiation persists there can be no let-up in South Africa's defences, any more than the American South of the ante-bellum era could permanently remain part of a union in which the majority believed that slavery was wrong.

For countries in such a situation all is subordinate to the defence of their primary positions, and in this cause they will be willing to make concessions that might seem to smack of opportunism. The Smith regime does not in fact embody the full philosophy of apartheid at present – whatever the future may hold in store – and even more paradoxical is South Africa's alignment with Portugal. For whatever the view one may hold about Portugal's handling of her overseas territories, no one can seriously maintain that on the basic question of race the two countries are anywhere near the same end of the spectrum.

It is only five years since the eminent Brazilian scholar Gilberto Freyre wrote of the Portuguese policy of close contact between

Europeans and non-Europeans as one that helped to bring east and west together on equal footing, and added that such a union was only possible, it would seem, 'by means of miscegenation and the blending of cultures' – and these are not sentiments that would evoke much sympathy in Pretoria.

But opportunism in the pursuit of an ideologically-based foreign policy should not be regarded as unusual; what is more unusual is the seriousness and consistency of the ideology itself. One reason why this is sometimes underrated is the tendency of polemics on the subject to wander off into side issues. No doubt this is in part the result of the natural wish of the regime's supporters to wage the battle of words on ground of their own choosing; but it is also the result of the restricted frame of reference accepted by some of its critics.

It is not a question of evil men wantonly causing pain to satisfy crude or ignoble desires of their own; it is rather that men convinced of their own disinterestedness and doing good according to their own lights, men with a sense of responsibility not limited to their own white community, nor even to the territory within their own frontiers, are, because of the falsity of their basic philosophy, unable to carry out their purposes except in ways which must rebound to their ultimate discomfiture.

They would wish South Africa to be a model of peace, prosperity and security for the entire continent; they would – many of them – even like to see South African skills play the part they ought to in raising material standards in the rest of Africa. But none of this can happen; and it is natural that those who in their hearts must know this should blame the outside world rather than their own shortcomings.

It is of course a fact that by material standards the Africans of South Africa are a good deal better off than those to the north of them; otherwise the pull that the country exerts upon labour far outside its borders would be inexplicable. But inroads upon primary poverty do not eliminate but rather sharpen the frustrations caused by political and social discrimination; all history bears witness to this fact. Again no one who has glimpsed the *favellas* of Latin America can dissent from the proposition that unfettered freedom of movement from rural to urban areas at a time of rapid industrialization creates social and human problems on an almost unmanageable scale.

But the pass laws which are the result of the attempt to control this influx in South Africa's case are emptied of the good they might do by the regime's refusal to accept the fact that the laws exist for people and not just for units in a labour force. Thus they operate to weaken the ties of family, when every consideration of prudence and humanity should wish to see them strengthened.

Most important of all, the concept of 'separate development', which might be thought of as representing a possible method for overcoming the weaknesses that all plural societies appear to show, it vitiated by the inherent inequality of the roles assigned to the several groups, and the hopelessly uneven allocation of resources between them. Nor is it possible to accept the easy rationalizations of so-called 'liberal' nationalists either when they point to an unfeasible geographical partition of the country as an ultimate solution or when they assert that economic progress itself will make it possible for the other communities to 'catch up' with the whites, since meanwhile everything that is being done is helping to make it certain that the non-white groups will be neither socially nor psychologically prepared for new responsibilities.

Indeed there is an inherent contradiction between the aims of the regime in economic development, upon which the future success of their policies, and the conversion of the world to a position of approval, is held to depend, and apartheid itself, which in respect of job-reservation and educational policy is a major obstacle to these same economic goals.

It is not the immorality of apartheid that worries South Africa's rulers – that it is in Gladstone's words about another tyranny 'the negation of God erected into a system of government', since the God of Nationalist South Africa has been pressed into the regime's own service – it is the fact that it is in practice, in the long run, untenable.

The eradication of political liberties may conceal this for a time, and the undercurrent of violence inseparable from a social order of this kind may be ascribed wholly to external instigation. The nature of the external threat is indeed often mistaken. Sophisticated arms are accumulated, more than are needed to cope with minor threats, irrelevant where great powers are concerned. Preparations are made to cope with sanctions that are unlikely to be more comprehensive or effective than those mobilized against Mr Smith.

What matters is that South Africa is set on a course which is viable neither internally nor externally. When the time comes to change course, as it must, South Africa will be in need of friends; the tragedy of her men of good will is that (Mr Smith apart) she has none.

26 *Isolation of South African Universities*

In a country dedicated to an ideology the role of higher educational institutions becomes of the first importance. In South Africa the policy of 'separate development' is held to apply to education at all levels including the university level.

The only exception made – and that a temporary one – is to permit non-whites to enter 'white universities' to follow courses still unavailable at the institutions in process of establishment for their own racial groups; and this permission is not easily obtained. (The large University of the Witwatersrand has under 200 non-white students, almost all of them drawn from the Indian community.)

Nevertheless the English-medium universities – notably Witwatersrand – are, along with a section of the English-language press, some religious circles, and one or two private institutions such as the invaluable South African Institute of Race Relations, the only elements in the South African complex which offer scope for criticism of the governing ideology itself.

If one believes – as one must – that apartheid will in the long run have to give way to a system more consonant with the realities of South Africa's position and the moral outlook of the rest of the world, then it should be a major concern of all those who would wish to see the inevitable transition a peaceful rather than a sanguinary one to do whatever is possible to strengthen the English universities as institutions, and to do what is possible to counteract the sense of isolation, and even despair, which is bound from time to time to afflict the most sensitive members of both their staffs and their student bodies.

For these reasons one can only regret both the unwillingness of the great American foundations to interest themselves in the problems of South African universities, and even more strongly the attempted boycott of them by members of the British academic community. Some of the sponsors of the 'boycott' are of course consistent in that they clearly believe that no change can come about in South Africa except through violence, and that everything should be done to harden the lines of division in the country so as to make the choice of revolution inescapable.

But for liberals who continue to hope for some evolutionary solution, such behaviour can only be explained by their ignorance of the situation with which they are dealing. The boycott is indeed deeply deplored by such a man as Alan Paton, who is more important as the conscience of South African liberalism than as the leader of the now tiny and powerless South African Liberal Party.

The misunderstanding arises from a completely false analogy with what went on, for instance, in the universities of Nazi Germany. If universities in any country adopt an ideology repugnant to the world academic community, and accept in its name the purging of their own body in the name of racial purity, or whatever it may be, then it is right that this should be taken into account in assessing what relations with them it is proper for the outside world to maintain. But this is not the South African situation.

The ideology of apartheid is not accepted by the English universities, but has indeed been specifically repudiated by them, notably by the University of the Witwatersrand, in a declaration of principle that deserves to be more widely known. So far as the law allows they have continued to admit students of all races – and it is worth adding that there is no question here of a duty to court martyrdom. The penalty for attending a university without governmental permission falls upon the student himself, not upon the university that admits him. So, to defy the law would be at the expense of the non-white student, not at that of the university authorities concerned.

Furthermore, while the Government controls admissions and is going to extend its powers still further into the internal workings of the universities in respect of their non-whites, what is actually taught or said on campus remains free. Lip service to the ideology is not demanded as yet.

One must not of course paint too glowing a picture of what the

English-speaking universities have managed to achieve. Their performance has been in this respect an uneven one, and has much depended upon the accident of personality. Not all university teachers by any means believe that their duties extend beyond the performance of their routine assignments – and these are much heavier than in British or American universities. Not all are willing to attract the limelight to themselves with the risk that this may entail not only for themselves but for their families.

The powers that the State possesses for dealing with recalcitrant individuals by purely administrative action without recourse to the courts are formidable – and they have been used and will be. It is not a matter for surprise that so much of the burden of public protest, and of the attempts to bridge the racial gap, falls upon student organizations, both within the universities themselves and through their national organization: NUSAS.

In order to understand the full implications of the responsibilities for intellectual leadership thus assumed – and sometimes inevitably prematurely assumed – by students, one has to remember some aspects of the South African university scene itself that have no connection (or at any rate no direct connection) with the racial situation.

There is little in the way of private endowment except for bricks and mortar *ad majorem benefactoris gloriam,* and public moneys are given on a formula based on the number of undergraduates registered for particular courses which makes innovation difficult. The amounts in themselves are also very limited so that the staff-student ratio is an unfavourable one, particularly in the non-scientific departments. The University of the Witwatersrand, which is about the size of Manchester University, has a history department with only three posts. In such circumstances any intensive teaching of a personal kind is difficult to allow for.

Furthermore, the very low standards demanded in secondary education mean that the first two or three years of university are essentially 'sixth form' work, and there is in addition, in the non-science departments, an almost complete absence of post-graduate work. The social sciences are particularly undeveloped – the only department of political studies in the country is at the Witwatersrand – and this also has its consequences. What it adds up to for the able and demanding student is the absence of the idea of the university as a community dedicated to the advancement of knowledge, and of himself as a member of it; it is largely an extension of school, and intellectual stimulus is sought outside the formal curriculum.

For these reasons, there is bound to be rather more than the usual lack of proportion that in all countries is the hall-mark of student politics, and it is no answer to say that those who pay for this are the students themselves. It is enormously important for their future and for the country's that they should be assisted to escape from their isolation. The American 'field service scholarships', looked at askance by authority, have done something for a number of those at the immediately pre-university level; British universities remain the Mecca of the intellectually ambitious, with Oxford's 'PPE' understandably the goal of most of them.

But not all can go abroad, and here again the argument for encouraging academic visitors from other English-speaking countries is an obvious one – particularly as for the foreigner the prudential argument does not apply. It would be a pity if South Africa's young people were left with the belief that Senator Robert Kennedy is the only person in the world who cares enough about liberalism to carry its torch among them.

Where the non-English-speaking universities are concerned the arguments are rather different but point the same way. One must disapprove not merely of the existence of separate institutions for non-whites but of the spirit in which they are now run – of the implied belief that 'Bantu' or 'Indian' education can be held to mean something different from education itself. But any voice from the external world may help to minimize the inevitable bitterness of those who have to accept these limitations, or get no higher education at all.

In the Afrikaans-speaking universities one is of course among those who accept the fundamentals of the ideology, and in institutions where some aspects of it are part of the curriculum itself. Nor is student dissent, except on a very small scale, to be expected.

Nevertheless the visitor will be welcomed and listened to with courtesy, even when the picture of the external world that he presents is one that his audience will largely find repugnant. And as the change, if it is to come, will have to involve Afrikaaner attitudes as well, it would be wrong to avoid this challenge.

27 *Indian Universities and Their Problems: Some First Impressions*[1]

I am going to say this evening to an audience which includes a number of friends some things which I fear they will not like. It would seem purposeless to come and give a talk in order to repeat cosy platitudes. I want to talk, since I have been asked to talk at this India International Centre, about the impressions which have been made upon me in the course of an all-too-brief stay, now nearly at the end, by the universities of this country, and do so in the light of the sole qualification which I have, which is that in the course of a probably rather ill-spent academic life, a good deal of it has been spent in foreign universities in Europe, in North and South America and Africa, so that if my knowledge here is not deep – it is not – I can at least say that the problems which strike me, or the questions I ask, are those which have been raised by one's immediate tendency to compare one's new experiences with one's old ones.

The experience I have had in India is of course wholly inadequate. I have only been to in all (on some occasions very briefly indeed) some half-a-dozen universities, though I have of course had the opportunity of discussing my experiences both at Delhi and at these other universities with people with longer experience and much deeper knowledge.

I would add a further difficulty that, although the universities that have been my hosts have been uniformly courteous and helpful in my travel, they have not shown themselves anxious to admit a foreigner to the midst of their own preoccupations and, as com-

[1] Lecture delivered at the India International Centre on 2 December 1966.

pared with almost all the other countries I have visited, I have found much less of these straight professional talks – or what we would call straight 'shop' talk – than I have elsewhere.

Far fewer university teachers seem anxious to make use of the experience of someone from another background for direct practical purposes. Furthermore, it does not appear to be in this country the custom, which the travellers certainly find very useful in visiting universities, to allow the visitor direct contact with the students by giving him the opportunity to take a class in his subject if he happens to turn up on an appropriate day. So that my knowledge of students and the student opinion is very restricted indeed.

The reasons, I think, for the difficulties which I have experienced in trying to form an impression spring in fact, no doubt, from what is one of the aspects of Indian universities that most strike one – someone who comes from a different, I think in particular from an Anglo-Saxon university background, and it is one of which I am aware Dr Deshmukh and others themselves feel is of considerable importance. That is to say, partly for physical reasons and also I think for many social and cultural reasons, there is the absence of what can be called an academic community at each university composed as a single unit of scholars at different levels, from professors to undergraduates, all united by propinquity in their daily life and in their social as well as their intellectual intercourse. The idea of a university as a place where one spends, at least in term time, most of one's waking hours, which is the centre of one's world, does not seem to me, in my brief observations, to have taken full root on Indian soil and has now been put out of reach, perhaps for financial reasons and for reasons of scale. This may be much less true of wholly residential institutions like the IITs, and generally speaking of course what I want to talk about relates primarily and inevitably to the arts and social sciences. I want to make this point not to disarm criticism, of which I am sure there will be a great deal and valuable to me, but to explain that the inquiry itself has not been an easy one.

The second point is that although the problems of the universities figure a great deal in the press and in public discussion, particularly at the moment – I can think of no other place which I have visited where universities and their affairs have occupied so many columns in the newspapers – much of the discussion obviously relates to immediate problems and avoids the fundamentals. There is of course available to the inquirer the recent report of the Education Commission and this document has of course been a useful guide in many ways, if only because the statistics it presents gives one some idea of the scale of the problems one is talking about. But if I may say so, I do not find this report, in as far as it relates to universities, a particularly enlightening document in its analysis. I think that there are certain features of it which suggest a certain timidity at facing the magnitude of the problems involved.

As I have pointed out in a talk I gave elsewhere – and I found it has been noticed by others – the authors of the report discuss India's educational needs within their conception of the country's future development and talk of the socialistic pattern of society which the nation desires to create. Now, whether such a phrase – socialistic pattern of society – is a concept which has sufficient meaning for a nation to make it a conscious objective, raises very wide questions. But I think it is curious to find so loose and sweeping a statement in a document which bears the stamp of authority.

The authors also declare that the educational system will need radical changes if it is to meet the purpose of modernizing democratic and socialistic society. I am not at all certain – and the discussion about this in the IIT seminar which is going on now rather confirms my doubt – I am not at all sure whether anyone knows whether a society can be modernizing, democratic and socialistic all at the same time. That may be a matter of opinion, but where I find the report hardest to follow is, for instance, where it also says that apart from her own resources, which must of course be fundamental, India also needs the cooperation and assistance of richer industrialised nations which 'share India's faith in democratic socialism'.

Now, I would assume the authors must be right in believing that India requires some external assistance. All nations do. But must this assistance really come from nations who share a faith in democratic socialism, however we define that phrase? If we take the nations who have hitherto given India assistance, particularly in the field of higher education – I am thinking of the foreign sponsorship of the IITs – we find something quite different. We have the United States, which presumably shares the faith in democracy, but certainly not in socialism. We have the Soviet Union, which has the faith in socialism, but which is not democratic in the Indian sense of the word. The United Kingdom no doubt has a government that professes both democracy and socialism, but this can hardly at the present time be called a part of the national faith.

So, if they seriously mean that India should not accept assistance in the field of higher education except from countries who fit into this ideological definition, then she is likely to get such aid only from countries much poorer in resources herself and far more likely to need India's help than to give it her! So that one might ask oneself what the learned authors of the report had really in mind? Are we doing more than inserting into a serious document, about a very serious and critical subject, a pinch of incense on the shrine of accepted ideologies?

Furthermore, the report – and it is not unique in the history of reports about education and higher education in particular – gives the impression of being based upon a bird's-eye view or, you will forgive me, a 'Vice-Chancellorial view' of universities. There does not appear to have been an attempt to see what are the realities of the problem in the lives of the individual students and teachers. And no one reading it when it came out just before last August could anticipate that the publication would herald a persistent bout of student agitation which has disrupted academic life in so many centres. One wonders whether the Commission ever found time to talk individually to, say, junior lecturers or to individual students and to ask how the university looks from the 'worm's-eye-view'.

Again, in the recommendations of the Commission – some of them I think in principle much to be applauded, for instance, their perception of the urgent need for a developing country of some major institutions working at international standards – the Commission still finds it hard to break away from the traditional structure of Indian universities. The whole division in the colleges and universities or, if you like, the gap between undergraduate and graduate teaching, which is in my view one of the most wasteful method of using educated manpower possible to conceive, is never even discussed in this report.

The Commission again associated with itself in a rather intermittent fashion some foreign experts, and this shows a commendable absence of national exclusiveness. But I cannot see in the report a great deal of evidence that foreign experience is taken seriously either for the warning or for the example it could give.

There is of course a prevalent view, not always consciously expressed, that Indian problems are unique and that therefore experience elsewhere is anyhow not a great deal of help. It would be presumptuous to argue whether or not this is true over the general range of India's problems. But with regard to universities and their problems it cannot be true for the simple reason that the tasks of universities are essentially identical the world over. Universities exist everywhere as a combination of teaching and research, a passing on of the knowledge of previous generations and an adding to the stock of knowledge. And indeed not only in its fundamental organization and methods but also in its subject matter in the arts and the social sciences as well as the natural sciences, there is after all only a single body of knowledge. Each different country may attach a different degree of importance to the national literature, the national history and so on, but basically knowledge is single and the approach is through similar mental concepts and a single set of methodology. Changes in either concept or methods in university activity may come from any quarter and the mark of a first-rate university, or one mark of it, is precisely its openness to external influence.

Nor, I think, except in scale, are the actual and immediate problems of Indian universities unique. The effects of the rapid inflation of university numbers in causing, for different reasons, an alienation of the student body from its teachers, bringing about indiscipline and even violence, are certainly not confined to this country. I have seen almost identical phenomena or read about them at other times in Latin America and many other parts of the world. Indeed, student unrest, as you know, has spread to the so-called developed countries. The students of the London School of Economics demonstrated outside their venerable building only a week ago against the appointment of a particular person as Principal – so that they and students of Hyderabad have something in common!

What is perhaps more singular in India, at least so far as my limited experience goes, is the lowly financial and consequently in part the lowly social status accorded to university teachers and an inevitable consequence – their tendency towards remoteness from the main currents of the nation's life and the main concerns of its intellectual elite. I had thought that the remoteness of the University of Delhi, Department of History, with which I have been concerned, from the very political and intellectual life of this capital city, from institutions such as this, was wholly an effect of geographical remoteness and a poor public transport. But I am beginning now, particularly since I have had a chance to see other universities and of comparing my experience with those who have been attached to other universities, to believe that the phenomenon is a much more general one with deep social reasons. Its effects are of course most obvious in the social sciences. In Political Science, for instance, teaching wholly divorced from practice or experience or the daily rubbing of elbows with people involved in the governing of great concerns, is bound to be in the worst sense 'academic' and where the conservatism, which all the university teachers suffer from, with regard to curriculum and methods, is sure to be reinforced if the community at large makes few or no demands upon the university to provide solutions for its problems.

Now, these impressions – as I said these are the random impressions of an academic vagabond – should lead one, I think, at least to the conclusion that one ought to try to do precisely the opposite of what I believe the Education Commission has done, not look at the universities in the light of an imaginary and perhaps utopian future, but to examine their predicament in the light of things as they are and of history as it has been. The university and indeed the whole idea of an integrated national educational system, of which the universities form the apex, has after all been borrowed from the industrial West. In the West the pre-existing systems were of course changed to meet the new needs of industrial society. But the course of development has gone differently in the developing countries and most differently perhaps in India, and this I am sure is the first point to grasp.

In the history of the Western world – and I include in the 'Western world' the developed countries such as the Soviet Union and Japan – three basic social processes have gone together: industrialization, education and the spread of political responsibilities to the population at large through the extension of the franchise. Industrialization, if you look, for instance, at nineteenth-century British or French history – British history is perhaps the classic example – promoted the need for literacy and made essential the provision of at least elementary education on a mass basis. Literacy permitted the extension of the franchise and its culmination in a democratic system. But here, in India, things have been entirely different. Instead of the three basic processes going alongside each other, politics at the time when the Constitution was made took a jump ahead of the others. Universal suffrage was adopted at a time when the country's economy was still basically agrarian and when education was very far from universal. It is easy to see why this is done and indeed you can pursue a fruitless inquiry, of interest perhaps only to historians, as to whether the decisions could have been otherwise. But it produced important consequences and they are the consequences that Indian education and particularly Indian universities are now living with.

After all, what was done was to give a largely rural and uneducated electorate an important and perhaps crucial voice in how the two other processes were going to be developed – industrialization and education itself. With regard to education it does seem that certain effects of giving the priority to political development are almost the same in any country. In the first place, it is likely that in such a situation any kind of education, however elementary, however incomplete will both confer tangible economic advantage and accord social prestige. We should therefore expect much greater pressure in favour of the expansion of facilities that exist to cover more and more of the population than we shall get towards the raising of their standards. And as a consequence of this, we may expect the cheaper forms of learning which, after the early years of schooling, means the arts rather than the social sciences to develop at a faster pace. Quite apart from the financial element, a system in which politics in the narrow party sense counts for so much, as it is bound to do with a mass electorate, will give to verbal facility a prestige which will make it desired for its own sake at the expense of some other aspects of education.

In other words, while the national interest in a developing country would demand a rather high concentration upon the preparation of persons of all ranges of technical skills, and not least in the highest skills – the technological skills – politically the selectivity in recruitment and the rigorousness of standards which institutions devoted to these purposes require, are bound to be unpopular. Dislike of elitism, which is the bane of education in the democracies even in advanced countries, is bound to be still more prevalent and more damaging in developing societies.

Allied with this is the concentration, long criticized and certainly not in India alone, upon the certificate of achievement rather than the achievement itself – there is a pressure, both social and political, to lower standards so that no one who has put in the required time at a course of study is deprived of its fruits by his mere inability to reach a level of performance. Political pressures which are recognized in societies still close to familial and neighbourhood notions of loyalties and obligations and where the sense of the *res publica* is weak, are thought of as proper means towards obtaining employment or contracts, so why not use them to try to reverse the views of the examiners? While these pressures can sometimes be resisted, it is more serious when the same false values are implicitly accepted by the student body itself. Students who strike in order to force down the level of the demands made upon them by their teachers are obviously damaging, in the long run, only themselves and their country, since at some point the reality behind their alleged achievements must catch up with them. In such an atmosphere, it is difficult to require the student to make the essential step in education of taking responsibility upon himself in his own mental development. A class that he misses for taking part in some strike or demonstration will seem to be something of which he has deprived the authorities, not a sacrifice of a part of his own education as a result of which he will be the first and major sufferer.

Indeed, I would go so far as to suggest that in any society in which high school or college students regard themselves as a class or a social group with demands of their own, rather than a collection of individuals working to prepare themselves for their role in the productive process, widely interpreted, such a country is one where educational policy has been misdirected. It almost certainly means that the rate of expansion has been higher than the society in question can handle and the failure to make this clear in speech and action is another of the unfortunate effects of premature political democratization.

Education in a developing country may also be saddled with tasks which are not essentially connected with the primary purposes of training individuals. If a society in question is an essentially pluralistic one, it may be expected to act as a solvent of the differences and the promoter of nationhood. Now, in certain conditions, this is not a task to which educational institutions are necessarily unsuited. In the United States the educational system has been the main vehicle in assimilating successive waves of immigration and in creating a common American nationhood.

But it is important to see what advantages it had and what price it has demanded. In the first place, the United States had the advantage of a given mould into which to pour the molten metal. It was assumed from the beginning that linguistic unity was essential and that young Americans, whatever their parents might talk, would be taught in English. In some cases, therefore, national unity was purchased at the price of tension between the generations. In the second place, it involved considerable attention to the promotion of a patriotic attitude towards the nation's history and institutions. American children have been encouraged to identify themselves with the founding fathers of the nation, even though their own ancestors in George Washington's time may have been far away on the East European plain or in some Mediterranean mountain village. To place such concentration on nation-building is bound to lead to some lack of attention to the history and institutions of other peoples, some neglect of foreign languages and, on the whole, a lower level of academic pressure, than in a fully formed nation such as Britain or France or Germany. In other words, by and large, the political and social objectives in education will then tend to prevail over its academic and professional purposes.

Naturally, this does not mean that the United States has at any time, and least of all today, failed to equip itself with sufficient specialists at high levels of skill and performance, but it has done this by prolonging in their case the process of formal education for much longer than in most other advanced countries, let alone developing countries. And this is because the United States is rich and therefore can afford to keep outside the productive process for longer time a high proportion of its young people.

It will be seen then that the American example certainly shows what could be done in the way of helping to make a viable political community through the medium of education, but its circumstances make it of little relevance to countries where a plurality of cultures exist and where there is no dominant structure of modernity into which to fit the new aspirants. Even the Soviet Union, which may have more to offer by way of example, had its task facilitated by the existence of one dominant national group and language with an established social and educational ideology.

The high priority the Soviet Union has been able to give educational expenditure, as compared, for instance, with its expenditure upon housing or medicine or general personal consumption, is partly because there is no need there to satisfy the pressures of a democratic electorate.

Again, for a variety of reasons inherent in the nature of the system, the Soviet Union is much freer to take the necessary disposition to see that its young people follow branches of study in numbers that are related to the development of the economy itself. There are of course imperfections, and the Soviet self-criticism makes it abundantly apparent. There does not seem to be any likelihood of the Soviet Union facing what much of the Western and Central Europe faced with disastrous results in inter-war years and what the developing world clearly faces today – the production of large numbers of graduates in humanities, social sciences or in law with no prospect of the kind of employment for which they believe themselves to be qualified by their university education. Such an output of unemployed graduates ought to be regarded as the most infallible of all recipes for political disorder and pointing the road to either chaos or fascism.

For the developing countries or many of them, these general problems are complicated by the problem of languages, except in those countries in Africa where the local languages never reached a written literary level and where therefore there is general agreement about the use of one of the world languages as a vehicle not only of government but for all the modernizing factors in society including education. The position is further complicated where, as in India, there is already a multilingual situation before the problems of external communication are reached at all. In such a situation, particularly where party attitudes have to be formed in conditions of universal franchise, political rather than educational considerations would tend to be the governing factor. It would be found difficult to exact standards in a foreign language as the price of admission to educational institutions without appearing to demote the national or regional languages. So during a period, when because of the absence of books and teachers a foreign language has to be used for instruction, classes are swelled by large numbers of students whose command of the language is inadequate and to whom the utility of instruction they are receiving is thereby much reduced, and such students by their presence force a slackening of the pace with which instruction could be absorbed by others and a lowering of the level of attainment for the whole of their age group. And those who are linguistically equipped to travel the road much faster would tend to suffer the most.

In such circumstances, it may well be that a rearguard action to hold the line in favour of a world language, generally speaking, is mistaken. It is, I believe, far too optimistic to believe that resources could be made available for teaching English or any other world language on the scale required if the present educational expansion in India continued. And those who believe that somehow English can be acquired, as it were, on the way, are simply blinding themselves to the difference between having a smattering of the language and a mastering of it for professional use in one's occupation or career. It is true that if the pace of academic life could be quickened, if students were made to work harder, and subjected to much more rigorous discipline, the situation could be improved perhaps. But it is hard to see universities tackling this problem at the time of such rapid expansion nor indeed would such a change be easily achieved.

In such circumstances, it seems necessary that something radical should be done. On the one hand, the clock's hands cannot be set back and some form of educational opportunity at all levels must now, I presume, be widely offered. The only hope, I think, as far as higher education is concerned, is to harden the distinction between different kinds of institutions. State universities and colleges, in my view – and I put this forward simply as a possible theme for discussion – should be encouraged as indeed I think the Commission encourages them, to make the switch to the regional language as soon as may be. Even if this means limiting their curriculum to the subjects where material in the regional languages exists it is surely better that the students should fully understand a narrow range of subjects than have a cloudy and incomplete view of a wider range. And the market opened, at any rate in the major languages of India, should call for in due course a more adequate supply of books, whether original or translations.

But for institutions of national status, whether universities or technological institutes, I fail to see a possible compromise. It is plain that without one of the world languages at the student's command much of what science and learning has to offer is simply out of reach and will remain out of reach at least until India has reached the level of achievement and power that will enable it to impose Hindi as a world language in the way in which the Soviet Union's scientific achievements as well as its political might have imposed Russian. With the demands that must be made on highly qualified manpower in a developing economy it is surely absurd that any part of the top institutions should be taken up in teaching students not fully equipped to benefit. A prerequisite of a high level in a world language for entry will in turn have its effect, lower down on the educational ladder, in the shape of foreign language medium or bi-lingual schools or special preparatory institutions for intensive language teaching on the lines pioneered by the armed services in a number of countries in the last war.

It will no doubt be argued that this solution and indeed the general trend of my argument suggests an elitist approach out of harmony with India's democratic aspirations, and that it would create greater divisions between a Western-educated elite and the mass of the people. This may well be true. But this is inevitable unless one accepts the fact that one is not going to try to get this very high level of achievement in India. One cannot defy the laws of social development by introducing democracy before the material and educational foundations exist and expect that no price will have to be paid for it. The alternative is surely, to go on pretending that such harsh decision need not be made and watching the outcome of this in increased disaffection and irresponsibility in the student body.

With this of course is involved a major change of outlook; and there are as well obviously changes that could be brought about even within the existing system, while the major ones are being hammered out. The difficulty is that many things which come to mind as a result of experience elsewhere are rendered peculiarly difficult here by the sheer size of the problem which the inflation of student numbers has brought about. One obvious example lies in the relationship between the printed word and the students. The reliance upon oral instruction and the textbook, appropriate for education at the primary and early stage of secondary education, is wholly out of place in colleges and universities and would indeed be regarded in many countries as inappropriate even in the latter years of secondary education. The reading of books, the use of libraries, should be the foundation of all education in the higher levels, at least in the arts and social sciences. But to think of what is needed to be done to make a really wide range of works available either in the Western languages or in the Indian languages is a terrifying thought and I have no solution to offer for it. Nor even at the present scale of university salaries does it seem possible to expect teachers, except the very fortunate ones, to build up the private libraries that should be the foundation of their own work and teaching.

But if there are to be, as the Commission says, five or six national universities, then there at least the effort should be made by a deliberate curtailment of numbers to make it possible to reform teaching: wide reading, regular written assignments and some degree of discussion are in the beginning a part of the natural order of things. This means a far more active and continuous presence of teachers and students on the university campus as well as bigger and better libraries. But unless the country is to rely wholly on a foreign trained elite, there is nothing to be done to avoid this expenditure.

To say this is not to condone the view that seems to exist in some quarters that the time ought to come when it would not be necessary to send students and teachers abroad to study. Nothing, I think, could be more short-sighted than cutting off funds for foreign study as a method of handling balance of payment problems. Study abroad is not a mark of a country's backwardness. Indeed, exchanges between the major industrial countries in the higher education field are constantly on the increase. I suspect that my own university is far more international now in its composition than it has been since the European Reformation.

But, once again a unique part of my experience in this country (the uniqueness compared with what has befallen me in many other countries), is that only once in these weeks has a university teacher taken the occasion of meeting me to ask about the courses and facilities available in the universities in the United Kingdom for Indian students.

Another point, which I referred to earlier and to which I attach increasing importance, is the current division of structure in most of the universities that I have seen between what are called universities and what are called colleges; that is to say, between institutions devoted to graduate instruction and research and those bodies devoted to education of the undergraduates. There are many reasons why I believe this division to be a great mistake. For one thing, the strength of any university faculty or department depends in part upon size – size alone permits specialization. As it is, departments seem to me to regard themselves as largely limited to their so-called 'university' teachers – a small permanent corps – and to treat college teachers as mere auxiliaries largely confined to what is thought to be the less important matter of undergraduate education. This may well be due to a historical misunderstanding of the college system in England, whether the Oxford and Cambridge variety or the London variety.

I do not think that it is fully appreciated that the teachers of the colleges in Oxford and Cambridge have a dual role to play – they are in the full sense members of the university and its faculty; nor that the colleges in London University are themselves (except for the examining function) full universities in that graduate and under-graduate instruction go alongside each other. I know there are geographical difficulties here and elsewhere where colleges are widely separated from the parent campus. But certainly I would have thought that more could be done to integrate universities and those colleges which have some kind of geographical proximity. Furthermore, I think it is improper, and it is difficult to see how this idea got about, that people should believe that university posts, professorships, should entitle their holders to ignore undergraduate teaching. This separation of the people who do the two jobs is unknown in the United Kingdom or Western Europe and unknown except to the most limited extent in the United States.

Other aspects of the university situation that might be thought to require improvement – the complexities of administration, the poorness of internal communication – have, of course, as have so many other things, a largely financial explanation, though I think in some respects these might be easily improved if people were more impatient of their consequences. The impatience of the Indian students seems to be matched only by the patience of the Indian teachers.

In the end, in all things it comes to a matter of attitude. To have effective universities one must want to have them. One must cease to think in terms of self-contained and mutually exclusive departments or colleges and one must have obviously leadership at the top. Nothing can be more depressing than your reported tendency to regard Vice-Chancellorships as political appointments. The running of a great university is a wholly professional job, no less professional, I would have thought, and I am still sure that Dr Deshmukh will agree with me, than that of a surgeon or a pilot. One would not like to fly a plane whose crew had been appointed on political grounds!

It should also, in my view, be the case that university teaching should be a full-time job. I think that at universities (and this has struck me so often when I have gone round that I cannot believe it is accidental) one far too often hears that the people one wants to meet are 'out of station'. I know that seminars and conferences are supposed to be intellectually refreshing and stimulating, but this should surely not be at the expense of the daily grind of research and teaching. These are things which in my view should take place only in vacation. Because if coercive discipline of the students is too difficult or out of harmony with the national genius, then the discipline of example is the only thing we have left.

As I said at the beginning, I have been, I am afraid, brutally frank and the role of a candid friend is notoriously unpopular the world over. I would myself have been much happier not to give this talk but to hear a talk on these lines from an Indian scholar. My impression is that quite a number of Indian academics I have met would in fact, if asked to give a talk of this kind, have said many, if not all, of the same things. My hope is that this exhibition of foolhardiness by someone who has only a week more in this country and no responsibility, will encourage them to speak out. Because, certainly I would be the first to agree that only those with the responsibility of making changes can really know what changes are possible.

What I have offered you at Dr Mazumdar's request this evening are merely, as I said, my impressions.

28 '*Imperial Sunset*'

To talk about an unwritten book is to offer hostages to fortune, but if *Imperial Sunset* is not written by me, it will have to be written by someone else. For the problem I am trying to tackle is one that stares us all in the face. Within the space of a single lifetime Britain has been transformed from being the centre and powerhouse of a world empire into an unsuccessful candidate for admission to an embryonic federation of Western Europe. How did this happen?

It would be relatively simple to account for such a reversal of fortunes in a number of different but complementary ways. One could dwell upon the scientific and technical developments that made a community of about 50,000,000 people inadequate to sustain the industrial panoply of a great power. One could point to the invention of the aircraft and to new developments in the weapons that were putting an end to the long superiority of sea power over land power. One could extol or deplore the spread of European education, technology, and ideas to non-European people that made it impossible for an island in the North Sea to perpetuate a dominion over palm and pine. One could examine, or speculate upon, the demographic, economic, and moral impact of massive involvement in two world wars separated by a mere twenty years' interval of uneasy peace. One could contrast the general acceptance of the desirability of those liberal ideas and institutions of which Britain was regarded as the exemplar at the end of the nineteenth century with their subsequent violent repudiation over most of the globe.

Historians have given and will give due weight to all these factors in the process by which Britain's position has been transformed. But the historical explanations that derive from a consideration of these elements have this in common: they are largely or wholly external to Britain itself. Britain is seen as the victim of history rather than as an active agent. And from the point of view of world history such an emphasis is no doubt correct.

The Britain of the 1960s is a Britain living with the consequences of these changes that have taken place in the world and in its own domestic arrangements, but still, and to a marked degree, the lineal successor of the Britain of the 1890s. Its political debates are carried on upon the assumption that what has changed has changed in large part in consequence of certain decisions that have been taken by rulers and people, acting through the accepted constitutional and governmental processes, and that the problems it now faces are similarly capable of solution. No doubt the decisions that have been or are being taken relate in large part to questions that have crystallized only recently. But the language in which they are discussed, the emotions they arouse, the ways in which groups and individuals align themselves on one side or the other are all understandable only in the light of past debates that have helped to mould the political consciousness of the nation as well as of its rulers, meaning by rulers both the overt spokesmen of political clans and the silent but not less important caste of anonymous public servants, which it was the glory of Victorian reformers to have bequeathed to the succeeding age.

We can divide those who thought positively about Britain's place in the world in the 1890s into the advocates of Anglo-Saxon partnership, of imperial federation and imperial preference, or of possible alignments with one or other grouping among the European powers. There were however important currents of opinion hostile to each of these conceptions, which were opposed indeed to the whole idea that Britain should adopt a new and more active role abroad. There were those who believed that Britain's invulnerability at home was still unimpaired, that imperial possessions were neither of practical utility nor compatible with the habits and moral outlook of a modern democracy, that the first duty of Britain's rulers was to promote the welfare of Britain's people. And the sentiment of the Little Englander was fortified by the intellectual conviction of the Free Trader.

In the 1890s it was the imperial school that had made the running; seventy years later it was the need to find an answer to what seemed to be the growing cohesion of Western Europe in a new European policy. The Commonwealth argument against Britain's entry into Europe was an entirely different argument from that used by the imperialists of the earlier decade; the Commonwealth itself was something very different from anything conceived of by Rhodes or Joseph Chamberlain. So too the 'special relationship' with the United States which played so important a part in British political thinking referred to a different strategic and political situation, and carried with it different connotations from any that can have been apparent in the 1890s. The 'Little Englander', again, could much less easily avoid identification with outright pacifism, or 'neutralism' in an age of ideological conflict and nuclear weapons. But in every case it is reasonable to assume that some degree of continuity in basic assumptions could be found underlying the changes in attitude and their practical applications.

The interaction of these and other schools of thought and their relative weight in British society and politics are subjects of study that transcend the ordinary divisions between domestic, imperial, and diplomatic history. For, in the event, the great decisions have been made by the same relatively small groups of men – our political and administrative elites – and we need to know how they came to be the men they were and to hold the ideas they held.

It was the men who were unable to enact imperial federation and imperial preference who brought Britain out of her diplomatic isolation; it was the peace-makers at Versailles who reluctantly bowed to the wishes of the older Dominions for a separate voice in international affairs; it was the 'appeasers' of the 1930s who had carried through the chief step towards Indian self-rule; it was the government that accepted the dissolution of Britain's Indian empire that rejected the original proposals for organic links with Western Europe and that made the continuing presence of the United States in Europe a prime objective of British policy; it was the Prime Minister who talked of the 'wind of change' in Africa and whose period of office saw the rapid multiplication of the independent members of the Commonwealth and the penultimate stages of decolonization in Africa who sought membership for Britain in the European Communities. And all the statesmen and civil servants concerned in these decisions played their part in the continuous reshaping of government and military institutions to suit the demands of new periods in the light of their own interpretation of past experience.

Even at the early stage of my inquiry some familiar questions are beginning to come up in an unfamiliar form. For instance, it begins to look as though the politics of the first world war and of its immediate aftermath will need to be looked at in a new light. It was not simply a question of winning the war and then trying to create an international framework that would make future wars unthinkable. It was also felt, by those who thought in terms of power, that Britain's influence in Europe and beyond could only be exercized if she could bring to bear on Europe the resources of the entire Empire. For this end, British statesmen were prepared to bid high, both in terms of sharing control of foreign policy with Dominion statesmen and in terms of reversing the historic policy of Free Trade in favour of far-reaching measures of imperial preference. Baldwin's decision in 1923 to come out for protection, and fight an election on it, has been ascribed to a wish to outbid Lloyd George, who was believed to be dallying with protectionist ideas. But in the proper imperial context it looks entirely different – and did so at the time to those who knew what had been going on behind the scenes.

Or again, let me consider the vexed question of 'appeasement'. The 'appeasers' set themselves two conditions that any European policy for Britain would have to meet. It would have to unite the broad mass of the Conservative Party upon which the government relied for support; and that made them wary of any close collaboration with Soviet Russia, and hence with Russia's ally, France. In the second place, it would have to be a policy which would receive the support of the entire Commonwealth. The time when Britain could bring the Dominions into a war by her own decision was over; now these great issues were for the Dominion governments and parliaments themselves to decide. And while the Pacific Dominions had been prepared to see more resistance to Mussolini, whose Mediterranean designs they feared, they were as opposed as were Canada and South Africa to any commitments which might drag Britain into war for purely European objectives. Before assessing the roles of Baldwin or Chamberlain or Halifax, we must reckon with the influence and prestige of those dominant Commonwealth figures of the inter-war period, the isolationist Prime Minister of Canada, Mackenzie King, and the South African protagonist of friendship towards Germany and suspicion of France, General Smuts.

But there is another aspect that should not be overlooked. The historian, overwhelmed by his material, over-abundant as it is despite the closure of the archives over the greater part of the period, faces an obvious temptation to divide the subject up into temporal segments of manageable proportions: before 1914, or between the wars. But these segments are often shorter than the active lives of the men with whom he is concerned. And each man carries forward the lessons of his apprenticeship and early experiences to his later tasks. The career of one British statesman, Winston Churchill, spans almost the entire period of my inquiry. But many others have adopted political attitudes at one juncture that they have sought to translate into action at another. How far in military and political thinking would one have to trace the consequences of the Boer war and of the subsequent settlement in South Africa? How can one classify the attitudes to foreign policy in the 1930s without reflecting on the different roles of those who had already been in political life during the first world war and of those whose earliest adult experience had been the parlous agony of trench warfare? Would Eden's policy over 'Suez' have been the same but for the identification of the new Egyptian dictator with the European dictators whom Britain had 'appeased' in vain?

Nor need experience of the past be direct or personal. Even in a period when the hereditary factor in British politics was diminishing, family ties may have played their part. The careers of Austen and Neville Chamberlain are perhaps to be studied in the light of their devotion to the memory and credo of their formidable father. Other statesmen may have the gift of forming around themselves a group of younger disciples whose influence may outlast their own; more than one writer has noted the persistent ramifications of the Milner 'kindergarten'. Even harder to assess is the influence of formal study, and personal reading and reflection; since one must remember that what may be relevant to understanding a statesmen's reactions to some event may be the ideas he acquired as a student thirty years earlier; and it is equally true that in assessing the influence of books one should remember the inevitable time-lag before they have their full effect.

But books are not everything. Personal acquaintance with another country or group of countries may powerfully affect political judgement. At some time in the latter half of the nineteenth century it became possible for the aspiring politician to contemplate a journey round the Empire as his predecessors when young might have made a Grand Tour of Europe. The fact that Milner's young men knew South Africa better than they did France or Germany was not without importance for their later judgements of men and events. Some men may have known European countries through business contacts; socialists like Ernest Bevin knew them through the international trade-union movement. But language can be an important barrier. The linguistic qualifications of professional diplomats are formal and open to scrutiny. But what of their political masters? Has there been a decline in the importance attached to the knowledge of modern languages in cultivated society? Was British insularity at times the product of an inability to enter into the intellectual world of Britain's neighbours? Was English thought to be as important an accomplishment for the foreign statesman and intellectual as it had been a century earlier?

What then of the materials for such a study? The diplomatic archives as they become available add to one's knowledge of the detail of events but are hardly likely now to reveal objectives or opinions not already known from some other source. The kind of material which we require to answer some of these questions is likely to be of a different sort. Few statesmen and fewer officials are of the contemplative turn of mind that would lead them formally to set out their appreciation of the political environment that has motivated their actions or advice. In memoirs written in later life, the temptation to give one's opinions greater consistency in retrospect must usually be too strong to be resisted. Something can be derived from a study of the public record – though the reasons used to support a policy in public are not necessarily those that have influenced one's private thinking. More is to be found in personal letters and in the contemporary record of conversations where these exist. How much the historians of the whole English-speaking world will eventually owe to Mackenzie King's obsessive habits as a diarist! When one tries to assess currents of feeling or opinion external to the ruling few, one needs the sense of period conveyed by the eye of the novelist or of the perceptive traveller. To such knowledge there are no short cuts; each decade passes into history more quickly than one often admits. The England of 1897 is as remote as the England of 1597 and can only be reconstructed imaginatively by the same methods. For recent years one can hope to check one's view directly against the views of some of the participants in the actual events; but those who have tried to use oral sources for the writing of contemporary history will be familiar with the inevitable pitfalls.

Now that I have outlined my project and its difficulties I feel appalled at my own temerity. My only hope lies in the fraternity of my profession. I hope to draw extensively on the wisdom of others and upon their knowledge of where one should look for answers to the kind of questions I have been raising. By talking of my project in advance I hope to get criticisms and suggestions when they can still affect the outcome.[1]

[1] Volume I of *Imperial Sunset*, entitled *Britain's Liberal Empire 1897–1921*, was published in 1969.

Index

ABOUT THE AUTHOR

MAX BELOFF is a Fellow of All Souls College and Professor of Government and Public Administration at the University of Oxford. He was born in London in 1913 and educated at St. Paul's School, and Corpus Christi College, Oxford, where he was awarded "first class honours" in the School of Modern History in 1935. His first book, a study of 17th-century politics, was published three years later. Among his subsequent books are: **Thomas Jefferson and American Democracy** *(1948);* **The Foreign Policy of the Soviet Union** *(2 volumes, 1947-49);* **The Age of Absolutism 1660-1815** *(1954);* **The Future of British Foreign Policy** *(1969)* and **American Federal Government** *(2nd ed., 1970).*

Professor Beloff has been at work in recent years on a trilogy which will describe and analyze the decline and fall of the British Empire. When the first installment was published - **Imperial Sunset,** *Vol. 1 (1969)* - it was acclaimed as a masterpiece of contemporary historical writing. As Anthony Hartley, editor of *Interplay* magazine wrote in *Book World: "A brilliant book, so suggestive in its insights and stimulating in its ideas that it is not easy for a reviewer to do it justice. . ."* The reader of **The Intellectual in Politics** will find these special qualities generously represented in this, his latest book.